Milestones in Neurotoxicity and Neuroprotection: A Lifelong Tribute to Toshiharu Nagatsu

Milestones in Neurotoxicity and Neuroprotection: A Lifelong Tribute to Toshiharu Nagatsu

Editors:

M. Naoi
Department of Brain Sciences
Institute of Applied Biochemistry
Yagi Memorial Park
Mitake Gifu, Japan

S.H. Parvez
Neuroendocrine Unit
CNRS
Gif Sur Yvette, France

W. Maruyama
Laboratory of Biochemistry and Metabolism
Department of Basic Gerontology
National Institute for Longevity Sciences
Obu, Japan

M.A. Collins
Department of Biochemistry
Loyola University Medical Center
Maywood, IL, USA

M.B.H. Youdim
Rappaport Medical Institute
Department of Pharmacology
Technion, Faculty of Medicine
Haifa, Israel

Reprinted from *Neurotoxicology and Teratology,* volume 24/5

ELSEVIER

2002

Amsterdam - London - New York - Oxford - Paris - Tokyo - Boston - San Diego
San Francisco - Singapore - Sydney

ELSEVIER SCIENCE B.V.

P.O. Box 211, 1000 AE Amsterdam, The Netherlands

First edition 2002

Reprinted from *Neurotoxicology and Teratology*, volume 24/5

Library of Congress Cataloging in Publication Data
A catalog record from the Library of Congress has been applied for.

British Library Cataloguing in Publication Data
A catalogue record from the British Library has been applied for.

ISBN: 0-444-51036-2

Transferred to digital print 2008
Printed and bound by CPI Antony Rowe, Eastbourne

SPECIAL ISSUE: MILESTONES IN RESEARCH ON NEUROTOXINS AND NEUROPROTECTION:

A TRIBUTE TO PROFESSOR TOSHIHARU NAGATSU

NEUROTOXICOLOGY AND TERATOLOGY

CONTENTS   VOLUME 24 NUMBER 5 2002

SPECIAL ISSUE: MILESTONES IN RESEARCH ON NEUROTOXINS AND NEUROPROTECTION: A TRIBUTE TO PROFESSOR TOSHIHARU NAGATSU

ELSEVIER

Neurotoxicology and Teratology 24 (2002) ix–xiv

NEUROTOXICOLOGY
AND
TERATOLOGY

www.elsevier.com/locate/neutera

Laudatio

Conformational diseases:
Misfolding mechanisms may pave the way to early therapy

C. Reiss[a], R. Ehrlich[b], T. Lesnick[a], S. Parvez[c], H Parvez[d]

[a]Alzheim' R&D-Vigicell, 2, rue de la Noue, F91190 Gif, France
[b]Seccion Bioquimica, Facultad de Ciencias, Universidad de la Republica, 11400 Montevideo, Uruguay
[c]Laboratoire de Neuroendocrinologie, Université Champagne Ardennes, Reims F51000, France
[d]IAF-CNRS, F91198 Gif, France

This contribution is dedicated to Professor Toshiharu NAGATSU. Since his discovery and identification of tyrosin hydroxylase, he introduced many novel concepts and provided milestones-setting results, which significantly contributed to establish the field of neurotransmitter biology, which particular emphasis on molecular aspects of neurological diseases.

1. Introduction

Since the early discovery that sickle-cell anaemia is a heritable disease related to specific DNA mutations, many pathologies were found to be linked to transition, transversion, deletion or insertion at single or multiple loci of the cellular genome. More recently, conditions were observed that result not from mutations, but from epigenetic modifications. Epigenetic events alter gene expression, in the absence of mutations in the nucleotide sequence of the gene and/or the elements in charge of controlling gene expression. An important class of phenotypes depending on epigenetic events involves reversible, irreversible, heritable etc. nuclear modifications, like chromatin organisation or DNA methylation for instance. These can deeply modify the life and fate of the cell, from global phenotypic traits like chromosome stability down to the regulation and control of gene expression, whether the modification is part of the normal biological activity of the cell (during development for instance), or comes in response to some external stimuli (action of various stressors or proliferators for instance).

Lately however, an entirely new class of epigenetic events was identified, which could not be linked to mutations or to some epigenetic alteration of the control or regulation of gene expression, as the genes involved are expressed in normal amounts and in the normal biological context. Although the proteins involved in these diseases had the normal primary sequences, they had not adopted the native conformation, hence the name "conformational" epigenetics.

It was recently observed that the production of misconformer is a very common event. Under physiologic conditions, up to 80% of all proteins produced in the cell may come as misconformers, often identified as such while the protein is still in the process of synthesis [21]. Conformation-defective, nascent or completed proteins are normally quickly tagged (ubiquinated in eukaryots) for degradation and directed to the proteolysis complex (proteasome for instance). However, the misconformers found in conformational diseases are characterized by their resistance to proteolysis, even when ubiquitinated. They can therefore accumulate in or around the cell, form aggregates, fibers, plaques or "bodies" which are either toxic to the cell, or else alter or abolish its normal biological role. If this happens in neurones, the outcome may be dementia. The "bodies", fibers, tangles or plaques found in the brain of patients with Alzheimer's, Parkinson's, Kreuzfeld-Jacob's (BSE, scrappy in cattle) and over 20 other "conformational" diseases identified so far, are built up from disease-characteristic proteins which have adopted some nonnative conformation often rich in β-structures (the deposits are called "amyloid" in case they display the tinctoral properties of starch (*amylum*) when treated with Congo red).

Formation and resistance to proteolysis (protein deposit) of misconformers have been observed in a variety of species, including yeast. In *S. cerevisiae* for instance, the transcription factor Ure2p involved in nitrogen uptake from a poor source is repressed in the presence of a rich nitrogen source. One out of 10^5 cells, however, carries a "prion" type misconformer of Ure2p, much more resistant to proteolysis than the native Ure2p and allowing uptake of nitrogen from a poor source even in the presence of a rich one. This misconformer and the associated phenotype can even be transmitted to daughter cells. Several examples of such a "cytoplasmic heritage" were found in fungi but have not been reported in mammalian cells so far, although this is conceivable.

Two rather heterodox biological processes characterize conformational epigenetic traits: production of proteins in some non-native conformation rich in β-structures, and their accompanying, or subsequent, resistance to proteolysis and degradation. Although the steady, intra- or intercellular accumulation of insoluble protein species seems to be responsible for the disease-specific toxicity, the initiating process is the production of misconformers. As already noticed, up to 80% of all proteins produced by the cell do not fold correctly, meaning that the production of misconformers is a rather common event. The very fact that a protein of given primary sequence can adopt, in addition to its native conformation, one (or more) different conformations indicates that its folding or refolding is not under thermodynamic, but under kinetic control. The folding or refolding of such a protein escapes the Anfinsen principle, which states that the native conformation corresponds to that of global-free energy minima. Rather, at points of its folding or refolding route, the protein meets bifurcations at which the route adopted is selected kinetically. Here, we examine the kinetic processes involved in this selection and their impairment, the initial step leading to misfolding and ultimately to conformational diseases. Having a clear view of the misfolding mechanisms, means to avoid misconformation can be developed and help develop possible cures of conformational diseases.

2. Local translation rate participates in folding

The vast majority of the misfolded proteins produced in the cell are identified as such (and tagged for degradation) while still attached to the ribosome, hence their misfolding has occurred co-translationally. The kinetic process involved would then be basically linked to the translation rate. In the following, we restrict to the study of co-translational misfolding, or misfolding during post-translational processing. Other routes to misfolding have been proposed, like for instance Prusiner's "protein only" hypothesis, which assumes that a pathologic prion protein is able to recruit through protein–protein contact a native prion protein to flip over to the pathologic conformation. Here, we focus on *de novo* synthesis of misconformers, which does not preclude their activity in recruiting sister proteins in the native conformation to flip to the odd conformation, as envisioned by Prusiner's hypothesis. Rather, our goal is to provide a rationale for the formation of the initial misconformer, which may then pass its conformation to other proteins.

How can the translation rate affect folding? Co-translational folding follows a vectorial pathway. On each amino acid addition, the nascent protein faces a choice among several (say 3 or 4) folding routes. Each route is characterized by a set of ϕ and ψ values (rotation angles around the two backbone bonds of the nascent peptide unit) accessible on the Ramachandran plot. The routes become populated according to the laws of thermodynamics, provided the protein folds freely and that, on each addition of the next peptide unit, the folding "reaction" reaches equilibrium.

The two conditions are not necessarily met in the cell. First, in the crowded cell, the folding space is restricted if

only by the bulky ribosome to which the nascent protein is attached [19]. Second, as a peptide unit is added to a polypeptide of $(n-1)$ amino acids, the conformational space available for the n-mer polypeptide may differ from that available for the $(n-1)$-mer. New conformers may have become accessible and some conformers accessible to the $(n-1)$-mer may now be forbidden. Kinetic partitioning of the n-mer conformation sets on and that of the $(n-1)$-mer is cancelled, whether or not the latter has reached equilibrium. Selection of the co-translational folding pathway could therefore be affected by the translation kinetics of the nth codon of the coding sequence. The local translation rate can indeed efficiently influence folding.

The next question is then: what controls the translation rate? It is well known that for a cell in a given state (for eukaryotic cells in G, S or M phases, or in the G_0 state), the global amount of protein synthesized varies linearly with time, hence proceeds at a constant, average rate. Therefore, per unit time, the average number of codons translated in the cell is constant also and corresponds to the average codon demand (Solomovici, et al., submitted). It follows that, at the level of an individual codon (i), the time t_i it takes to translate that codon is proportional to the inverse of its demand, the codon usage 5; in all the genes of the species under consideration. Hence, 5.

The genetic code is degenerate, as 61 codons code for 20 amino acids, some of which have up to six "synonymous" codons. The cell is free to select anyone of them for specifying a given amino acid (Met and Trp excepted). In all genomes sequenced so far, each synonymous codon appears to be used at a specific, species-characteristic rate [18]. Species families, and even species within families, appear to have their own strategy in using synonymous codons, different of, and sometimes opposite to, that in other families. Furthermore, within a given cell, the usage of synonymous codons may vary hundredfold. Hence in this cell, the time taken to add a particular peptide unit to the nascent protein depends on the synonymous codon selected, and can vary up to hundredfold depending on this selection [22].

It follows that the local translation rate of a messenger RNA is far from uniform and changes from codon to codon. Synthesis of the nascent peptide can be fast over mRNA sequences constraint in frequent synonymous codons, and slow over sequences with the opposite constraint. In all species searched so far, coding frames frequently bear sequences usually 5 to 15 codons long, constraint to various degrees in frequently (fast) or infrequently (slow) used codons in that species [8,22].

3. Local translation rates involved in local conformation

Local translation rates might modify folding at the level of secondary structures. Alpha helices tend to fold rather fast, as mainly neighbouring amino acids participate [6].

The corresponding messenger sequence is expected to bear frequent synonymous codons, corresponding to abundant cognate tRNAs, which could therefore be rapidly translated. Notice that slow translation (messenger sequence constrained in rare synonymous codons) of a sequence element destined to be part of an α helix, could favour its interaction with another sequence and thereby produce misfolding. β structures in contrast build up much more slowly, since amino acids in some cases rather distant along the primary sequence may be involved (Finkelstein, 1991). β structures formation could therefore demand kinetic conditions favouring distant interactions, granting enough time for retrieval and adequate chaperoning of the structure elements. Slow translation of the corresponding messenger sequences, warranted by a bias favouring rare codons for instance, would provide the time necessary for this process to be accomplished, fast translation would not [16].

The translation rate might further influence cotranslational folding by yet another mechanism. For many proteins, folding must be assisted or catalysed by a variety of factors. DnaK for instance, an *E. coli* chaperone of the Hsp70 family, can transiently bind to sequences of 10–15 amino acids with a hydrophobic core, and differentiate between native and non-native conformers [20]. Proline cis/trans isomerization is catalysed by specific enzymes [24], and disulfide bonds formation and isomerization are also catalysed by a variety of enzymes [3], etc. These so-called "foldases" are crucial for many protein species to reach their native conformation. The "reaction" of a foldase with its specific target in the nascent protein, or during post-translational folding steps, requires that the target peptide sequence remains available for sufficient time before becoming buried inside the protein. It is again the translation rate, which is critical for this reaction to occur and permit— or cancel—the corresponding folding step.

Experimental evidence in support of the role of foldases in misconformation is provided by Komar et al. [14]. They translated the same messenger RNA in two cell-free translation mixtures, which differentiate mainly in their tRNA and foldase contents. The messenger was transcribed from the *URE2* gene of *S. cereviciae*, known to code for the yeast Ure2p prion, which can appear in a native or a misfolded conformation (these conformers differ by their resistance to proteinase K, PK). According to the analysis of the misfolding mechanisms made above, it should be possible to synthesize each conformer of Ure2p by the proper adjustment of translation conditions. This possibility was tested by translating the mRNA of the *URE2* gene in two classical cell-free systems, derived from rabbit reticulocyte lysate (RRL) and wheat germ extract (WGE). The former is known to have higher tRNA and foldase concentrations and to translate on average faster than the latter. The protein produced in RRL was readily proteolyzed by PK, whereas that produced in WGE was much more resistant. Digestion patterns and kinetics of conformers produced in RRL and WGE were respectively identical with those of Ure2p isolated from normal yeast cell and Ure2p extracted from cells exhibiting the prion phenotype. Digestions were made in all cases with an identical ratio of PK over total protein, and controls showed that both extracts have no inhibitory activity on PK.

The experiment proves that the prion isoform can be produced *de novo* at will, and in almost 100% amounts, by selecting the proper translation conditions. In this experiment, any contamination of native Ure2p by a misfolded version is excluded. Assuming this observation holds *in situ*, phenotypes corresponding to normal or misfolded prions could result from a slight modification of the translation apparatus, involving one or several of the misfolding factors mentioned above. It is plausible that similar mechanisms *might* produce misfolded proteins involved in some conformational diseases.

The local translation rate, constitutively encoded in the messenger sequence of the protein by an appropriate choice of synonymous codons, appears therefore as an important folding factor which supplements the information encoded in primary sequence to this end [2]. Translation rates at places critical for bending of the protein, for secondary structure formation (β in particular) or for interaction with foldases, are expected to be marked by specific selections of synonymous codons [16]. It is worth mentioning that it is not the absolute rate of local translation, which is critical for folding, but the relative rates enabling the "reactions" leading to secondary, tertiary and quaternary structures.

The dependence of protein folding on its translation rate has been considered previously at several places, based mostly on theoretical arguments [1,16,19,23]. Experimental proof has been obtained in our lab *in vitro* by Komar et al. [15], who showed that the translation in a cell-free system of a CAT mRNA in which a sequence of 16 frequent codons found in the wild-type gene was replaced by a sequence of 16 less frequent synonymous codons (silent mutations). The translation yielded a protein with significantly reduced specific activity, compared to that of the wild-type protein. Proof *in situ* of the dependence of protein folding on its translation rate may be more difficult to obtain, as the misconformer would be readily proteolyzed, but see Cortazzo et al. [7]. This difficulty could be overcome, provided the protein investigated could adopt a proteolysis-resistant misconformation, i.e. expression of its gene would occur in a cellular environment allowing for conformational epigenetics.

4. Modication of the local translation rate and misfolding

Can intra- or extra cellular factors induce co-translational misfolding? Since spontaneous, silent or effective mutations are rare, one would expect the local translation rate of a given gene to remain essentially invariant. Consequently co-translational misfolding should be a rather rare event, the associated conformational pathologies would be mainly restricted to familial cases linked to inherited mutations in the disease-specific gene.

This is obviously not the case, as demonstrated by the large production of misconformers in the healthy cell and the frequent occurrence of sporadic (i.e., devoid of mutation) conformational diseases, for instance. Examination of the actual translation elongation process allows solving this apparent discrepancy.

In species for which pertinent biometric data are available, a fair correlation is observed between codon usage and the cellular concentration of cognate tRNA [11], likely to hold as well for ternary elongation complexes (amino-acylated tRNA, elongation factor and GTP). It was further shown that the concentration of ternary complexes limits the rate of transcription elongation [9]. The elongation reaction can be considered pseudo-first order with respect to ternary complex concentration and for each codon, the translation rate constant raises with the codon–anticodon stability [12], determined by the modifications of the tRNA bases at, and in the vicinity of, the anticodon [22].

Despite that its coding sequences are almost immutable, the cell keeps nevertheless full control over the translation rates of its genes. Indeed, the cell can set the pace of its translation machinery by tuning the activity of several independent or coupled factors involved in translation.

The most prominent of these factors is the population of its ternary complex pool. This pool (and consequently the translation rate) can be modified by: (i) changes in the absolute and/or relative concentrations of tRNAs, by controlling the transcription of the corresponding genes; (ii) changes in the nature and the degree of base modification, by controlling the expression and activity of the enzymes in charge of base modifications; (iii) modulation of the activities of tRNA synthases (control of their expression); (iv) control of the amount of amino-acylated tRNAs (dependence on expression of synthases and tRNAs, status of tRNA base modification, amino-acid supply); (v) changes in the expression level of elongation factors, etc. The cell can also change its total population of active ribosomes, the population of ribosomes engaged in translation (change of transcription activity, activation of internal translation initiation sites), or modulate translation-regulation factors like ppGpp, etc.

For folding steps requiring foldases, the total cellular concentration (expression level) of the latter, and the concentration of those free to engage in folding processes (depending on cellular concentration of proteins in need of foldases) are critical. Shortage in any of the foldases, whether in the cytoplasm or in the endoplasmic reticulum (for proteins getting their final conformation in the ER), would produce misconformers.

5. Factors involved in conformational epigenetics

The fact that a majority of proteins, produced by a cell under normal conditions, are identified as misconformer while still attached to the ribosome, is a proof that to a large extend, the translation apparatus is under normal conditions, constitutively defective in the production of native conformations. According to the arguments developed above, the misconformations would result from a deregulation of the relative local translation rates. For some optimal setting of the translation apparatus and associated folding factors, the local translation rate of the mRNA would enable the native folding of the protein. Upon shifting away from this optimum, the translation machinery would progressively modify the local translation rate and start producing misconformers. Proteins may be more or less prone to misfolding, depending on primary sequence (constraints in amino acid types for instance), on multimerization, on need of assistance by foldases, etc.

Which epigenetic (to which we restrict here) events could force the translation apparatus to change the codon translation rate? As noticed above, it would be enough for these events to modify the expression of one or more of the various genes involved in the production of the translation apparatus. In response to a change of its metabolic state or its position in the cell cycle, the cell deeply modifies the relative and absolute composition of its translation equipment. For instance, as the growth rate of E. coli increases from 0.4 to 2.5 doublings/hr, the total tRNA population increases by 80%. However, the ribosome population increases by 250%, meaning that the total tRNA to ribosome ratio drops by 50%. The decrease affects preferentially rare tRNAs [10]. Many tRNAs are encoded by more than one gene and the corresponding promoters are subject to complex regulations in relation to growth [13]. The kind and level of modifications of tRNA bases also depend on the cell growth status [5]. Rapidly dividing cells tend to get rid of sophisticated base modifications, probably favouring high translation rates rather than high accuracy. Change of growth rate also affects the expression level of synthases and their efficiency in amino-acylating tRNAs [17].

Stimuli from its environment may force the cell to change its rate of protein production, sometimes extensively. For instance, hormone-dependent secretion can induce the production of massive amounts of proteins, which the cell must face by producing the translation machinery and supplies adapted for this burden.

The changes in the cell status just mentioned, and others, have the potential to affect co-translational folding and to lead to misconformation. We believe this to be the major contribution to the spontaneous, massive production of cotranslational misconformers [21].

In addition, if the change corresponds to an increase in protein production, larger amounts of foldases are required, and the more so as the amount of misfolded proteins increases concomitantly. It is likely that above a critical load of misfolded proteins, misfolding enters an avalanche regime, which could force the cell to store misfolded proteins in inclusion bodies. As an example, for identical cells growing under similar conditions, those expressing a heterologous gene at low rate and in moderate amounts may produce the soluble protein in its native conformation, whereas those expressing the same gene but at high rate and/ or in large amounts often produce inclusion bodies [4]. Similar observations were made upon tuning the expression of a homologous gene, which produced inclusion bodies above certain rate of protein production.

6. Can misfolded protein synthesis and accumulation be stopped?

There are three main ways to prevent conformational diseases: (i) avoid formation of misfolded versions of the disease-specific protein(s); (ii) in case this cannot be achieved, avoid that the protein adopts a misfolded conformation that can resist proteolysis; (iii) avoid high intra- or extra-cellular concentration of the misfolded protein, as it can favour the formation of insoluble, proteolysis-resistant aggregates.

As discussed above, misfolded proteins are produced either because of unscheduled, excessive gene expression which exceeds the cell's resources (aas, tRNAs, elongation factors, etc.), or to the fact some components of the translation apparatus (foldases) are missing or not available as required, or to changes in the metabolic state of the cell or its environment, which modify significantly the translation rate of the cell. In each case, molecular biology provides methods allowing identifying the culprit and quantitating its effects. At the cell level, transcriptome and proteasome techniques identify genes expressed abnormally. They can reveal also insufficient production of parts of the translation machinery or of the resources needed for translation. 2-D protein gels can be used to identify proteolysis-resistant components. Inducible expression of various foldases, born by adapted vectors, can allow probing the threshold level of a given foldase, needed by the cell for synthesis of native proteins. The results could be used to establish the cell's "phase diagram", showing for each of the foldases the threshold concentration below which protein misconformation sets on. Techniques of gene therapy could be used to supply the cell with, for instance, the defective or missing foldase(s).

Since misconformers can result from local modifications of the translation rate, a given protein can show up in multiple misfolded conformations, corresponding to alterations of the translation rate at various places of the mRNA. It would be of interest to sort out and identify varieties of the misconformers, which actually resist proteolysis, in order to identify the local mRNA or protein sequence involved. If for instance the misconformation were due to too high a local translation rate, the corresponding mRNA sequence would be uncovered for a relative long time and could therefore be available for tagging by RNAi and destroyed. RNAi techniques could also be used to control the level of expression of genes, which have a propensity to aggregate, thereby reducing the risk of aggregation, which follows the mass action law.

References

[1] I.A. Adzhubei, A.A. Adzhubei, S. Neidle, An integrated sequence-structure database incorporating matching mRNA sequence, amino acid sequence and protein three-dimensional structure, Nucleic Acids Res. 26 (1998) 327–331.

[2] C.B. Anfinsen, Principles that govern the folding of protein chains, Science 181 (1973) 223–230.

[3] F. Aslund, J. Beckwith, The thioredoxin superfamilly: redundancy, specificity and gray area genomics, J. Bacteriol. 181 (1999) 1375–1379.

[4] F. Banneyx, Recombinant protein expression in E. coli, Curr. Opin. Biotechnol. 10 (1999) 411–421.

[5] G.R. Björk, T. Rasmuson, Links between tRNA modification and metabolism, and modified nucleosides as tumour markers, in: H. Grosjean, R. Benne (Eds.), Modification and editing of RNA, ASM press, Washington, DC, 1998, pp. 471–492.

[6] D.T. Clarke, A.J. Doid, B.J. Stapley, G.R. Jones, The alpha helix folds in the millisecond time scale, Proc. Natl. Acad. Sci. 96 (1999) 7232–7237.

[7] P. Cortazzo, C. Cervenansky, M. Marin, C. Reiss, R. Ehrlich, A. Deana, Silent mutations affect in vivo protein folding in E. coli, Biochem. Biophys. Res. Commun. 293 (2002) 537–541.

[8] A. Deana, R. Ehrlich, C. Reiss, Silent mutations in the E. coli ompA leader peptide region strongly affect transcription and translation in vivo, Nucleic Acids Res. 26 (1998) 4778–4782.

[9] V. Emilsson, C.G. Kurland, Growth rate dependence of transfer RNA abundance in E. coli, EMBO J. 9 (1990) 4359–4366.

[10] V. Emilsson, A.K. Naslund, C.G. Kurland, Growth-rate dependent accumulation of twelf tRNA species in E. coli, J. Mol. Biol. 230 (1993) 483–491.

[11] T. Ikemura, Correlation between the abundance of E. coli tRNA and the occurrence of the respective codons in its protein genes, J. Mol. Biol. 146 (1981) 1–21.

[12] T. Ikemura, Correlation between the abundance of yeast tRNA and the occurrence of the respective codons in protein genes. Differences in synonymous codon choice patterns of yeast and E. coli with reference to the abundance of isoaccepting tRNAs, J. Mol. Biol. 158 (1982) 573–594.

[13] H. Inokuchi, F. Yamao, Structure and expression of prokaryotic tRNA genes, in: D. Söll, U.L. RajBhandary (Eds.), tRNA Structure, Biosynthesis and Function, ASM press, Washington, DC, 1994, pp. 17–30.

[14] A.A. Komar, T. Lesnik, C. Cullin, E. Guillemet, R. Ehrlich, C. Reiss, Differential resistance to proteinase K digestion of the yeast prion-like (Ure2p) protein synthesized in vitro in wheat germ extract and rabbit reticulocyte cell-free translation systems, FEBS Lett. 415 (1997) 6–10.

[15] A.A. Komar, T. Lesnik, C. Reiss, Synonymous codon substitution affect ribosome traffic and protein folding during in vitro translation, FEBS Lett. 462 (1999) 387–391.

[16] T. Lesnik, J. Solomovici, A. Deana, R. Ehrlich, C. Reiss, Ribosome traffic in E. coli and regulation of gene expression, J. Theor. Biol. 202 (2000) 175–185.

[17] T. Meinnel, Y. Mechulam, S. Blanquet, Aminoacyl tRNA synthetases: genes and regulation of expression, in: D. Söll, U.L. RajBhandary (Eds.), tRNA Structure, Biosynthesis and Function, ASM press, Washington, DC, 1994, pp. 251–292.

[18] Y. Nakamura, T. Gojobori, T. Ikemura, Codon usage tabulated from international DNA sequence database status for the year 2000, Nucleic Acids Res. 28 (2000) 292.

[19] C. Reiss, Selected topics in molecular biology, in need of "hard" science, in: M. Peyrard (Ed.), Non-Linear Excitations in Biomolecules, Springer-Verlag, 1995, pp. 29–54.

[20] S. Rüdiger, L. Germeroth, J. Schneider-Mergener, B. Bukau, Substrate specificity of the DnaK chaperone determined by screening cellulose-bound peptide libraries, EMBO J. 16 (1997) 1501–1507.

[21] U. Schubert, L.C. Anton, J. Gibbs, C.C. Norbury, J.W. Yewdell, J.R. Bennink, Rapid degradation of a large fraction of newly synthesized proteins by proteasomes, Nature 404 (2000) 770–773.

[22] J. Solomovici, T. Lesnik, C. Reiss, Does *E. coli* optimize the economics of the translation process, J. Theor. Biol. 185 (1997) 511–521.

[23] T.A. Thanaraj, P. Argos, Protein secondary structural types are differentially encoded in mRNA, Protein Sci. 5 (1996) 1973–1983.

[24] Y. Zhan, D.R. Olsen, K.B. Nguyen, E.T. Rhodes, D. Mascarenhas, Expression of eukaryotic proteins in soluble form, Protein Exp. Purif. 12 (1998) 159–165.

Neurotoxicology and Teratology 24 (2002) 563–564

NEUROTOXICOLOGY

AND

TERATOLOGY

www.elsevier.com/locate/neutera

Editorial

Tribute to Professor Toshiharu Nagatsu

The 9th International Catecholamine Symposium took place in Kyoto, Japan, as a joint symposium with the 5th International Conference on Progress in Alzheimer's and Parkinson's disease from March 31 to April 5, 2001. Professor Toshiharu Nagatsu, as the president of Catecholamine Symposium, organized the joint symposium with Professor Yoshikuni Mizuno, Juntendo University School of Medicine, Tokyo, Japan. In this symposium, the friends of Professor Nagatsu dedicated him a symposium, "Milestones in Catecholamine Research", and symposiums on "Catecholaminergic Neurotoxins and Neuroprotection" and "The Role of Neuromelanin in the Substantia Nigra in Parkinson's Disease" were organized. We, the speakers in these symposiums, his friends, colleagues and previous students decided to pay homage to Professor Nagatsu in the form of a collection of papers written by the entire scientific community well known in catecholamine research.

The lifelong devotion of Professor Nagatsu to the studies on neurotransmitter regulation carries an important impact to basic and clinical research. His discovery of tyrosine hydroxylase with Professor Sydney Udenfriend at early 1960s opened the golden days in catecholamine research. He and his colleagues in NIH and Japan successfully accomplished the purification and characterization of enzymes in catecholamine synthesis and metabolism in the 1970s and the 1980s. Then, he dared again to enter the novel field of molecular neuroscience. His research groups at the Nagoya University School of Medicine (1984–1991) and the Institute for Comprehensive Medical Science, Fujita Health University (1991–2000) succeeded in the cloning of tyrosine hydroxylase, aromatic L-amino acid decarboxylase, phenylethanolamine N-methyltransferase, dopamine β-hydroxylase and many important enzymes in the biosynthesis of biopterin, the cofactor of tyrosine, tryptophan and phenylalanine hydroxylases. His contribution to neurosciences was not limited to the basic sciences, it also extends to the studies of human diseases, such as Parkinson's disease and other neurological and psychiatric disorders. His overall scientific contributions are beyond administration, and he will always be considered as the main architect of catecholamine research.

We should emphasize that during the 1980s, Professor Nagatsu also started the studies on endogenous neurotoxins to cause Parkinsonism in humans, after the discovery of 1-methyl-4-phenyl-1,2,3,6-tetrahydropyridine (MPTP). He organized the research groups with the late Professor Mitsuo Yoshida, which brought about the findings on catechol and simple isoquinolines as endogenous dopaminergic neurotoxins. The studies on the intracellular mechanism of the neurotoxicity clarified the molecular mechanism underlying the neuroprotection in Parkinson's disease and other neurodegenerative disorders.

In this special issue of Neurotoxicology and Teratology, "Milestones in Research on Neurotoxins and Neuroprotection: A Tribute to Professor Toshiharu Nagatsu", the most active researchers in neurochemistry presented their recent advances in the research on neurotoxins and neuroprotection. The paper by Professor Nagatsu will give us an overview of the neurotoxins and the pathogenic factors in Parkinson's disease. This issue presents the new results on mechanisms of neurotoxicity by endogenous and xenobiotic toxins to cause selective cell death in Parkinson's disease, such as MPTP, isoquinolines, β-carbolines and L-DOPA. The possible role of N-methyltransferase and other enzymes in the biosynthesis and metabolism of the toxins in the pathogenesis of Parkinson's disease was also presented. In addition, the involvement of neuromelanin and manganese in the cell death of dopamine neurons was also reviewed. A quite new finding on the role of salsolinol in pituitary prolactin secretion suggested the physiological function of the dopamine-derived isoquinoline in the brain.

Recent research advances on the molecular mechanisms of neuroprotection were also included in this issue, and cyclic and aliphatic propargylamines were proven to suppress the death process in the animal and cellular models of Parkinson's disease. The neuroprotection by GDNF and other small molecules in neurodegenerative disorders was also reviewed.

It is our great pleasure to edit a review issue on these very important subjects as a Tribute to Professor Nagatsu. His more than 40 years of contributions to neuroscience constitute as "True Milestones of Catecholamine Research". We hope that this issue will increase our knowledge on neurodegeneration and neuroprotection, and that the basic and clinical researchers will find how profoundly we owe to Professor Nagatsu for the advances in these subjects.

Makoto Naoi[*],
Hasan Parvez,
Wakako Maruyama,
Michel A. Collins,
Moussa B.H. Youdim
(Editors)

---

[*] Corresponding author. Department of Brain Sciences, Institute of Applied Biochemistry, Mitake, Kani-gun, Gifu, Japan. Tel.: +81-574675500; fax: +81-574675310.

E-mail address: mnaoi@giib.or.jp

NEUROTOXICOLOGY

AND

TERATOLOGY

Neurotoxicology and Teratology 24 (2002) 565–569

www.elsevier.com/locate/neutera

Review article

Amine-related neurotoxins in Parkinson's disease
Past, present, and future

Toshiharu Nagatsu*

Institute for Comprehensive Medical Science, Fujita Health University, Toyoake, Aichi, 470-1192, Japan

Received 8 December 2001; accepted 24 January 2002

Abstract

Parkinson's disease (PD) is an aging-related movement disorder caused by a deficiency of the neurotransmitter dopamine (DA) in the striatum of the brain as a result of selective degeneration of nigrostriatal DA neurons. The molecular basis of the cell death of DA neurons is unknown, but one hypothesis is the presence of some amine-related neurotoxins that kill specifically nigrostriatal DA neurons over a long period of time. This neurotoxin hypothesis of PD started in the 1980s when 1-methyl-4-phenyl-1,2,3,6-tetrahydropyridine (MPTP) was discovered to produce acutely PD-like symptoms. Two groups of natural MPTP-like and amine-related neurotoxins have been investigated as endogenous candidate compounds: isoquinolines (IQs) and beta-carbolines. These neurotoxins are speculated to cause oxidative stress, mitochondrial dysfunction, apoptotic cell death, and PD symptoms. However, since PD is a neurodegenerative disorder that progresses slowly over a period of many years, a long-term study may be required to elucidate the neurotoxicity of such neurotoxins in relation to PD.
© 2002 Elsevier Science Inc. All rights reserved.

Keywords: Parkinson's disease; Neurotoxins; MPTP; Tetrahydroisoguinolines; Beta-carbolines; Rotenone; TaClo

1. Introduction

Parkinson's disease (PD) is a movement disorder caused by a deficiency of the neurotransmitter dopamine (DA) in the striatum of the brain, which occurs as a result of degeneration of DA neurons in the substantia nigra pars compacta (SNc) that project to the striatum (A9 neurons). PD is the second most common aging-related neurodegenerative disorder after Alzheimer's disease. The characteristic clinical symptoms of PD are movement disorders called parkinsonism, i.e., rigidity, bradykinesia, and resting tremor. Cell death of A9 DA neurons in idiopathic PD is believed to be caused both by genetic factors and by environmental factors. As a group of environmental factors, some neurotoxins are speculated to be related to PD. This neurotoxin hypothesis of PD started from the discovery of 1-methyl-4-phenyl-1,2,3,6-tetrahydropyridine (MPTP) in the 1980s [5,12]. Since then, natural MPTP-like neurotoxins in the brain and/or cerebrospinal fluid (CSF) of PD patients have been investigated, and

two groups of amine-related compounds, i.e., isoquinolines (IQs) and beta-carbolines, have been suggested as candidates of such neurotoxins producing PD [4,16].

2. MPTP-PD

MPTP is the only recognized synthetic neurotoxin that is capable of inducing PD symptoms in humans and nonhuman primates.

The first case of PD that appeared after intravenous injection of MPTP as a contaminant of 1-methy-4-phenyl-piperidine-4-carboxylic acid ethyl ester (meperidine), which is a synthetic heroin, was a 23-year-old chemistry student at Bethesda, MD. He synthesized that meperidine containing MPTP as a by-product and injected it intravenously into himself. L-3,4-Dihydroxyphenylalanine (DOPA) to supplement DA was effective in that patient. Kopin's group at the National Institutes of Health (NIH) identified MPTP in that meperidine preparation and reported the case in 1979 [5]. At that time, the NIH group examined MPTP neurotoxicity to produce PD symptoms in experimental animals including rats, but found them to be resistant to MPTP neurotoxicity.

---

* Tel.: +81-562-93-9392; fax: +81-562-93-8831.

E-mail address: tnagatsu@fujita-hu.ac.jp (T. Nagatsu).

Then, in 1983 in California, a group of young drug addicts acutely showed PD-like symptoms after self-administration of street batches of meperidine contaminated by MPTP. Like idiopathic PD, L-DOPA, which supplements DA in the brain, was an effective cure for the symptoms. These cases were reported in 1983 by Langston et al. [12], and since then the molecular mechanism of MPTP-elicited PD has been extensively studied.

Humans and monkeys are the species most sensitive to MPTP, and some nonprimate animals such as C57/BL mice are also susceptible to the neurotoxin. The mechanism of species differences in susceptibility to MPTP still remains unknown. One possibility is a species difference in the glial cells that produce cytokines and neurotrophins to regulate the pathway of apoptosis (programmed cell death), which may be triggered by MPTP. Another important factor may be the species difference in the DA-transporter (DA-T).

MPTP is highly lipophilic, and after its systemic administration, it rapidly crosses the blood–brain barrier to enter the brain. Once in the brain, MPTP, which is a proneurotoxin, is metabolized to 1-methyl-4-phenyl-2,3-dihydropyridinium (MPDP+) by the enzyme monoamine oxidase B (MAO-B), which is localized in the outer membrane of mitochondria, within non-DA cells such as glial cells or serotonin neurons. MPDP+ is then probably spontaneously oxidized to 1-methyl-4-phenylpyridinium (MPP+), the active neurotoxin. MPP+ is then taken up across the cell membrane to enter into the DA neurons via DA-T mostly at the nerve terminals in the striatum. As acute reactions, MPP+ is taken up into synaptic vesicles from the cytoplasm by vesicular monoamine transporter type 2 (VMAT2) to release DA from the nerve terminals; it also inhibits and inactivates tyrosine hydroxylase (TH) to decrease DA synthesis. In the chronic phase, MPP+ is transported from the nerve terminals of A9 DA neurons in the striatum to the cell bodies in the SNc by retrograde axonal flow. MPP+ is also accumulated within the inner mitochondrial membrane, where it inhibits Complex I (NADH ubiquinone reductase), one of the five enzyme complexes of the inner mitochondrial membrane involved in oxidative phosphorylation, interrupts electron transport, releases reactive oxygen species (ROS), and depletes ATP. Inhibition of mitochondrial Complex I opens mitochondrial permeability transition pore, and may trigger apoptotic cell death. Thus, MPP+ decreases DA acutely and chronically to produce PD-like symptoms.

3. Environmental or endogenous MPTP-like and amine-related neurotoxins

Assuming that some MPTP-like neurotoxins may trigger idiopathic PD, endogenous MPTP-like compounds have been examined in postmortem brains and in the CSF from patients with PD. Two MPTP-like compounds, IQs and beta-carbolines, were found in the brain and CSF from patients with PD and have been extensively examined.

3.1. IQs in PD

We found that MPP+, the active metabolite of MPTP, acutely inhibits the TH system in tissue slices of rat striatum. In screening for various MPTP-like compounds that inhibit the striatal TH system, we found tetrahydroisoquinoline (TIQ) and its derivatives to be active inhibitors [6]; and TIQ produced parkinsonism in monkeys [25]. A controversy has existed as to whether or not TIQs exist in the brain in vivo. Caution is always necessary to demonstrate that TIQs are not analytical artifacts generated during sample work-up and/or analysis. Gas chromatography–mass spectrometry is the most suitable analytical method for TIQs. TIQ and 1-methyl-TIQ (1-Me-TIQ) were found in both normal and PD brains as endogenous amines by gas chromatography–mass spectrometry [19,20]. Various TIQs were subsequently found in the brains of patients with PD and in those of control patients: TIQ, 1-Me-TIQ, N-Me-TIQ, N-Me-6,7-(OH)2-TIQ (N-Me-norsalsolinol), 1,N-(Me)2-6,7-(OH)2-TIQ (N-Me-salsolinol), 1-phenyl-TIQ, N-Me-1-phenyl-TIQ, and 1-benzyl-TIQ (1-Bn-TIQ) [16]. TIQs were also found in various foods in small concentrations. Exogenously administered TIQ easily crosses the blood–brain barrier and passes into the brain, although TIQs are metabolized in the liver to 4-OH-TIQs by debrisoquine hydroxylase (P-450 CYP2D6). However, TIQs in the brain are believed to be synthesized from precursor endogenous amines such as phenylethylamine or DA by enzymes. Only the (R) enantiomer, (R)-N-Me-6,7-(OH)2-TIQ (R-N-Me-salsolinol), is speculated to be synthesized enzymatically in the brain [18]. Among these TIQs in the brain, 1,N-(Me)2-6,7-(OH)2-TIQ (N-Me-salsolinol) [18], N-Me-6,7-(OH)2-TIQ (N-Me-norsalsolinol) [15], and 1-Bn-TIQ [10,11] have been extensively investigated as probable neurotoxins to cause PD. The levels of these TIQs were found to be increased in the brain and/or the CSF of patients with PD. Naoi et al. [18] proposed that R-N-Me-salsolinol may be an enzymatically formed PD-producing neurotoxin. R-N-Me-salsolinol produced PD symptoms in rats by stereotaxic injection into the striatum. After the injection, behavioral changes similar to those of PD were induced; and R-N-Me-salsolinol and its oxidation product, 1,N-(Me)2-6,7-(OH)2-isoquinolinium ion (IQ+), were accumulated in the striatum. Also, a definite amount of 1,N-(Me)2-6,7-(OH)2-IQ+ was detected in the SN. After continuous administration of R-N-Me-salsolinol to the rat striatum, selective destruction of TH-containing neurons was observed in the SN. This may suggest that R-N-Me-salsolinol, like MPP+, is taken up into the nerve endings of the nigrostriatal DA neurons in the striatum, and is oxidized to the N-Me-salsolinium ion, which is then transported to the cell bodies in the SNc to cause cell death, and subsequently, PD symptoms. N-Me-norsalsolinol and salsolinlol were identified in the CSF from PD patients, but not in that from normal controls. *N*-Me-salsolinol, but not salsolinol, was oxidized to *N*-Me-6, 7-(OH)2-IQ+ by MAO. The level of *R*-*N*-Me-salsolinol in PD patients was significantly higher than that in control subjects [15]. These results may suggest that endogenous, probably enzymatically synthesized, *R*-*N*-Me-salsolinol might be involved in the pathogenesis of PD. 1-Bn-TIQ was detected in mouse brain and also in the CSF from patients with PD [11]. The level of 1-Bn-TIQ was three times higher in the CSF of PD patients than in that of patients with other neurological diseases. Repeated administration of 1-Bn-TIQ induced PD-like behavior abnormalities in mice. It was also suggested that some cases of atypical PD in the French West Indies may have a link with the consumption as food of tropical plants that contain Bn-TIQs [3].

Among endogenous TIQs, 1-Me-TIQ, which was specifically distributed in the striatum of the human brain, is unique in that the level was not increased as those of other neurotoxic TIQs but was decreased in PD. 1-Me-TIQ also prevented PD-like behavior abnormalities produced by MPTP, TIQ, and 1-Bn-TIQ. These results suggest that 1-Me-TIQ may be preventive against PD and that the observed decrease in its level could be related to the pathogenesis of PD [10,20,24].

3.2. Beta-carbolines in PD

Beta-carbolines have structures similar to those of MPTP/MPP+, and may be synthesized in vivo from tryptophan via tryptamine [4,13]. Like MPTP, beta-carbolines may be precursor neurotoxins that are *N*-methylated and oxidized by MAO-B to form, in their case, beta-carbolinium ions, which may produce neuronal death and PD symptoms. Norharman, harman, 2-Me-norharmanium, and 2,9-(Me)2-norharmanium were found in the human brain and/or the CSF. Beta-carbolines are methylated to 2-mono-*N*-methylated beta-carbolines and then to neurotoxic 2,9-di-methylated beta-carbolinum ions in the brain. Although simple beta-carbolines and 2-monomethylated beta-carbolines, which were present in almost all subjects, showed slightly higher levels in PD patients than in controls, the difference was not significant. In contrast, a neurotoxic 2,9-dimethylated beta-carbolinium ion, 2,9-dimethylated norharman, was found in half of the PD patients examined, but was not at all found in non-PD patients. These results suggest that beta-carbonium compounds like IQ+ could also be candidate neurotoxins to produce PD [4,13].

1-Trichloromethyl-1,2,3,4-tetrahydro-beta-carboline (TaClo) is another neurotoxic beta-carboline [2]. TaClo can be synthesized in vivo from tryptamine and the synthetic aldehyde chloral after application of the hypnotic chloral hydrate or after exposure to the widely used industrial solvent trichloroethylene, which is metabolized to chloral. TaClo was reported to cause degeneration of DA neurons in vivo. However, since TaClo and the *N*-methylated derivative

had no DA-T-mediated neurotoxicity in cultured cells transfected with the human DA-T gene, they may not cause neurotoxicity by a mechanism analogous to that of MPTP/MPP+ involving the uptake into DA neurons by DA-T.

4. 6-Hydroxydopamine

6-Hydroxydopamine (6-OHDA) is a neurotoxin specific to DA neurons in vitro and in vivo [9]. It is a very unstable compound and is easily oxidized to produce ROS, which may kill DA neurons via apoptosis. 6-OHDA is formed from DA in vitro, but is believed not to be produced in vivo. However, since 6-OHDA is a neurotoxin selective for DA neurons in vivo, it is widely used to produce hemiparkinsonian animal models by stereotaxic injection of it into the brain. Since 6-OHDA does not cross the blood–brain barrier, it is directly injected stereotaxically into the striatal region of animals to produce hemiparkinsonian animal models. The resultant cell death of the DA neurons on the injected side causes a decrease in the DA concentration in the presynaptic nerve terminals and also an increase in the number of DA receptors, i.e., supersensitivity, in the postsynaptic neurons in the striatum. Thus, the 6-OHDA-induced hemiparkinsonian rats show turnings to the injured side (ipsilateral rotations) by injection of DA-releasing drugs such as methoamphetamine due to decreased DA, but to the normal side (contralateral rotations) by injection of DA receptor agonists such as apomorphine due to this supersensitivity. 6-OHDA-induced hemiparkinsonian rats are useful experimental models for pharmacological studies of DA agonists and antagonists. However, since 6-OHDA may not be formed in vivo, it cannot be considered a causative neurotoxin producing idiopathic PD.

5. Rotenone-induced PD in rats

Epidemiological studies have suggested that insecticide exposure is associated with an increased risk of developing PD. Rotenone is a naturally occurring, lipophilic compound from the roots of certain plants (Derris species), and is used as the main component of many insecticides. Rotenone is not structurally related to amines, but is a specific inhibitor of Complex I of the inner mitochondrial membrane involved in oxidative phosphorylation. Interestingly, a selective defect in Complex I of the mitochondria in the nigrostriatal DA neurons has been reported from several laboratories. Betarbet et al. [1] reported that in Lewis rats, chronic, systemic inhibition of Complex I by the lipophilic insecticide, rotenone, caused highly selective degeneration of nigrostriatal DA neurons with behavioral PD symptoms of hypokinesia and rigidity. Important morphological findings in these rotenone-treated rats were that the nigrostriatal DA neurons had accumulated fibrillar cytoplasmic inclusions containing ubiquitin and alpha-synuclein, similar to Lewy bodies in idiopathic PD. This finding is in contrast to MPTP-induced animal models of PD, in which typical cytoplasmic inclusions that closely resemble Lewy bodies are not observed. One reason for the differences between MPTP-induced PD and rotenone-induced PD may be that rotenone produces a systemic Complex I defect in contrast to the defect selective for DA neurons by MPTP, because the extremely lipophilic rotenone, in contrast to MPTP, crosses the plasma membrane easily without depending on DA-T. The production of Lewy body-like cytoplasmic inclusions in the rotenone-induced PD in animals is interesting, since recent discoveries of the two causative gene products of familial PD, i.e., Parkin, which is a ubiquitin ligase (E3) in autosomal recessive juvenile PD [8,22], and alpha-synuclein [21], which is a substrate of Parkin [23] in autosomal dominant PD, indicate that failure of the ubiquitin–proteasome system may be common in both idiopathic PD and familial PD [14].

6. Neurotoxins and apoptosis in PD

Recent reports by us and others suggest that the neuronal death in idiopathic PD may be due to apoptosis (programmed cell death) [17]. In vitro experiments using neurotoxins such as MPTP, 6-OHDA, and R-N-Me-salsolinol in cell culture systems indicate that the death of DA neurons are apoptotic in nature. Such neurotoxins may also produce apoptotic cell death of the nigrostriatal DA neurons in vivo. In the nigrostriatal region of the postmortem brain from patients with idiopathic PD, the levels of proapoptotic cytokines such as TNF-alpha and IL-1beta were increased. On the contrary, the levels of antiapoptotic neurotrophins, such as brain-derived neurotrophic factor (BDNF) and nerve growth factor (NGF), were decreased. Furthermore, various factors from the upstream to downstream pathway of apoptosis were found to be activated specifically in the nigrostriatal region, i.e., soluble Fas (sFas), TNF-receptor R1 (TNF-R1, p55), Bcl-2, caspase 1, and caspase 3. Similar results as obtained from postmortem human PD brains were also obtained from MPTP-PD mice and 6-OHDA-PD rats. These results suggest that in idiopathic PD, some unknown environmental factors such as neurotoxins may probably trigger microglia to release proinflammatory cytokines, which may then subsequently activate the apoptotic pathway in the nigrostriatal DA neurons. The decrease in the level of DA in the striatum to less than 20% of the normal level as the result of DA neuron death may produce PD symptoms. Another important concept in the neuronal death in familial PD is failure of the ubiquitin–proteasome system [14]. Parkin protein is ubiquitin ligase (E3) [22], and alpha-synuclein and unfolded Parkin-associated endothelin receptor-like receptor (Pael receptor) are substrates of Parkin [7]. Mutations and loss of activities of Parkin and the resultant failure of the ubiquitin–proteasome pathway may cause the accumulation of abnormal insoluble proteins and may also

cause apoptotic cell death. The final part of the pathway of apoptotic cell death could be common to both idiopathic PD and familial PD. The differences between them may be the initial part of this pathway where apoptosis starts.

7. Conclusions

As a cause of idiopathic PD, MPTP-like neurotoxins have been implicated. Several IQs and beta-carbolines similar to MPTP have been identified in postmortem brain and in the CSF from patients with idiopathic PD. These neurotoxins cause mitochondrial dysfunction, inhibition of Complex I, oxidative stress, and apoptosis in vitro in cultured DA neurons. In vivo experiments also indicate that such MPTP-like neurotoxins may cause apoptotic death of DA neurons. However, MPTP is the only neurotoxin that has definitely been proven to cause PD in humans, with its finding occurring by serendipity. The toxicity of these neurotoxins that were identified in the brain or in the CSF is weak compared with that of MPTP, which may explain the very slow progression of PD. Most experimental evidences suggest that such MPTP-like neurotoxins may be synthesized in the brain. Like MPTP, such neurotoxins may be activated first by N-methyltransferase and then by MAO. If this hypothesis is true, the activity of N-methyltransferase may be important in PD. Deprenyl, which is a MAO inhibitor specific for MAO-B, is assumed to be effective for PD, not only by preventing the degradation of DA due to inhibition of MAO, but also by neuroprotective action preventing apoptosis. In fact, deprenyl was reported to decrease the level of neurotoxic 1-Bn-TIQ and to increase the content of neuroprotective 1-Me-TIQ in MPTP-elicited PD in mice [10]. Furthermore, deprenyl was reported to increase the level of neurotrophins with antiapoptotic activity such as BDNF, NGF, and GDNF, which levels in the nigrostriatum in PD were found to be decreased.

The neurotoxin hypothesis for PD has contributed to the elucidation of neuronal death in PD, especially by the studies on the action of MPTP. Since PD is a slowly progressing neurodegenerative disorder, further studies are needed to elucidate the relationship between MPTP-like neurotoxins and PD.

References

[1] R. Betarbet, T.B. Sherer, G. Mackenzie, M. Garcia-Osuna, A.V. Panov, J.T. Greenamyre, Chronic systemic pesticide exposure reproduces features of Parkinson's disease, Nat. Neurosci. 3 (2000) 1301–1306.

[2] G. Bringmann, R. Brückner, M. Münchbach, D. Feineis, R. God, W. Wesemann, C. Grote, M. Herderich, S. Diem, K.-P. Lesch, R. Mössner, A. Storch, 'TaClo', a chloral-derived mammalian alkaloid with neurotoxic properties, in: A. Storch, M.A. Collins (Eds.), Neurotoxic Factors in Parkinson's Disease and Related Disorders, Kluwer Academic Publishing/Plenum, New York, 2000, pp. 145–149.

[3] D. Caparros-Lefebvre, A. Elbaz, The Caribbean Parkinsonism Study Group, Possible relation of atypical parkinsonism in the French West Indies with consumption of tropical plants: A case–control study, Lancet 354 (1999) 281–285.

[4] M.A. Collins, E.J. Neafsey, β-Carboline analogues of MPP⁺ as environmental neurotoxins, in: A. Storch, M.A. Collins (Eds.), Neurotoxic Factors in Parkinson's Disease and Related Disorders, Kluwer Academic Publishing/Plenum, New York, 2000, pp. 115–130.

[5] G.C.B. Davis, A.C. Williams, S.P. Markey, M.H. Ebert, E.D. Caine, C.M. Reichert, I.J. Kopin, Chronic parkinsonism secondary to intravenous injection of meperidine analogues, Psychiatry Res. 1 (1979) 249–254.

[6] Y. Hirata, H. Sugimura, H. Takei, T. Nagatsu, The effects of pyridinium salts, structurally related compounds of 1-methyl-4-phenylpyridinium ion (MPP⁺), on tyrosine hydroxylation in rat striatal tissue slices, Brain Res. 397 (1986) 341–344.

[7] Y. Imai, M. Soda, H. Inoue, N. Hattori, Y. Mizuno, R. Takahashi, An unfolded putative transmembrane polypeptide, which can lead to endoplasmic reticulum stress, is a substrate of parkin, Cell 105 (2001) 891–902.

[8] T. Kitada, S. Asakawa, N. Hattori, H. Matsumine, Y. Yamamura, S. Minoshima, M. Yokochi, Y. Mizuno, N. Shimizu, Mutations in the *parkin* gene cause autosomal recessive juvenile parkinsonism, Nature 392 (1998) 605–608.

[9] R.M. Kostrzewa, D.M. Jacobowitz, Pharmacological actions of 6-hydroxydopamine, Pharm. Rev. 26 (1974) 199–288.

[10] Y. Kotake, Y. Tasaki, M. Hirobe, S. Ohta, Deprenyl decreases an endogenous parkinsonism-inducing compound, 1-benzyl-1,2,3,4-tetrahydroisoquinoline, in mice: In vivo and in vitro studies, Brain Res. 787 (1998) 341–343.

[11] Y. Kotake, Y. Tasaki, Y. Makino, S. Ohta, M. Hirobe, 1-Benzyl-1,2,3,4-tetrahydroisoquinoline as a parkinsonism-inducing agent: A novel endogenous amine in mouse brain and parkinsonian CSF, J. Neurochem. 65 (1995) 2633–2638.

[12] J.W. Langston, P. Ballard, J.W. Tetrud, I. Irwin, Chronic parkinsonism in humans due to a product of meperidine-analog synthesis, Science 219 (1983) 979–980.

[13] K. Matsubara, N-Methyl-β-carbolinium neurotoxins in Parkinson's disease, in: A. Storch, M.A. Collins (Eds.), Neurotoxic Factors in Parkinson's Disease and Related Disorders, Kluwer Academic Publishing/Plenum, New York, 2000, pp. 131–143.

[14] K.S.P. McNaught, C.W. Olanow, B. Halliwell, O. Isacson, P. Jenner, Failure of the ubiquitin–proteasome system in Parkinson's disease, Nat. Rev. Neurosci. 2 (2001) 589–594.

[15] A. Moser, D. Kömpf, Presence of methyl-6,7-dihydroxy-1,2,3,6-tetrahydroisoquinolines, derivatives of the neurotoxin isoquinoline, in parkinsonian lumber CSF, Life Sci. 50 (1992) 1885–1891.

[16] T. Nagatsu, Isoquinoline neurotoxins in the brain and Parkinson's disease, Neurosci. Res. 29 (1997) 99–111.

[17] T. Nagatsu, M. Mogi, H. Ichinose, A. Togari, Changes in cytokines and neurotrophins in Parkinson's disease, J. Neural Transm., Suppl. 60 (2000) 277–290.

[18] M. Naoi, Y. Maruyama, P. Dostert, Y. Hashizume, D. Nakahara, T. Takahashi, M. Ota, Dopamine-derived endogenous 1(R), 2(N)-dimethyl-6,7-dihydroxy-1,2,3,4-tetrahydroisoquinoline, N-methyl-(R)-salsolinol, induced parkinsonism in rats: Biochemical, pathological and behavioral studies, Brain Res. 709 (1996) 285–295.

[19] T. Niwa, N. Takeda, N. Kaneda, Y. Hashizume, T. Nagatsu, Presence of tetrahydroisoquinoline and 2-methyl-tetrahydroisoquinoline in parkinsonian and normal human brains, Biochem. Biophys. Res. Commun. 144 (1987) 1084–1089.

[20] S. Ohta, M. Kohno, Y. Makino, O. Tachikawa, O. Hirobe, Tetrahydroisoquinoline and 1-methyl-tetrahydroisoquinoline are present in the human brain, Biomed. Res. 8 (1987) 453–456.

[21] M.H. Polymeropoulos, C. Lavedan, E. Leroy, S.E. Ide, A. Dehejia, A. Dutra, B. Pike, H. Root, J. Rubenstein, R. Boyer, E.S. Stenroos, S. Chandrasekharappa, A. Athanassiadoce, T. Papapetropoulos, W.G. Johnson, A.M. Lazzarini, R.C. Duvoisin, G. DiIoro, L.I. Golbe, R.L. Nussbaum, Mutation in the α-synuclein gene identified in families with Parkinson's disease, Science 276 (1997) 2045–2047.

[22] H. Shimura, N. Hattori, S. Kubo, Y. Mizuno, A. Asakawa, S. Minoshima, N. Shimizu, K. Iwai, T. Chiba, K. Tanaka, T. Suzuki, Familial Parkinson disease gene product, parkin, is a ubiquitin-protein ligase, Nat. Genet. 25 (2000) 302–305.

[23] H. Shimura, M.G. Schlossmacher, N. Hattori, M.P. Frosch, A. Trockenbacher, R. Schneider, Y. Mizuno, K.S. Kosik, D.J. Selkoe, Ubiquitination of a new form of α-synuclein by Parkin from human brain: implications for Parkinson's disease, Science 293 (2001) 263–269.

[24] Y. Tasaki, Y. Makino, S. Ohta, M. Hirobe, 1-Methyl-1,2,3,4-tetrahydroisoquinoline, decreased in 1-methyl-4-phenyl-1,2,3,6-tetrahydropyridine-treated mouse, prevents parkinsonism-like behavior abnormalities, J. Neurochem. 57 (1991) 1940–1943.

[25] M. Yoshida, T. Niwa, T. Nagatsu, Parkinsonism in monkeys produced by chronic administration of an endogenous substance of the brain, tetrahydroisoquinoline: The behavioral and biochemical changes, Neurosci. Lett. 119 (1990) 109–113.

NEUROTOXICOLOGY

AND

TERATOLOGY

Neurotoxicology and Teratology 24 (2002) 571–577

www.elsevier.com/locate/neutera

Review article

Potential neurotoxic "agents provocateurs" in Parkinson's disease

Michael A. Collins*, Edward J. Neafsey

*Department of Cell Biology, Neurobiology, and Anatomy, Division of Biochemistry, Loyola University School of Medicine,
2160 South First Avenue, Maywood, IL 60153, USA*

Received 16 January 2002; accepted 24 January 2002

Dedicated to Professor Toshiharu Nagatsu on the occasion of his 70th birthday

Abstract

Idiopathic Parkinson's disease (PD), one of the most common neurodegenerative disorders associated with aging, is characterized neurochemically by abnormal and profound loss of nigrostriatal dopamine (DA) neurons. A prominent current view is that the excessive degeneration of the dopaminergic system is the outcome of extended insults by environmental neurotoxins or endogenous neurotoxic factors in genetically vulnerable or susceptible individuals. Recent insights into the identities and mechanisms of potential neurotoxic species, which span pesticides, environmental contaminants including heterocyclic amines with β-carboline (βC) and isoquinoline (IQ) structures, endogenous DA metabolites or intermediates, neuromelanin, metals, and infectious agents, are presented. © 2002 Elsevier Science Inc. All rights reserved.

Keywords: Parkinsonism; Pesticides; Dopamine; Carbolines; Isoquinolines

1. Introduction

Considerable progress has been made toward identifying neurotoxic agents possibly associated with the etiology and/or progression of idiopathic, aging-related Parkinson's disease (PD). The research that spurred this progress is both neurobiological and epidemiological. First and foremost, the discovery of 1-methyl-4-phenyl-1,2,3,6-tetrahydropyridine (MPTP) and its monoamine oxidase (MAO)-dependent activation to MPP^+, a pyridinium cation with a predominantly mitochondrial-based toxic mechanism in the nigrostriatal neurons, greatly helped to focus attention on pesticides and contaminants in the environment that could chronically imitate this process and on underlying mitochondrial deficiencies or predisposition. The key research regarding MPTP is summarized in Speciale's article (this issue).

Second, epidemiological findings in the past decade have supported the view, perhaps to a greater extent than with other neurodegenerative disorders, that environmental factors are at least as important as genetics in the etiology of late onset PD. Notably, an oft-cited large twin study concluded that heredity apparently does not have a major role in the etiology of typical disease, leaving nongenetic factors (e.g., pesticide exposure) as the primary antecedents [85]. Most recently, a study of 310 orchardists indicated that parkinsonism may be associated with long-term occupational exposure to pesticides [28]. A meta-analysis of peer-reviewed studies on PD in the past decade concluded that rural environments and pesticides constitute significant risk factors [70].

A number of potential agents emerge as candidate neurotoxins, some with poorly understood preference for the nigrostriatal dopaminergic system (Table 1). Most attention has focused on exogenous (environmental) compounds or metals such as iron or manganese, either naturally present in or introduced into our environments, and these are initially discussed. Second, a substantial part of the repertoire is constituted by endogenous and primarily dopamine (DA)-derived molecules. In some cases, the above exogenous or endogenous factors require important metabolic steps for activation, brain accumulation, and selective uptake, e.g., oxidation (including nitrogen or sulfur), N-methylation (quaternization), and/or conjugation. Last, there is the

---

* Corresponding author. Tel.: +1-708-216-4560; fax: +1-708-216-8523.
E-mail address: mcollin@lumc.edu (M.A. Collins).

Table 1
Potential neurotoxic factors linked to PD

Pesticides/ herbicides	Lipophilic industrial compounds	Endogenous DA-related compounds	Heterocyclic alkaloidal compounds (environmental and endogenous)	Infectious agents	Metal ions
Paraquat DTCs including maneb Rotenone Dieldrin	polychlorinated biphenyl heptachlor Tinuvin 123	6-HODA thiocatechol adducts of DA, DOPA, and DOPAC benzthiazoles + mercapturates 3-nitroso-tyrosine melanin dopaminachrome dopaldehyde	βCs: 2,9-di-N-methyl-norharman and harman cations TaClo TaBro IQs: TIQ 1-benzyl-TIQ N-methyl-salsolinols THP reticuline or other Annonaceae alkaloids	Gram-negative bacteria (N. asteroides) BV (G. vaginalis)	iron manganese (copper)

long-held idea that, in some instances, infectious agents are the culprits in PD.

2. Environmental neurotoxic factors

Pesticides/herbicides were associated with idiopathic PD shortly after the MPTP discovery [3,7]. One of the most acknowledged is paraquat, an N-methylated dipyridinium herbicide that is in the same chemical group as MPP^+ and has been linked to PD in epidemiological studies [46]. Unexpectedly and unlike MPP^+, the bication penetrates the BBB [10,22]. However, while paraquat can cause neuronal degeneration, by directly or indirectly promoting oxidative stress [68,83] and possibly by stimulating neuronal synuclein aggregation [54], its actions are reported to lack specificity for the basal ganglia [22].

Of importance is that paraquat's toxicity toward nigrostriatal neurons appears to be selectively enhanced in mice when given with maneb, a manganese-containing dithiocarbamate (DTC) pesticide that at the dose used is not neurotoxic [87,88]. In a more recent study, neonatal exposure to a paraquat–maneb combination enhanced susceptibility to subsequent treatments in adulthood [51]. Considerably earlier, a DTC was found to unmask the selective neurotoxicity of MPTP in mice [24,44], and DTC-based pesticides combined with other moderately toxic environmental agents were suggested to cause synergistic induction of nigrostriatal degeneration [7]. The potentiating effect of DTC analogs correlates with their inhibition of mitochondrial respiration at complex IV [2], but also may be related to their inhibition of enzymes such as cytochrome P450s that deactivate neurotoxins [93]. However, to be noted is that DTCs alone, either containing (the case with maneb) or lacking manganese, were relatively neurotoxic in mesencephalic–striatal cocultures [78]. There was little selectivity, since both DA and GABA terminals were equally affected, and the manganese component apparently did not contribute to the toxicity.

The plant-derived pesticide rotenone has emerged as a likely environmental candidate only in the past several years, although its highly potent inhibition of mitochondrial complex I respiration has been known for decades. Rotenone parallels MPP^+ and 6-hydroxy-DA (6-HODA) in damaging rat striatal DA terminals after stereotaxic injection into the median forebrain bundle [42] and is neurotoxic to neuroblastoma cell cultures [76]. Rotenone perfusion with intravenous minipumps in rats at a dose of 10–18 mg/kg/day for a week resulted in damage to the striatum but not to the substantia nigra — a pattern that is more analogous to manganese and carbon monoxide than to MPTP [29]. However, a lower rotenone dose (2–3 mg/kg/day) with essentially the same delivery mode, duration, and species induced selective nigral DA cell loss, nigral cytoplasmic inclusions resembling Lewy bodies (pathologic hallmarks of PD), and parkinsonian motor deficits along with the striatal lesions [6,46]. Although Lewis rats were stated to be more reproducible than Sprague–Dawley rats [6], a recent review nevertheless concluded that the high variability in damage would preclude the rotenone rat model's usefulness in anti-PD drug testing [4]. It is also curious that a similar rotenone dose given every other day for 3 weeks to C57Bl/6J mice, a strain more susceptible to MPTP neurotoxicity than are rats or most other murine strains [41,80], failed to provoke degeneration of nigrostriatal neurons [86]. Despite these concerns, the rotenone model is gaining favor for experimental studies with other potential environmental neurotoxins [38,46].

Among chlorinated pesticides, dieldrin surfaces prominently as a potential parkinsonian neurotoxin. Dieldrin was reported to be elevated in PD brain [23,30], despite the fact that its agricultural use has been banned in the US for several decades. It is also a relatively potent and selective dopaminergic toxin in mesencephalic cultures, with EC50 values in the low micromolar range [79].

Although not a pesticide, ortho-polychlorinated biphenyl (o-PCB) is a widespread hydrophobic contaminant that has been largely unappreciated by PD researchers. It displays selective toxicity toward DA neurons in culture, particularly in combination with methylmercury, and also in utero, causing DA deficits into adulthood [74]. Parenthetically, combinations of any of the above pesticides with the also ubiquitous heptachlor may be important to consider; studies indicate that heptachlor alone can exert selective striatal neurotoxic effects in mice [47] and it could potentiate the actions of other environmental lipophilic contaminants by increasing DA transporters and inhibiting mitochondrial respiration [61].

A piperidine derivative used as a light stabilizer in plastics manufacturing, Tinuvin 123 (bis-[1-octyloxy-2,2,6, 6-tetramethyl-4-piperidinyl]sebacate), has been proposed as an industrial parkinsonian neurotoxin [57], since the compound displays selective toxicity toward dopaminergic neurons in culture and in vivo following acute central administration into rat mesencephalon. However, the selectivity of Tinuvin-induced neurodegeneration was not confirmed in the in vivo model [45]. The possibility remains that, much like the case with rotenone (above), chronic, systemic low-dose exposure to Tinuvin 123 might nevertheless selectively endanger nigrostriatal neurons.

Certain environmental pyrido-indoles (carbolines), non-pesticidal in nature and, like paraquat, remarkably similar in structure to MPP$^+$ when N-methylated, are potential parkinsonian agents [20,21]. The carboline class receiving the most interest, β-carbolines (βCs)—specifically, norharman and harman—are ubiquitous. They form pyrolytically during grilling or broiling of tryptophan-containing foodstuffs and in tobacco smoking; they are also produced by industrial pyrolysis processes, arising in smoke, industrial wastes, and even groundwater [21,92]. There is an appreciable literature concerned with the (co)mutagenic and sometimes carcinogenic actions of these and related "heterocyclic aromatic amines." In terms of neurodegeneration, βCs are viewed as pro-toxins, activated by one or possibly two N-methylation steps in the CNS to form MPP$^+$-like N-methyl quaternary compounds.

Two of these βCs, the 2,9-*N,N*-dimethylated-norharman and 2,9-*N,N*-dimethylated-harman cations, are reported to be elevated in PD CSF [60]; also, an N-methylating enzyme activity that forms them is higher in the brain from PD subjects than controls [34]. Depending on preexposure interval, the above dimethylated βC cations are somewhat more effective than MPP$^+$ as mitochondrial respiratory inhibitors, and they are relatively potent, albeit less selective, dopaminergic neurotoxins in vitro (several orders of magnitude less effective than MPP$^+$) [21,58]. A relatively unrecognized parkinsonian model was produced by subchronic intraperitoneal treatment of C57/Bl mice with norharman or 2-methylated norharman [59]. In a human brain phage display library study of potential cellular targets for the N-methylated norharman cations, the βCs bound avidly to tubulin, also a binding partner of α-synuclein, and to paraoxonase, a pesticide-catabolizing enzyme, raising questions about interactions with pesticides [35]. In light of this and the aforementioned paraquat–maneb experiments, it would be enlightening to determine the effects of N-methylated βC cations in mice cotreated with maneb. Whether epidemiological relationships exist between envir-

onmental βC and pesticide exposure/intake and PD risk is an unexplored question.

Other βCs (and tetrahydro-βCs) that might have roles in PD include 3-carboxylated derivatives of tryptophan and DOPA that, like norharman, are formed in brewing and fermentation processes and are found in smoked foods and cheeses [21]. There has been little if any research on the neurotoxic potencies of these compounds or their N-methylated forms. A further interesting 3-carboxylated βC warranting further study, originally developed as an anxiogenic GABA receptor inverse agonist but found to be an industrially derived xenobiotic with neurotoxic effects in vitro, is *N*-methyl-βC-3-carboxamide or FG 7142 [53,96].

TaClo, a trichloromethyl-1,2,3,4-tetrahydro-βC reported to be produced from Pictet–Spengler condensations of tryptamine with chloral, is found in chloral-exposed individuals and exerts DA neurotoxicity in the absence of N-methylation [9,39]. Also, even greater neurotoxic effects were observed with the analogous tribromomethyl derivative, TaBro, which results from condensations of the tribromoaldehyde, bromal [8]. Relative to norharman-related βCs with their environmental ubiquity, the involvement of TaClo or TaBro in the etiology of PD appears limited in scope. Only a small fraction of PD sufferers could possibly have been or are being exposed to chloral, its potential precursors such as trichloroethylene [40] or bromal.

3. Endogenous DA-related neurotoxic compounds

DA itself, although greatly reduced in advanced PD because of neurodegeneration, has been implicated as an active player in the early neurotoxic process (e.g., Ref. [31]). Initial loss of nigrostriatal neurons (perhaps involving environmental agents discussed herein) would result in a homeostatic response of augmented DA turnover; i.e., increased DA formation by tyrosine hydroxylase and metabolism primarily by MAO. Both tyrosine hydroxylase and MAO generate reactive oxygen species (ROS) that promote oxidative stress and further endanger nigrostriatal neurons [1,16,97]. Also not to be discounted is that excess nonvesicular DA, as a catechol, can generate ROS nonenzymatically, particularly in the presence of ferrous ion or other metals. In addition, oxidized and/or cyclic derivatives of DA have been postulated as endogenous PD toxins because of their neurotoxic actions. Most prominent is 6-HODA, a well-known catecholamine neurotoxin that potentially arises in brain under oxidative stress conditions [52,69]. During its facile autoxidation, 6-HODA forms ROS and quinones that covalently modify neuronal proteins [17,69].

Oxidation of DA by tyrosinase, a key enzyme in neuromelanin formation, also produces ROS [1]. One of the organic intermediates, dopaminachrome, a cyclized precursor of neuromelanin, is toxic to neuroblastoma cells [33], but the mechanism remains obscure. Furthermore, excess neuromelanin itself is neurotoxic particularly in the presence of iron, perhaps through ROS generation [27]. However, DA-derived melanin initially was considered to be neuroprotective in the PD process, and recently has been shown to scavenge peroxynitrite in vitro, attenuating its nitrating and oxidizing actions [82]. This would argue against the notion that polymeric neuromelanin has a deleterious role in PD.

Thiocatechol adducts derived from quinoidal DA, DOPA, and DOPAC, elevated in PD brain [81], have received increased attention; they and their derived benzothiazines and mercapturates are neurotoxic [63,75]. MAO-mediated deamination of DA generates the aldehyde intermediate, dopaldehyde, which has had a modest history in PD research. Its importance initially was related to the fact that it is a coreactant in the cellular formation of the neurotoxic benzyl-isoquinoline (IQ), tetrahydropapaveroline (THP), during L-DOPA therapy or treatment [15,73] (see THP discussion below). Dopaldehyde is now being reinvestigated because of its direct neurotoxic effects in rats and in neural cultures, as well as its potentiation of respiratory inhibition [11,48]. The DA precursor, tyrosine, when 3-nitrosylated via nitric oxide mechanisms (usually considered a biomarker for peroxynitrite reactions), has been proposed as a possible neurotoxin in PD, since it caused nigrostriatal damage and behavioral deficits after intrastriatal injection [62]. As is the case to some extent with the thiol adducts and catechol aldehydes, the specificity of the neurotoxicity requires clarification, as does the underlying mechanism(s).

IQs constitute a structurally interesting group arising primarily from endogenous DA condensations and they are also plant alkaloids along with βCs [18,65]. The relevant IQs derived from DA are N-methyl-norsalsolinol, elevated in PD CSF [64], and N-methyl-salsolinol, and in particular, the R stereoisomer. The latter IQ, formed centrally from salsolinol by brain N-methyltransferase activity, is neurotoxic in vivo and induces apoptosis in vitro [56,66]. A third trace DA-derived (tetrahydro) IQ found in human CNS, THP, can induce apoptotic/necrotic neuronal loss, possibly via oxidation products [55,84]. The view that THP is an endogenous neurotoxin must be reconciled with the fact that brain THP levels are enhanced by L-DOPA treatment [15].

Reticuline, a phenolic 1-benzylated tetrahydro-IQ constituent of Annonaceae fruits and derived teas (as well as a likely intermediate in the morphinan biosynthesis pathway in poppies), biosynthesized in theory via THP and therefore DA, has been suggested as a possible cause of atypical parkinsonism occurring in the French West Indies [13]. However, while reticuline does induce neurotoxicity in vitro at high micromolar concentrations [50], it is more reasonable that the alkaloid primarily exerts antioxidant and neuroprotectant effects and that other Annonaceae components might be the responsible toxic agents [19]. Two further IQs derived not from DA but from phenethylamine, namely, N-methyl-IQ cation and 1-benzyl-tetrahydro-IQ, have been proposed as possible parkinsonian toxins, since they might be present in PD brain and induce parkinsonism

in animal models [65,67]. More thorough summaries of IQs and their possible roles in neurodegeneration are elsewhere in this volume.

4. Possible infectious agents and PD

Although the apparent viral epidemic of *encephalitis lethargica* (von Economo's disease) in the first decades of the past century caused thousands of cases of parkinsonism [72], the idea that CNS infections could underlie idiopathic PD has not been widely accepted. One interesting research direction is the possible role of pervasive environmental bacteria in predisposing an individual to eventual onset of PD. Mice inoculated with sublethal doses of a Gram-negative bacteria pervasive throughout the soil, *Nocardia asteroides*, gradually developed slowed movement and motor disorders, and showed selective loss of nigrostriatal dopaminergic neurons [5,43].

Another infectious trigger for PD under consideration is bacterial vaginosis (BV), which is relatively common in pregnancies (incidence approaching 15%) and produces a Gram-variable bacteria, *Gardinerella vaginalis*, as well as Gram-negative bacteria [26,91]. While it can be either symptomatic or asymptomatic, BV is associated with elevated levels of pro-inflammatory cytokines and endotoxin in the chorioamniotic environment. BV infection could lead to abnormal development of the nigrostriatal system and a predisposition toward the later onset of PD, perhaps in concert with the environmental neurotoxins discussed herein [14].

5. Environmental metal ions

There has been vigorous investigation into the possibility that selected essential metal ions play roles in abnormal nigrostriatal degeneration. Iron and manganese have received the most attention because of their ability to stimulate destructive hydroxyl radical formation via the Fenton reaction, and the former metal is discussed in greater detail elsewhere in this volume. Total brain iron content is largely established during development; however, some redistribution or dysregulation must occur leading up to PD because the brains of afflicted individuals show increased levels of ferric iron in the basal ganglia, and most of the basal ganglia iron increase is manifested in the substantia nigra [36,77]. Accordingly, of great interest is the possible association between iron and nigral melanin as a source of free radicals [27]. Also, iron is postulated to interact with DA or derived catecholic chrome derivatives discussed above, specifically by reacting with hydrogen peroxide that can be generated to produce short-lived hydroxyl radicals [95].

Manganese, a very early risk factor for PD [25], has received attention of late because of environmental concern arising from increased use of a manganese tricarbonyl antiknock compound, MMT, in gasoline. However, it is clear that while manganese can induce a parkinsonian syndrome that is sometimes reversible, it differs biochemically and clinically from PD [12]. No association of PD with 20 years of exposure to metals including manganese and iron was found in a recent population-based case-control study [37]. Nonetheless, studies of the neurotoxicity of manganese as well as iron have contributed to the notion that oxidative stress is a fundamental underlying mechanism in PD. Furthermore, there are indications that early manganese and iron intake might have profound effects on later-life susceptibility to neurotoxins and subsequent degeneration [32,49,94]. Of related interest are results suggesting that welding, which exposes career welders to a range of metals including manganese, iron, and gases such as carbon monoxide, is associated with a form of parkinsonism that is reported to be distinguished from idiopathic PD only by an earlier onset [71].

6. Summary and speculations

Overall, the emerging picture is that persistent, long-term exposure to diverse xenobiotics—including but not restricted to pesticides, industrial pollutants, metals, infectious agents, and naturally formed heterocyclics such as IQs and βCs—either directly or via bioactivation and in critical association with inherent neurocellular aging changes, can promote oxidative/peroxidative processes leading to a common endpoint of nigrostriatal neuronal loss. If this occurs in individuals with particular genetic susceptibilities (e.g., deficient mitochondrial function or catabolic enzyme activity, or impairment of heat shock proteins to stabilize protein secondary structure), it would contribute to the development of idiopathic PD.

Mechanistically, considerable effort has been expended in ascertaining the role of mitochondrial respiratory inhibition by xenobiotics. However, it is possible that aside from (or linked to) the respiratory inhibition, energy depletion, and oxidative stress, protein aggregation phenomena induced by xenobiotics and metals could be important in PD-related neurodegeneration [54,89,90,97]. It would be informative to determine the potencies of an inclusive range of xenobiotic and endogenous agents in terms of their abilities to induce aggregation, oligomerization, and fibrillation of α-synuclein and other proteins localized in Lewy bodies.

References

[1] J.D. Adams, M.L. Chang, L. Klaidman, Parkinson's disease — redox mechanisms, Curr. Med. Chem. 8 (2001) 809–814.

[2] S.O. Bachurin, E.P. Shevtzova, N.N. Lermontova, T.P. Serkova, R.R. Ramsay, The effect of dithiocarbamates on neurotoxic action of 1-methyl-4-phenyl-1,2,3,6 tetrahydropyridine (MPTP) and on mitochondrial respiration chain, Neurotoxicology 17 (1996) 897–903.

[3] A. Barbeau, M. Roy, G. Bernier, G. Campanella, S. Paris, Ecogenetics of Parkinson's disease: Prevalence and environmental aspects in rural areas, Can. J. Neurol. Sci. 14 (1987) 36–41.

[4] M.F. Beal, Experimental models of Parkinson's disease, Nat. Rev. Neurosci. 2 (2001) 325–334.

[5] B.L. Beaman, D. Canfield, J. Anderson, B. Pate, D. Calne, Site-specific invasion of the basal ganglia by *Nocardia asteroides* GUH-2, Med. Microbiol. Immunol. 188 (2000) 161–168.

[6] R. Betarbet, T.B. Sherer, G. MacKenzie, M. Garcia-Osuna, A.V. Panov, J.T. Greenamyre, Chronic systemic pesticide exposure reproduces features of Parkinson's disease, Nat. Neurosci. 3 (2000) 1301–1306.

[7] A. Bocchetta, G.U. Corsini, Parkinson's disease and pesticides, Lancet 2 (1986) 1163.

[8] G. Bringmann, D. Feineis, R. Bruckner, M. Blank, K. Peters, E.M. Peters, H. Reichmann, B. Janetzky, C. Grote, H.W. Clement, W. Wesemann, Bromal-derived tetrahydro-beta-carbolines as neurotoxic agents: Chemistry, impairment of the dopamine metabolism, and inhibitory effects on mitochondrial respiration, Bioorg. Med. Chem. 8 (2000) 1467–1478.

[9] G. Bringmann, R. God, S. Fahr, D. Feineis, K. Fornadi, F. Fornadi, Identification of the dopaminergic neurotoxin 1-trichloromethyl-1,2,3,4-tetrahydro-beta-carboline in human blood after intake of the hypnotic chloral hydrate, Anal. Biochem. 270 (1999) 167–175.

[10] A.I. Brooks, C.A. Chadwick, H.A. Gelbard, D.A. Cory-Slechta, H.J. Federoff, Paraquat elicited neurobehavioral syndrome caused by dopaminergic neuron loss, Brain Res. 823 (1999) 1–10.

[11] W.J. Burke, Catecholamine-derived aldehyde neurotoxins, in: A. Storch, M.A. Collins (Eds.), Neurotoxic Factors in Parkinson's Disease and Related Disorders, Kluwer–Plenum Publishers, New York, 2001, pp. 167–180.

[12] D.B. Calne, N.S. Chu, C.C. Huang, C.S. Lu, W. Olanow, Manganism and idiopathic parkinsonism: Similarities and differences, Neurology 44 (1994) 1583–1586.

[13] D. Caparros-Lefebvre, A. Elbaz, Possible relation of atypical parkinsonism in the French West Indies with consumption of tropical plants: A case-control study. Caribbean Parkinsonism Study Group, Lancet 354 (1999) 281–286.

[14] P.M. Carvey, J.W. Lipton, D.A. Gayle, T. Landers, C. Tong, S. Gyawali, M. Ptaszny, Q. Chang, Z.D. Ling, In utero lipopolysaccharide (LPS) followed by post-natal toxin exposure produces additive dopamine neuron losses in rats, Abstr. Soc. Neurosci. 27 (2001), #194.16.

[15] J.L. Cashaw, C.A. Geraghty, B.R. McLaughlin, V.E. Davis, Effect of acute ethanol administration on brain levels of tetrahydropapaveroline in L-dopa-treated rats, J. Neurosci. Res. 18 (1987) 497–503.

[16] G. Cohen, R. Farooqui, N. Kesler, Parkinson disease: A new link between monoamine oxidase and mitochondrial electron flow, Proc. Natl. Acad. Sci. U. S. A. 94 (1997) 4890–4894.

[17] G. Cohen, R.E. Heikkila, Mechanisms of action of hydroxylated phenylethylamine and indoleamine neurotoxins, Ann. N. Y. Acad. Sci. 305 (1978) 74–84.

[18] M.A. Collins, Mammalian alkaloids, in: A. Brossi (Ed.), Alkaloids XXI, Academic Press, New York, 1983, pp. 321–350.

[19] M.A. Collins, Atypical parkinsonism in the French West Indies, Lancet 354 (1999) 1473–1474.

[20] M.A. Collins, E.J. Neafsey, β-Carboline analogues of MPTP: Endogenous factors underlying idiopathic parkinsonism? Neurosci. Lett. 55 (1885) 179–184.

[21] M.A. Collins, E.J. Neafsey, β-Carboline analogues of MPP$^+$ as environmental neurotoxins, in: A. Storch, M.A. Collins (Eds.), Neurotoxic Factors in Parkinson's Disease and Related Disorders, Kluwer–Plenum Publishers, New York, 2001, pp. 115–130.

[22] M.T. Corasaniti, G. Bagetta, P. Rodino, S. Gratteri, G. Nistico, Neurotoxic effects induced by intracerebral and systemic injection of paraquat in rats, Hum. Exp. Toxicol. 11 (1992) 535–539.

[23] F.M. Corrigan, L. Murray, C.L. Wyatt, R.F. Shore, Diorthosubstituted polychlorinated biphenyls in caudate nucleus in Parkinson's disease, Exp. Neurol. 150 (1998) 339–342.

[24] G.U. Corsini, S. Pintus, C.C. Chiueh, J.F. Weiss, I.J. Kopin, MPTP

neurotoxicity in mice is enhanced by pretreatment with diethyldithio-carbamate, Eur. J. Pharmacol. 119 (1985) 127–128.

[25] J. Couper, On the effects of black oxide of manganese when inhaled in the lungs, Br. Ann. Med. Pharm. 1 (1837) 41–42.

[26] O. Dammann, A. Leviton, Does prepregnancy bacterial vaginosis increase a mother's risk of having a preterm infant with cerebral palsy? Dev. Med. Child Neurol. 39 (1997) 836–840.

[27] K.L. Double, L. Zecca, D. Ben-Shachar, M.B. Youdim, P. Riederer, M. Gerlach, Neuromelanin may mediate neurotoxicity via its interaction with iron, in: A. Storch, M.A. Collins (Eds.), Neurotoxic Factors in Parkinson's Disease and Related Disorders, Kluwer–Plenum Publishers, New York, 2001, pp. 211–218.

[28] L.S. Engel, H. Checkoway, M.C. Keifer, N.S. Seixas, W.T. Longstreth Jr., K.C. Scott, K. Hudnell, W.K. Anger, R. Camicioli, Parkinsonism and occupational exposure to pesticides, Occup. Environ. Med. 58 (2001) 582–589.

[29] R.J. Ferrante, J.B. Schulz, N.W. Kowall, M.F. Beal, Systemic administration of rotenone produces selective damage in the striatum and globus pallidus, but not in the substantia nigra, Brain Res. 753 (1997) 157–162.

[30] L. Fleming, J.B. Mann, J. Bean, T. Briggle, J.R. Sanchez-Ramos, Parkinson's disease and brain levels of organochlorine pesticides, Ann. Neurol. 36 (1994) 100–103.

[31] B. Fornstedt, A. Brun, E. Rosengren, A. Carlsson, The apparent autoxidation rate of catechols in dopamine-rich regions of human brains increases with the degree of depigmentation of substantia nigra, J. Neural Transm. 1 (1989) 279–295.

[32] A. Fredriksson, N. Schroder, P. Eriksson, I. Izquierdo, T. Archer, Neonatal iron exposure induces neurobehavioural dysfunctions in adult mice, Toxicol. Appl. Pharmacol. 159 (1999) 25–30.

[33] L. Galzigna, A. De Iuliis, L. Zanatta, Enzymatic dopamine peroxidation in substantia nigra of human brain, Clin. Chim. Acta 300 (2000) 131–138.

[34] D.A. Gearhart, M.A. Collins, J.M. Lee, E.J. Neafsey, Increased β-carboline 9N-methyltransferase activity in the frontal cortex in Parkinson's disease, Neurobiol. Dis. 7 (2000) 201–211.

[35] D.A. Gearhart, P.F. Toole, Brain proteins that bind to potential parkinsonian-inducing toxicants, N-methylated beta-carbolines, Abstr. Soc. Neurosci. 27 (2001), #654.19.

[36] M. Gerlach, M.B.H. Youdim, P. Riederer, Brain iron and other trace metals in neurodegenerative diseases, in: A. Storch, M.A. Collins (Eds.), Neurotoxic Factors in Parkinson's Disease and Related Disorders, Kluwer–Plenum Publishers, New York, 2001, pp. 259–276.

[37] J.M. Gorell, C.C. Johnson, B.A. Rybicki, E.L. Peterson, G.X. Kortsha, G.G. Brown, R.J. Richardson, Occupational exposure to manganese, copper, lead, iron, mercury and zinc and the risk of PD, Neurotoxicology 20 (1999) 239–247.

[38] J.T. Greenamyre, R. Betarbet, T. Sherer, A. Panov, Response: Parkinson's disease, pesticides and mitochondrial dysfunction, Trends Neurosci. 24 (2001) 247.

[39] C. Grote, H.W. Clement, W. Wesemann, G. Bringmann, D. Feineis, P. Riederer, K.H. Sontag, Biochemical lesions of the nigrostriatal system by TaClo (1-trichloromethyl-1,2,3,4-tetrahydro-β-carboline) and derivatives, J. Neural Transm. Suppl. 46 (1995) 275–281.

[40] D. Guehl, E. Bezard, S. Dovero, T. Boraud, B. Bioulac, C. Gross, Trichloroethylene and parkinsonism: A human and experimental observation, Eur. J. Neurol. 6 (1999) 609–611.

[41] K. Hamre, R. Tharp, K. Poon, X. Xiong, R.J. Smeyne, Differential strain susceptibility following MPTP administration acts in an autosomal dominant fashion: Quantitative analysis in seven strains of Mus musculus, Brain Res. 828 (1999) 91–103.

[42] R.E. Heikkila, W.J. Nicklas, I. Vyas, R.C. Duvoisin, Dopaminergic toxicity of rotenone and the 1-methyl-4-phenylpyridinium ion after their stereotaxic administration to rats: Implication for the mechanism of MPTP toxicity, Neurosci. Lett. 62 (1985) 389–394.

[43] K. Hyland, B.L. Beaman, P.A. LeWitt, A.J. DeMaggio, Monoamine changes in the brain of BALB/c mice following sub-lethal infection

with Nocardia asteroides (GUH-2), Neurochem. Res. 25 (2000) 443–448.

[44] I. Irwin, E.Y. Wu, L.E. DeLanney, A. Trevor, J.W. Langston, The effect of diethyldithiocarbamate on the biodisposition of MPTP: An explanation for enhanced neurotoxicity, Eur. J. Pharmacol. 141 (1987) 209–217.

[45] V. Jackson-Lewis, G. Liberatore, Effects of a unilateral stereotaxic injection of Tinuvin 123 into the substantia nigra on the nigrostriatal dopaminergic pathway in the rat, Brain Res. 866 (2000) 197–210.

[46] P. Jenner, Parkinson's disease, pesticides and mitochondrial dysfunction, Trends Neurosci. 24 (2001) 245–246.

[47] M.L. Kirby, R.L. Barlow, J.R. Bloomquist, Neurotoxicity of the organochlorine insecticide heptachlor to murine striatal dopaminergic pathways, Toxicol. Sci. 61 (2001) 100–106.

[48] I. Lamensdorf, G. Eisenhofer, J. Harvey-White, Y. Hayakawa, K. Kirk, I.J. Kopin, Metabolic stress in PC12 cells induces the formation of the endogenous dopaminergic neurotoxin, 3,4-dihydroxyphenyl-acetaldehyde, J. Neurosci. Res. 60 (2000) 552–558.

[49] J. Lan, D.H. Jiang, Excessive iron accumulation in the brain: A possible potential risk of neurodegeneration in Parkinson's disease, J. Neural Transm. 104 (1997) 649–660.

[50] A. Lannuzel, P.P. Michel, J. Abaul, Y. Agid, M. Ruberg, D. Caparros-Lefebre, Alkaloids from Annona muricata (soursop) are toxic to dopaminergic neurons: Potential role in the etiology of atypical parkinsonism in the French West Indies, Abstr. Soc. Neurosci. 25 (1999) 2018.

[51] V.G. Laties, M. Thiruchelvam, E.K. Richfield, A.W. Tank, R.B. Baggs, D.A. Cory-Slechta, Neonatal exposure to combined paraquat and maneb produces progressive locomotor deficits and enhances susceptibility to subsequent exposures, Abstr. Soc. Neurosci. 27 (2001), #653.11.

[52] W. Linert, E. Herlinger, R.F. Jameson, E. Kienzl, K. Jellinger, M.B. Youdim, Dopamine, 6-hydroxydopamine, iron, and dioxygen—their mutual interactions and possible implication in the development of Parkinson's disease, Biochim. Biophys. Acta 1316 (1996) 160–168.

[53] B. Malgrange, J.M. Rigo, P. Coucke, S. Belachew, B. Rogister, G. Moonen, Beta-carbolines induce apoptotic death of cerebellar granule neurones in culture, NeuroReport 7 (1996) 3041–3045.

[54] A.B. Manning-Bog, A.L. McCormack, J. Li, V.N. Uversky, A.L. Fink, D.A. Di Monte, The herbicide paraquat causes upregulation and aggregation of alpha-synuclein in mice, J. Biol. Chem. 276 (2002) 1641–1644.

[55] W. Maruyama, K. Sango, K. Iwasa, C. Minami, P. Dostert, M. Kawai, M. Moriyasu, M. Naoi, Dopaminergic neurotoxins, 6,7-dihydroxy-1-(3′,4′-dihydroxy-benzyl)-isoquinolines, cause different types of cell death in SH-SY5Y cells: Apoptosis was induced by oxidized papaverolines and necrosis by reduced tetrahydropapaverolines, Neurosci. Lett. 291 (2000) 89–92.

[56] W. Maruyama, M. Strolin-Benedetti, M. Naoi, N-Methyl (R) salsolinol and a neutral N-methyltransferase as pathogenic factors in Parkinson's disease, Neurobiology 8 (2000) 55–68.

[57] R. Masalha, Y. Herishanu, A. Alfahel-Kakunda, W.F. Silverman, Selective dopamine neurotoxicity by an industrial chemical: An environmental cause of Parkinson's disease? Brain Res. 774 (1997) 260–264.

[58] K. Matsubara, N-Methyl-β-carbolinium neurotoxins in Parkinson's disease, in: A. Storch, M.A. Collins (Eds.), Neurotoxic Factors in Parkinson's Disease and Related Disorders, Kluwer–Plenum Publishers, New York, 2001, pp. 131–143.

[59] K. Matsubara, T. Gonda, H. Sawada, T. Uezono, Y. Kobayashi, T. Kawamura, K. Ohtaki, K. Kimura, A. Akaike, Endogenously occurring β-carboline induces parkinsonism in nonprimate animals: A possible causative protoxin in idiopathic Parkinson's disease, J. Neurochem. 70 (1998) 727–735.

[60] K. Matsubara, S. Kobayashi, Y. Kobayashi, K. Yamashita, H. Koide, M. Hatta, K. Iwamoto, O. Tanaka, K. Kimura, β-Carbolinium cations, endogenous MPP+ analogs, in the lumbar cerebrospinal fluid of patients with Parkinson's disease, Neurology 45 (1995) 2240–2245.

[61] M.J. Mihm, B.L. Schanbacher, B.L. Wallace, L.J. Wallace, N.J. Uretsky, J.A. Bauer, Free 3-nitrotyrosine causes striatal neurodegeneration in vivo, J. Neurosci. 21 (2001) RC149 (1–5).

[62] G.W. Miller, M.L. Kirby, A.I. Levey, J.R. Bloomquist, Heptachlor alters expression and function of dopamine transporters, Neurotoxicology 20 (1999) 631–637.

[63] T.J. Montine, V. Amarnath, M.J. Picklo, K.R. Sidell, J. Zhang, D.G. Graham, Endogenous brain catechol thioethers in dopaminergic neurodegeneration, in: A. Storch, M.A. Collins (Eds.), Neurotoxic Factors in Parkinson's Disease and Related Disorders, Kluwer–Plenum Publishers, New York, 2001, pp. 155–166.

[64] A. Moser, D. Kompf, Presence of methyl-6,7-dihydroxy-1,2,3,4-tetrahydro-isoquinolines, derivatives of the neurotoxin isoquinoline, in parkinsonian lumbar CSF, Life Sci. 50 (1992) 1885–1891.

[65] T. Nagatsu, Isoquinoline neurotoxins, in: A. Storch, M.A. Collins (Eds.), Neurotoxic Factors in Parkinson's Disease and Related Disorders, Kluwer–Plenum Publishers, New York, 2001, pp. 69–76.

[66] M. Naoi, W. Maruyama, Y. Akako, Y. Nakagawa, T. Takahashi, Apoptosis by an endogenous neurotoxin, N-methyl [R] salsolinol: Relevance to Parkinson's disease, in: A. Storch, M.A. Collins (Eds.), Neurotoxic Factors in Parkinson's Disease and Related Disorders, Kluwer–Plenum Publishers, New York, 2001, pp. 77–89.

[67] S. Ohta, Isoquinolines in Parkinson's disease, in: A. Storch, M.A. Collins (Eds.), Neurotoxic Factors in Parkinson's Disease and Related Disorders, Kluwer–Plenum Publishers, New York, 2001, pp. 91–100.

[68] D.E. Pellegrini-Giampietro, G. Cherici, M. Alesian, V. Carla, F. Moroni, Excitatory amino acid release and free radical formation may cooperate in the genesis of ischemia-induced neuronal damage, J. Neurosci. 10 (1990) 1035–1041.

[69] A. Pezzella, M. d'Ischia, A. Napolitano, G. Misuraca, G. Prota, Iron-mediated generation of the neurotoxin 6-HODA quinone by reaction of fatty acid hydroperoxides with dopamine: A possible contributory mechanism for neuronal degeneration in Parkinson's disease, J. Med. Chem. 40 (1997) 2211–2216.

[70] A. Priyadarshi, S.A. Khuder, E.A. Schaub, S.S. Priyadarshi, Environmental risk factors and PD: A meta-analysis, Environ. Res. 86 (2001) 122–127.

[71] B.A. Racette, L. McGee-Minnich, S.M. Moerlein, J.W. Mink, T.O. Videen, J.S. Perlmutter, Welding-related parkinsonism: Clinical features, treatment, and pathophysiology, Neurology 56 (2001) 8–13.

[72] O.W. Sacks, Awakenings, Vintage Books, New York, 1999, 464 pp.

[73] M. Sandler, S.B. Carter, K.R. Hunter, G.M. Stern, Tetrahydroisoquinoline alkaloids: In vivo metabolites of L-dopa in man, Nature 241 (1973) 439–443.

[74] R.F. Seegal, Polychlorinated biphenyls and methylmercury act synergistically to reduce rat brain dopamine content in vitro, Environ. Health Perspect. 107 (1999) 879–885.

[75] X.M. Shen, H. Li, G. Dryhurst, Oxidative metabolites of 5-S-cysteinyldopamine inhibit the α-ketoglutarate dehydrogenase complex: Possible relevance to the pathogenesis of Parkinson's disease, J. Neural Transm. 107 (2000) 959–978.

[76] T.B. Sherer, P.A. Trimmer, K. Borland, J.K. Parks, J.P. Bennett Jr., J.B. Tuttle, Chronic reduction in complex I function alters calcium signaling in SH-SY5Y neuroblastoma cells, Brain Res. 891 (2001) 94–105.

[77] E. Sofic, W. Paulus, K. Jellinger, P. Riederer, M.B. Youdim, Selective increase of iron in substantia nigra zona compacta of parkinsonian brains, J. Neurochem. 56 (1991) 978–982.

[78] L. Soleo, G. Defazio, R. Scarselli, R. Zefferino, P. Livrea, V. Foa, Toxicity of fungicides containing ethylene-bis-dithiocarbamate in serumless dissociated mesencephalic–striatal primary coculture, Arch. Toxicol. 70 (1996) 678–682.

[79] S. Song, F. Cardozo-Pelaez, J. Sanchez-Ramos, Relationship of organochlorine pesticides to parkinsonism, in: A. Storch, M.A. Collins (Eds.), Neurotoxic Factors in Parkinson's Disease and Related Disorders, Kluwer–Plenum Publishers, New York, 2001, pp. 237–245.

[80] P.K. Sonsalla, R.E. Heikkila, The influence of dose and dosing interval on MPTP-induced dopaminergic neurotoxicity in mice, Eur. J. Pharmacol. 129 (1986) 339–345.

[81] J.P. Spencer, P. Jenner, S.E. Daniel, A.J. Lees, D.C. Marsden, B. Halliwell, Conjugates of catecholamines with cysteine and GSH in Parkinson's disease: Possible mechanisms of formation involving reactive oxygen species, J. Neurochem. 71 (1998) 2112–2122.

[82] K. Stepien, A. Zajdel, A. Wilczok, T. Wilczok, A. Grzelak, A. Mateja, M. Soszynski, G. Bartosz, Dopamine-melanin protects against tyrosine nitration, tryptophan oxidation and Ca2+-ATPase inactivation induced by peroxynitrite, Biochim. Biophys. Acta 1523 (2000) 189–195.

[83] A.Y. Sun, X. Li, Paraquat is a model environmental neurotoxin for studying Parkinson's disease, in: A. Storch, M.A. Collins (Eds.), Neurotoxic Factors in Parkinson's Disease and Related Disorders, Kluwer–Plenum Publishers, New York, 2001, pp. 247–257.

[84] Y.J. Surh, Tetrahydropapaveroline, a dopamine-derived isoquinoline alkaloid, undergoes oxidation: Implications for DNA damage and neuronal cell death, Eur. J. Clin. Invest. 29 (1999) 650–651.

[85] C.M. Tanner, R. Ottman, S.M. Goldman, J. Ellenberg, P. Chan, R. Mayeux, J.W. Langston, Parkinson disease in twins: An etiologic study, JAMA, J. Am. Med. Assoc. 281 (1999) 341–346.

[86] C. Thiffault, J.W. Langston, D.A. Di Monte, Increased striatal dopamine turnover following acute administration of rotenone to mice, Brain Res. 885 (2000) 283–288.

[87] M. Thiruchelvam, B.J. Brockel, E.K. Richfield, R.B. Baggs, D.A. Cory-Slechta, Potentiated and preferential effects of combined paraquat and maneb on nigrostriatal dopamine systems: Environmental risk factors for PD? Brain Res. 873 (2000) 225–234.

[88] M. Thiruchelvam, E.K. Richfield, R.B. Baggs, A.W. Tank, D.A. Cory-Slechta, The nigrostriatal dopaminergic system as a preferential target of repeated exposures to combined paraquat and maneb: Implications for Parkinson's disease, J. Neurosci. 20 (2000) 9207–9214.

[89] V.N. Uversky, J. Li, A.L. Fink, Pesticides directly accelerate the rate of alpha-synuclein fibril formation: A possible factor in Parkinson's disease, FEBS Lett. 500 (2001) 105–108.

[90] V.N. Uversky, J. Li, A.L. Fink, Metal-triggered structural transformations, aggregation, and fibrillation of human alpha-synuclein. A possible molecular link between Parkinson's disease and heavy metal exposure, J. Biol. Chem. 276 (2001) 44284–44296.

[91] R. Vigneswaran, Infection and preterm birth: Evidence of a common causal relationship with bronchopulmonary dysplasia and cerebral palsy, J. Paediatr. Child Health 36 (2000) 293–296.

[92] K. Wakabayashi, Y. Totsuka, K. Fukutome, A. Oguri, H. Ushiyama, T. Sugimura, Human exposure to mutagenic/carcinogenic heterocyclic amines and comutagenic carbolines, Mutat. Res. 376 (1997) 253–259.

[93] T.L. Walters, I. Irwin, K. Delfani, J.W. Langston, A.M. Janson, Diethyldithiocarbamate causes nigral cell loss and dopamine depletion with nontoxic doses of MPTP, Exp. Neurol. 156 (1999) 62–70.

[94] R. Witholt, R.H. Gwiazda, D.R. Smith, The neurobehavioral effects of subchronic manganese exposure in the presence and absence of preparkinsonism, Neurotoxicol. Teratol. 22 (2000) 851–861.

[95] F. Yantiri, J.K. Andersen, The role of iron in Parkinson disease and MPTP toxicity, IUBMB Life 48 (1999) 139–141.

[96] J. Yuan, S. Manabe, N-Methyl-β-carboline-3-carboxamide (FG 7142), an anxiogenic agent in airborne particles and cigarette smoke-polluted indoor air, Environ. Pollut. 90 (1995) 349–355.

[97] Y. Zhang, V.L. Dawson, T.M. Dawson, Oxidative stress and genetics in the pathogenesis of Parkinson's disease, Neurobiol. Dis. 7 (2000) 240–250.

ELSEVIER

Neurotoxicology and Teratology 24 (2002) 579–591

NEUROTOXICOLOGY

AND

TERATOLOGY

www.elsevier.com/locate/neutera

Review article

Dopamine-derived endogenous *N*-methyl-(*R*)-salsolinol
Its role in Parkinson's disease

Makoto Naoi[a],*, Wakako Maruyama[b], Yukihiro Akao[c], Hong Yi[c]

[a]*Department of Brain Sciences, Institute of Applied Biochemistry, Yagi Memorial Park, Mitake, 505-0116 Gifu, Japan*
[b]*Department of Basic Gerontology, National Institute of Longevity Sciences, Obu, 474-8522 Aichi, Japan*
[c]*Gifu International Institute of Biotechnology, Mitake, 505-0116 Gifu, Japan*

Received 8 December 2001; accepted 24 January 2002

Abstract

A dopamine-derived alkaloid, *N*-methyl-(*R*)-salsolinol [*N*M(*R*)Sal], enantioselectively occurs in human brains and accumulates in the nigrostriatal system. It increases in the cerebrospinal fluid (CSF) of parkinsonian patients and the activity of a neutral (*R*)-salsolinol [(*R*)Sal] *N*-methyltransferase, a key enzyme in the biosynthesis of this toxin, increases in the lymphocytes from parkinsonian patients, suggesting its involvement in the pathogenesis of Parkinson's disease (PD). The studies of animal and cellular models of PD proved that this isoquinoline is selectively cytotoxic to dopamine neurons. Using human dopaminergic SH-SY5Y cells, *N*M(*R*)Sal induces apoptosis by the activation of the apoptotic cascade initiated in mitochondria. In this article, we review the recent advance in proving our hypothesis that the dopamine-derived neurotoxin causes the selective depletion of dopamine neurons in PD. © 2002 Elsevier Science Inc. All rights reserved.

Keywords: *N*-Methyl-(*R*)-salsolinol; Dopaminergic neurotoxin; Parkinson's disease; Apoptosis; Catechol isoquinoline

1. Introduction

Dopamine neurons in the substantia nigra of human brains are known to be selectively vulnerable, and neuronal loss with advancing age was estimated to be more than one-third between the age of 20 and 90 years [34]. There is linear fallout of dopamine neurons with aging at a rate of 5–10% per decade [8], and the limited number of the cells causes dysfunction in cognition and motor movement. Similarities have been drawn between senility and Parkinson's disease (PD), based on the similar degeneration of dopamine neurons [19]. In addition, aging has been considered to play a role in the pathogenesis of PD. In PD, the clinical signs are detected when 50% of nigral neurons and 80% of striatal dopamine are lost. In PD, the velocity and intensity of the neuronal loss are more marked than in physiological aging, and dopamine neurons are estimated to decline at a rate of about 30 neurons daily. PD was once considered to be a form of accelerated aging, but it is now believed that pathological process other than aging is involved in PD, because the regional selectivity of dopamine depletion in the substantia nigra is different in PD from aging [14].

The pathogenesis of PD remains to be clarified, and various genetic and environmental factors have been proposed to cause PD in humans. Genes responsible for autosomal dominant and recessive juvenile parkinsonism were found to be α-synuclein [55] and parkin gene [15], respectively. However, genes for sporadic PD have never been identified. On the other hand, there has been an increasing body of evidence indicating the involvement of neurotoxins in the deterioration of nigrostriatal dopamine system in sporadic cases of PD. The finding that 1-methyl-4-phenyl-1,2,3,6-tetrahydropyridine (MPTP) [5] caused parkinsonism in humans stimulated search for endogenous and xenobiotic toxins as pathogenic agents. Three endogenous alkaloids have been proposed as candidates of dopaminergic neurotoxins: derivatives of dopamine-derived 1-methyl-4-phenyl-1,2,3,4-tetrahydroisoquinoline (salsolinol, Sal) [52] and 1-benzyl catechol isoquinoline [6,7-dihydroxy-1-(3′,4′-dihydroxybenzyl)-1,2,3,4-tetrahydroisoquinoline, tetrahydropapaveroline, THP] [29], simple 1,2,3,4-tetrahydroisoquinolines [43], and β-carbolines [3,4]. Among these alkaloids, *N*-methyl-(*R*)-salsolinol [1(*R*),2(*N*)-dimethyl-6,7-dihydroxy-1,2,3,4-tetrahydroisoquinoline, *N*M(*R*)Sal] is

* Corresponding author. Tel.: +81-574-67-5500; fax: +81-574-67-5310.
E-mail address: mnaoi@giib.or.jp (M. Naoi).

Table 1
Concentration of salsolinol derivatives in human brains

Salsolinol derivatives	Tissue	Content	Reference
(R)Sal[a]	frontal cortex	134 ± 125 pmol/g wet tissue	[27]
	caudate	73.3 ± 79.9	
	putamen	37.8 ± 23.0	
	substantia nigra	94.5 ± 78.7	
	IVF	0.39 ± 0.21 nM	[25]
$NM(R)$Sal[a]	caudate	65.7 ± 88.3 pmol/g wet tissue	[27]
	putamen	110 ± 126	
	substantia nigra	76.6 ± 23.0	
	CSF		
	parkinsonian	8.32 ± 2.89 nM	[24]
	control	4.53 ± 2.08 nM	
	IVF	9.15 ± 9.08 nM	[25]
DMDHIQ$^+$	substantia nigra	254 ± 59.0 pmol/g wet tissue	[27]
	IVF	16.0 ± 16.5 nM	[25]
(S)THP[b]	gray matter	$0.12 – 0.28$ pmol/g wet tissue	[58]

[a] The (S) enantiomers were not detected.
[b] The (R)THP was not detected.

now considered to be a most possible neurotoxin candidate involved in the pathogenesis of PD [51–54].

This review summarizes our data on the synthesis and selective occurrence of this neurotoxin in the human brain, preparation of a PD model in rats, and the analyses of clinical samples from parkinsonian patients. $NM(R)$Sal was

found to induce apoptosis in dopaminergic cells and the mechanism of the cell death was studied by use of human dopaminergic SH-SY5Y cells. The possible involvement of $NM(R)$Sal and related catechol isoquinolines in the pathogenesis of PD is discussed.

2. $NM(R)$Sal in human brain

2.1. Occurrence of NM(R)Sal in human brain

Sal was identified for the first time in urine from patients treated with L-DOPA [57] and then in the cerebrospinal fluid (CSF) and brain tissues [4]. Sal synthesis in the human body was once considered to be nonenzymatic by a Pictet–Spengler reaction of dopamine with acetaldehyde yielding the racemic form [56]. However, predominant occurrence of the (R)-salsolinol [(R)Sal] in human urine suggests that the (R) enantiomer may be synthesized enzymatically [7,59]. Recently, a sensitive enantiomeric analysis of Sal derivatives has been devised by use of a chiral column and high-performance liquid chromatography (HPLC) with electrochemical detection (ECD) [53]. Using this method, tissues from various human brain regions [27], CSF [24], and intraventricular fluid (IVF) [25] were analyzed for the enantiomeric character of

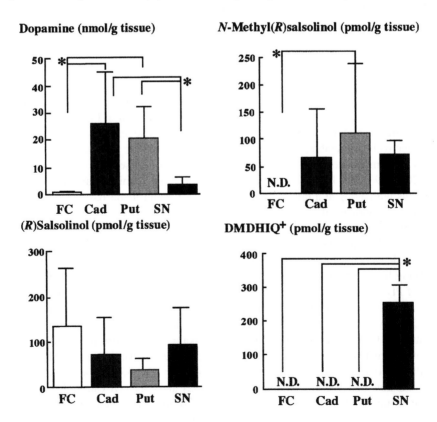

Fig. 1. Distribution of $NM(R)$Sal and related compounds in human brain regions. Tissues were punched out from frontal cortex (FC), caudate (Cad), putamen (Put), and substantia nigra (SN) of nine control brains. Dopamine, Sal, and NMSal were quantitatively analyzed by HPLC–ECD and the isoquinolinium ion by HPLC–fluorometric detection. Only the (R) enantiomer of Sal and NMSal were detected and selective distribution of $NM(R)$Sal and DMDHIQ$^+$ in the nigrostriatum was confirmed. * Significantly different from the frontal, $P < .01$. N.D. = not detected.

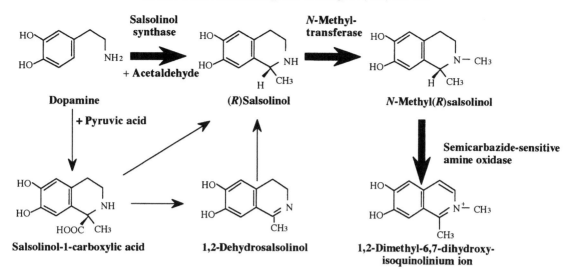

Fig. 2. Biosynthesis pathway of NM(R)Sal and its metabolism in human brain. (R)Sal synthase catalyses the reaction of dopamine with acetaldehyde or pyruvic acid to produce (R)Sal or (R)Sal 1-carboxylic acid, respectively. (R)Sal N-methyltransferase catalyses N-methylation of (R)Sal, but not (S)Sal, into NM(R)Sal, a neurotoxin. Oxidation of NM(R)Sal is nonenzymatic and enzymatic by an oxidase, sensitive to semicarbazide, but not to the inhibitors of mitochondrial MAO.

Sal derivatives. As summarized in Table 1, in human brains, CSF, and IVF, only the (R) enantiomers were detected, while the (S) enantiomer of Sal and NMSal [NM(S)Sal] were under the detectable limit.

In control human brains, the distribution of Sal derivatives was examined in four regions, frontal cortex, caudate, putamen, and substantia nigra [27]. NM(R)Sal was found to occur selectively in the nigrostriatum, whereas (R)Sal distributes uniformly among the brain regions, as shown in Fig. 1. On the other hand, 1,2-dimethyl-6,7-dihydroxyisoquinolinium ion (DMDHIQ$^+$) with a chemical structure similar to 1-methyl-4-phenylpyridinium ion (MPP$^+$) was detected only in the substantia nigra. These results suggest that the N-methylation of (R)Sal by an N-methyltransferase brings about the selective distribution in dopamine neurons, because of the uptake by a dopamine uptake system [60]. Binding of DMDHIQ$^+$ to melanin may account for the selective accumulation of the isoquinolinium ion in the substantia nigra [45]. On the other hand, THP is synthesized by a Pictet–Spengler reaction of dopamine with 3,4-dihydroxyphenyl-acetaldehyde, an oxidation product of dopamine by monoamine oxidase (MAO) [12]. However, recently, we detected only (S)THP in human brains, suggesting its

enzymatic synthesis, but the enzymes related to the enantioselective biosynthesis remain to be characterized [58].

2.2. Biosynthesis and metabolism of NM(R)Sal in human brain

The enzymatic biosynthesis pathway of (R)Sal and NM(R)Sal in the human brain has been clarified, as shown in Fig. 2. The enzymatic properties of related enzymes, (R)Sal synthase, (R)Sal N-methyltransferase, and NM(R)Sal oxidase, are summarized in Table 2. We purified a novel (R)Sal synthase from human brain [48], which condenses dopamine with acetaldehyde enantioselectively to (R)Sal. (R)Sal synthase catalyses the reaction of dopamine with pyruvic acid to produce (R)Sal 1-carboxylic acid, but the oxidative decarboxylation directly to (R)Sal was not confirmed, and it was nonenzymatically converted to 1,2-dehydrosalsolinol. However, the enantioselective synthesis of (R)Sal from 1,2-dehydrosalsolinol was not confirmed in human brain sample. Adrenaline, noradrenaline, and N-methyldopamine (epinine) are not the substrates, suggesting that NM(R)Sal is not synthesized directly from N-methyldopamine. At present, it is not clear whether the

Table 2
Characteristics of enzymes related to the metabolism of NM(R)Sal in the human brain

Enzymes	Some characteristics	Substrate	Reference
(R)Sal synthase	molecular weight: 34.3 kDa optimum pH: 7.4	dopamine, acetaldehyde, pyruvic acid	[48]
Neutral (R)Sal N-methyltransferase	methyl donor: SAM[a] molecular weight: 35.6 kDa optimum pH: 7.0	(R)Sal, norsalsolinol, but not (S)Sal	[49]
NM(R)Sal oxidase	sensitive to semicarbazide not sensitive to clorgyline, (−)deprenyl	NM(R)Sal, NM(S)Sal, N-methyl-norsalsolinol	[46]

These enzymes we isolated from the cytosol soluble fraction of human brain gray matter.

[a] S-adnosyl-l-methionine.

DMDHIQ⁺ concentration in the substantia nigra
(pmol/g tissue)

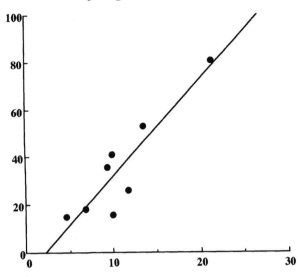

Activity of neutral *N*-methyltransferase in the caudate
(pmol/min/mg protein)

Fig. 3. Correlation of DMDHIQ $^+$ concentration in the substantia nigra with the activity of neutral (*R*)Sal *N*-methyltransferase in the caudate. In eight control human brains, the activity of the *N*-methyltransferase and the levels of DMDHIQ $^+$ and *N*M(*R*)Sal were quantitatively analyzed. The enzymatic activity in the caudate ($P < .01$) and putamen ($P < .05$) were well correlated with the level of the isoquinolinium ion in the substantia nigra. In other brain regions, there was no correlation between the enzyme activity and the levels of the isoquinolinium ion or *N*M(*R*)Sal.

(*R*)Sal synthase catalyses the enantioselective biosynthesis of (*S*)THP in the brain.

N-Methylation of (*R*)Sal into *N*M(*R*)Sal was proven in the rat striatum by in vivo microdialysis method [20], and the enzyme was isolated from human brains and partially characterized [49]. In human brain, there are two types of *N*-methyltransferase to (*R*)Sal as a substrate, with the optimal pH at 7.0 and 8.5. The substrate specificity of neutral *N*-methyltransferase activity measured at pH 7.0 was restricted to (*R*)Sal and (*S*)Sal was not *N*-methylated by this enzyme. This neutral (*R*)Sal *N*-methyltransferase has been suggested to play a major role in the biosynthesis of *N*M(*R*)Sal, as described below.

*N*M(*R*)Sal is oxidized into DMDHIQ $^+$ enzymatically by an oxidase sensitive to semicarbazide, but not to MAO inhibitors [46], or nonenzymatically by autoxidation [22,23]. The concentration of DMDHIQ $^+$ in the substantia nigra is determined by a neutral (*R*)Sal *N*-methyltransferase activity in the striatum, as shown in Fig. 3 [49]. No correlation between other isoquinolines and the activities of any related enzymes could be confirmed.

3. Preparation of a rat PD model by *N*M(*R*)Sal

To find neurotoxins capable of inducing parkinsonism in rats, the catechol isoquinolines were injected into the stria-

tum of male Wistar rats, and the behavioral, biochemical, and histopathological effects were examined [47]. Among catechol isoquinolines, only *N*M(*R*)Sal proved to be cytotoxic to dopamine neurons. This is the first endogenous isoquinoline capable of preparing a rat PD model with selective depletion of dopamine neurons in the substantia nigra.

3.1. Behavior changes due to perturbation of dopaminergic system

After a single injection of *N*M(*R*)Sal into the left striatum, rats exhibited postural abnormality with the head and trunk deviating toward the lesion side, lateral extension of the right hind limb, and stiffness of the tail elevated above the ground. The blinking of the right eye was markedly reduced. During spontaneous activity, the rats showed ipsilateral circulation toward the injection site. Injection of *N*M(*R*)Sal into both sides of the striatum induced akinesia. Some rats exhibited fine regular twitching of right limbs at rest with a cycle of about 60/min. Twitching lasted for several hours and the circulation was detected for 24–48 h after a single injection. An injection of DMDHIQ $^+$ in the left striatum induced hypokinesia in the right limbs, but no involuntary movement was detected. Other catechol isoquinolines, *N*M(*S*)Sal, (*R*)Sal and (*S*)Sal, norsalsolinol, and 2(*N*)-methyl-6,7-dihydroxy-1,2,3,4-tetrahydroisoquinoline (*N*-methyl-norsalsolinol), did not induce any behavioral changes to rats. By continuous infusion of *N*M(*R*)Sal with an osmotic minipump, rats showed ipsilateral rotation for the first 48 h and further akinesia for 4 or 5 days.

3.2. Biochemical changes after infusion of *N*M(*R*)Sal

Three days after a single injection, a rat was sacrificed and the brain was cut in 2-mm-thick slices by coronal section. Catecholamines, indoleamines, and their metabolites and catechol isoquinolines were analyzed by use of HPLC with a multi-ECD system or with a Coulochem-II ECD (ESA, Chelmsford, MA, USA) [44].

*N*M(*R*)Sal and DMDHIQ $^+$ accumulated in rat brain. *N*M(*R*)Sal were 1.03 ± 0.17 and 1.71 ± 0.73 pmol/mg wet weight in Slices I and II (the striatum), respectively. In the Slices V–VII, no significant amounts of *N*M(*R*)Sal could be detected. However, large amounts of DMDHIQ $^+$, an oxidation product of injected *N*M(*R*)Sal, were identified in Slices I and II. In addition, definite amounts of DMDHIQ $^+$ were found in Slices VI and VII containing the substantia nigra: 0.45 ± 0.16 and 0.48 ± 0.09 pmol/mg wet weight, respectively. After injection of DMDHIQ $^+$, remarkable amounts of DMDHIQ $^+$ were detected in Slices I–III and the amount was the highest in Slice I, but it was not detected in the substantia nigra. These results clearly show that *N*M(*R*)Sal injected into the striatum is transported by retrograde axonal flow to the substantia nigra, where it is oxidized into the isoquinolinium ion and accumulates in dopamine neurons. This is relevant with the result from human brains, where DMDHIQ$^+$ was detected only in the substantia nigra [27].

In the brain slices containing a portion of the striatum and the substantia nigra (Slices II and VII), marked reduction in dopamine was observed after injection of NM(R)Sal. The reduction was more manifest with NM(R)Sal than with DMDHIQ$^+$, and in the substantia nigra (Slice VII), dopamine was lower than the detection limit. Noradrenaline and its metabolite, 3-methoxy-4-hydroxyphenylethylene glycol, were reduced in the striatum and substantia nigra after the injection of NM(R)Sal ($P < .05$). By contrast, serotonin and its metabolite, 5-hydroxyindole-acetic acid, did not change after injection of NM(R)Sal. In other brain regions, the contents of dopamine, noradrenaline, serotonin, and their metabolites were not affected after injection of NM(R)Sal or DMDHIQ$^+$. These results demonstrate that NM(R)Sal is selectively cytotoxic to dopamine and noradrenaline neurons, but not to serotonin neurons.

The activity of a rate-limiting enzyme of dopamine synthesis, tyrosine hydroxylase (TH), was measured as reported previously [21] in the brain slices. As shown in Fig. 4, TH activity in the striatum and substantia nigra was significantly reduced after a single injection of NM(R)Sal in the striatum, while DMDHIQ$^+$ did not affect TH activity. In other brain regions, no significant reduction of TH activity was detected after injection of NM(R)Sal or DMDHIQ$^+$. The results of the in vivo experiments show that NM(R)Sal depleted the dopamine content and TH activity more profoundly than the more cytotoxic DMDHIQ$^+$ shown by in vitro experiments [61].

3.3. Histopathological changes after NM(R)Sal infusion

After 1-week continuous infusion of catechol isoquinolines in the left side of the striatum, the rat was perfused and fixed with a formaldehyde solution. The brain was cut into section of 8-μm thickness and stained by the hematoxylin–eosin (H–E) and Klüver–Barrera (K–B) method. Immunohistochemical examination was performed by the streptavidin–biotin complex method with rabbit anti-TH antibody (Chemicon International, Temecula, CA, USA). After NM(R)Sal administration, mild necrosis was detected around the injected site, whereas DMDHIQ$^+$ administration caused massive necrosis around the injected spot and numerous macrophages were observed. K–B staining showed also massive destruction of myelin structure after injection of

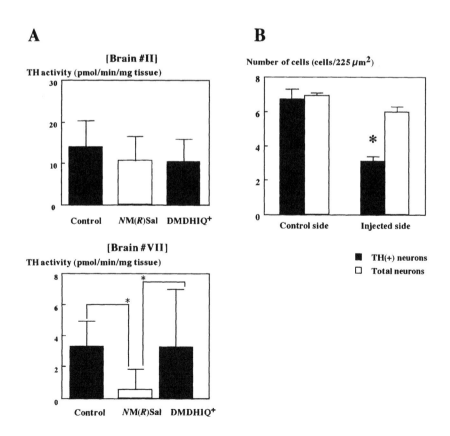

Fig. 4. Reduction of TH activity and dopamine neurons in the substantia nigra after isoquinoline injection. (A) TH activity in Slices II (the striatum) and VII (the substantia nigra). Three days after a single injection of NM(R)Sal or DMDHIQ$^+$, the activity of TH was measured in rat brain slices. Significant reduction of the activity was confirmed in the substantia nigra after NM(R)Sal injection ($P < .05$), whereas the activity did not change in the striatum significantly. (B) Number of dopamine neurons in the substantia nigra after continuous NM(R)Sal infusion in the striatum. NM(R)Sal was infused in the striatum for 1 week and the brain tissues were strained with anti-TH antibody to identify dopamine neurons. At the injected site, significant reduction of dopamine neurons was observed compared with the total neurons or control site. * Significantly different from control side, $P < .05$.

DMDHIQ$^+$, whereas NM(R)Sal did not cause significant changes in the structure as compared with control. The density of neurons stained with anti-TH antibody was reduced markedly in the substantia nigra of the treated side as compared with the control side, as shown in Fig. 4. The number of K–B stained cells was almost the same in either site of the substantia nigra. TH-positive neurons in the ventral tegmental area were not affected by the injection of NM(R)Sal. These results suggest that NM(R)Sal causes cell death selectively in dopamine neurons by a mechanism other than necrosis, namely apoptosis.

4. Analyses of NM(R)Sal and the related enzymes in clinical samples from parkinsonian patients

The involvement of NM(R)Sal in the pathogenesis of PD was proven by analyses of the CSF from patients with PD. The NM(R)Sal levels in CSF from parkinsonian patients were significantly higher than in control [24], suggesting that the metabolism of NM(R)Sal in the brain may be altered

in PD. To confirm this point, the activities of enzymes related to the metabolism of NM(R)Sal were examined in lymphocytes prepared from parkinsonian patients [50]. The results support our view that neutral (R)Sal N-methyltransferase may be one of the endogenous factors in the pathogenesis of PD.

4.1. Analyses of NM(R)Sal in CSF from parkinsonian patients

The lumbar CSF samples from 16 patients with newly diagnosed and untreated PD and from 29 control subjects without neurological disorders were used for the analysis. As a disease control, CSF from five patients with multiple system atrophy (MSA) was analyzed. All the patients were fully informed on the risks and potential benefits of the CSF examination, and the Ethical Committee approved the protocol of this study. Enantiomers of Sal and NMSal, dopamine, and 3-methoxy-4-hydroxyphenylacetic acid (homovanillic acid, HVA) were analyzed by HPLC as reported previously [44].

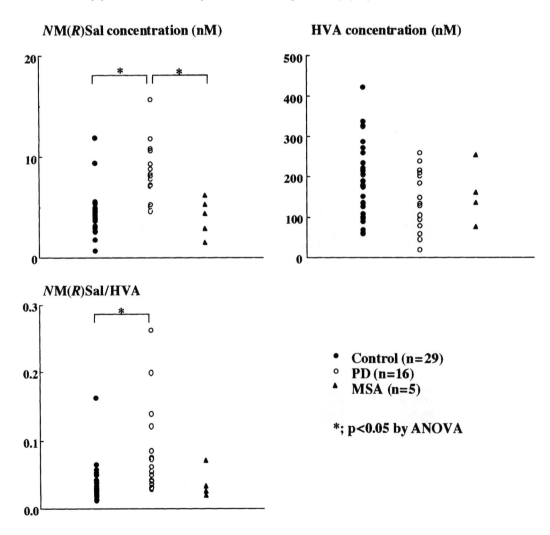

Fig. 5. NM(R)Sal concentrations in parkinsonian CSF. CSF samples from untreated patients with PD ($n = 16$), control ($n = 29$), and patients with a disease control (MSA, $n = 5$) were analyzed for NM(R)Sal and a dopamine product, HVA. * The difference between the group was statistically significant ($P < .05$).

NM(R)Sal was detected in CSF from control and patients with PD and MSA. The (S) enantiomer of NMSal, was under detection limit (<0.01 nM). These results are supported by other reports that (R)Sal and (S)Sal did not increase in parkinsonian plasma [40] or CSF [41]. NM(R)Sal concentration in the control group was not affected by the age from 2 to 76 years (r=.141) or by the sex (mean±S.D.: male: .39±1.73 nM; female: 4.89±2.79 nM). As shown in the Fig. 5, in all patients with MSA and almost all control, except out of 29, NM(R)Sal level was lower than 6 nM. On the other hand, in 12 out of 16 PD patients, the level was higher than 6 nM and significantly higher than that either in control or patients with MSA. The concentration of (R)Sal and free dopamine, the metabolic precursors of NM(R)Sal, were under 0.1 nM in CSF. To estimate the biosynthesis rate of the isoquinoline from dopamine, the NM(R)Sal concentration was compared with that of HVA, a major metabolite of dopamine. The ratio also significantly increased in PD patients compared with control or MSA patients. The HVA level was reduced in the CSF of PD and MSA, but the difference from control was not statistically significant. The results suggest that the levels of this neurotoxin in the CSF and maybe also in the brain of PD patients are significantly higher than in control, suggesting the altered activity of enzymes related to the metabolism.

.2. (R)Sal N-methyltransferase activity in PD lymphocytes

As described above, NM(R)Sal is synthesized in human brains by stepwise enzymatic reactions from dopamine with (R)Sal synthase [48] and (R)Sal N-methyltransferase [49]. NM(R)Sal oxidase [46] catalyze the metabolism of NM(R)Sal. Patients with PD and normal control subjects without neurological diseases were recruited after the Ethical Committee approved. The effect of pH on the activity of (R)Sal N-methyltransferase was examined. In control lymphocytes, the activity showed the two peaks around .0 and 8.0 with almost the same values. In parkinsonian lymphocytes, the highest activity was detected around pH .0, which was much higher than in control subjects,

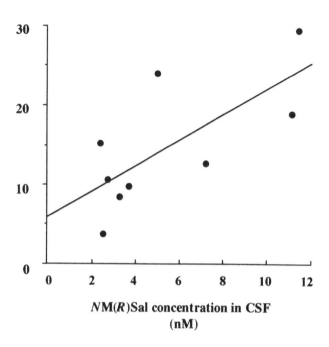

**(R)Sal N-methyltransferase activity in lymphocytes
(pmol/mg protein)**

**NM(R)Sal concentration in CSF
(nM)**

Fig. 6. The correlation of the activity of neutral (R)-salsolinol N-methyltransferase in the lymphocyte and the level of NM(R)Sal in CSF of untreated parkinsonian patients. There was significant positive correlation between the activity of (R)Sal N-methyltransferase and the level of NM(R)Sal in samples from the same patients (P<.05).

whereas the activity peak at alkaline site was almost the same as in control.

The activity levels of the neutral (R)Sal N-methyltransferase were measured in the lymphocytes from 24 control subjects and 56 parkinsonian patients. As summarized in Table 3, the activity in parkinsonian lymphocytes was significantly higher than in the control and disease control. There was no difference between male and female subjects in either the parkinsonian or control lymphocytes. There was no correlation of the activity with the age of PD patients (r=0) or control subjects (r=.078) or with the age of the onset of PD (r=.3). The relation of Hoehn–Yahr stages to the activity in the lymphocytes was not confirmed. The activities of other enzymes related to the metabolism of NM(R)Sal were examined (Table 3). (R)Sal synthase activity was not detected in lymphocytes, the activity of an alkaline (R)Sal N-methyltransferase measured at pH 8.0 and that of NM(R)Sal oxidase were almost the same as in control and parkinsonian lymphocytes.

Recently, the correlation was confirmed between neutral (R)Sal N-methyltransferase activity in lymphocytes and the NM(R)Sal concentration in CSF from untreated newly diagnosed PD patients, as shown in Fig. 6. The results support our view that neutral (R)Sal N-methyltransferase in the brain may increase the neurotoxin in the substantia nigra and the activity in the lymphocytes may be applicable as a peripheral

Table 3

Activity of a neutral (R)Sal N-methyltransferase and related enzymes in lymphocytes

	Enzyme activity in lymphocytes prepared from	
	Parkinsonian patients (n=56)	Control (n=24)
Neutral N-methyltransferase[a] (pmol/min/mg protein)	100.2±81.8	18.9±15.0
Alkaline N-methyltransferase[a] (pmol/min/mg protein)	41.8±17.3	25.0±23.0
NM(R)Sal oxidase (pmol/min/mg protein)	2.15±2.43	1.38±2.23

[a] The activities of neutral and alkaline (R)Sal N-methyltransferase were measured at pH 7.0 and 8.0, respectively. The value represents the

5. Mechanism of cell death induced by $NM(R)$Sal

There has been increasing evidence suggesting that apoptosis is a major form of cell death in development of the brain system [13] and neurodegenerative disorders [63]. A potent dopaminergic neurotoxin, MPP^+, was reported to induce apoptosis in cultured cerebellar granule neurons [6], cultured rat ventral mesencephalic and striatal cells [35], and in pheochromocytoma PC12 cells [42]. In addition, recently, apoptotic cell death was confirmed in the substantia nigra of

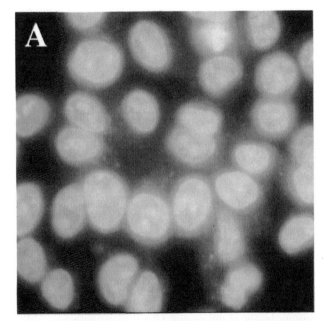

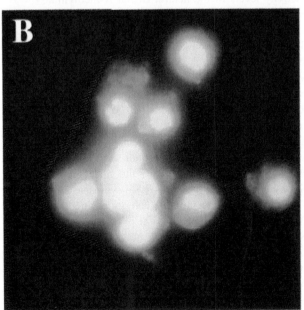

Fig. 7. Morphological aspects of $NM(R)$Sal-treated SH-SY5Y cells stained with Hoechst 33342. (A) Control. (B) Cells treated with 0.5 mM $NM(R)$Sal for 72 h. In some $NM(R)$Sal-treated cells, condensed and fragmented

Table 4
The cytotoxicity of catechol isoquinolines to dopaminergic SH-SY5Y cells

Isoquinoline	Apoptotic calls	Necrotic cells (% of the total)
Control	4.86 ± 4.34	1.21 ± 1.68
(R)Sal	9.55 ± 2.66	1.70 ± 1.41
(S)Sal	10.71 ± 4.54	1.33 ± 1.43
$NM(R)$Sal	100	
$NM(R)$Sal[a]	26.73 ± 4.57	0.85 ± 0.68
$NM(S)$Sal	10.90 ± 3.04	0.28 ± 0.44
$DMDHIQ^+$	28.02 ± 9.09	1.58 ± 1.82
(R)-1-Carboxy-salsolinol	3.73 ± 1.75	1.08 ± 1.10
(S)-1-Carboxy-salsolinol	5.72 ± 2.41	1.01 ± 0.78
Norsalsolinol	34.06 ± 5.05	1.63 ± 1.68
N-Methyl-norsalsolinol	8.87 ± 0.76	0.13 ± 0.32
(R)THP	0.25 ± 0.62	42.00 ± 7.45
(S)THP	0.07 ± 0.28	41.10 ± 8.10
Tetrahydropapaverine	10.13 ± 4.57	1.79 ± 0.50

SH-SY5Y cells were incubated with 500 μM isoquinolines for 24 h and stained with Hoechst 33342 and apoptotic and necrotic cells were assessed from morphological observation. The value represents mean ± S.D. of three experiments.

[a] 250 μM isoquinoline was used.

parkinsonian brains [2,36] and the activation of apoptotic cascade, such as the activated form of caspase-3 and Bax, and the nuclear translocation of glyceraldehyde-3-phosphate dehydrogenase (GAPDH) were detected in the parkinsonian substantia nigra [11,37,62].

We found that $NM(R)$Sal induces apoptosis in human dopaminergic neuroblastoma SH-SY5Y cells [1,26,28]. Among catechol isoquinolines, NMSal was the most potent to induce DNA damage, whereas Sal and $DMDHIQ^+$ were less potent. The studies on the intracellular mechanism underlying apoptosis suggest that mitochondria are the site to decide the cell death induced by this neurotoxin.

5.1. Apoptotic DNA damage by NM(R)Sal

After incubation with the catechol isoquinolines, DNA damage was assessed from shrinkage and fragmentation of cells and nuclei, loss of microvilli, and extensive degradation of chromosomal DNA. Fig. 7 shows apoptotic cells after incubation with $NM(R)$Sal. By staining with Hoechst 33342 and incubating with $NM(R)$Sal for 72 h, typical nuclear condensation and fragmentation were detected in almost all the cells. In control cells, such morphological changes were not detected. DNA damage was confirmed to be apoptotic by detection of DNA ladder formation [1], a hallmark of apoptosis [17].

Table 4 also shows the structure–activity relationship of cytotoxicity among catechol isoquinolines. Catechol isoquinoline induced apoptosis in SH-SY5Y cells rather than necrosis by morphological observation of the cells after staining with Hoechst 33342 and propidium iodide. $NM(R)$Sal markedly induced apoptotic DNA damage and the (S) enantiomer was much less cytotoxic. The results suggest that an intracellular mechanism capable of distin-

Phase contrast

JC-1 staining

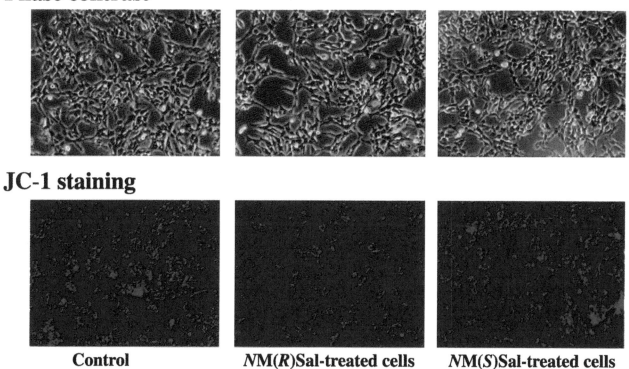

Control   **NM(R)Sal-treated cells**   **NM(S)Sal-treated cells**

Fig. 8. Changes of ΔΨm in the cells treated with NMSal. SH-SY5Y cells were incubated in the absence (control) or presence of 250 μM of NM(R)Sal or NM(S)Sal for 3 h. Red fluorescence represents J-aggregates insides intact mitochondria. The fluorescence decreased in the cells treated with NM(R)Sal without morphological change shown by phase-contrast microscopy. Red fluorescence of J-aggregate did not change in the cells treated with NM(S)Sal compared to control.

the death process in the cells. DMDHIQ$^+$ and norsalsolinol induced apoptosis, whereas (R)THP and (S)THP induced necrosis rather than apoptosis.

2. Intracellular death process induced by NM(R)Sal

Apoptosis is a death process tightly regulated and characterized by a series of stereotypic morphological features. Mitochondria are considered to control cell life and death. Three mechanisms are known to initiate and execute cell death [10,17]: (1) mitochondrial permeability transition (PT), (2) disruption of mitochondrial function, and (3) release of proteins activating caspase family proteases. A large conductance channel known as the mitochondrial PT pore occurs at the contact site between the outer and inner membrane of mitochondria. The structure and composition of the PT pore remain only partially defined, but it consists of adenine nucleotide translocator (ANT), voltage-dependent anion channel, and peripheral benzodiazepine receptor [65]. Mitochondrial PT is a sudden increase of the inner mem-

I   **II**   **III**

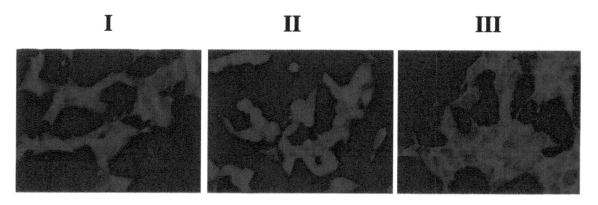

Fig. 9. Nuclear translocation of GAPDH by NM(R)Sal. SH-SY5Y cells were treated with 200 μM of NM(R)Sal for 24 h, and localization of GAPDH was determined by immunostaining with anti-GAPDH antibody. Fluorescence microscopy of GAPDH immunostaining of control cells without treatment (I),

brane permeability to solutes with molecular mass below 1500 Da. Apoptogenic proteins, cytochrome c, and apoptosis inducing factor, which are normally sequestered in mitochondria, are released through the outer membrane into cytosol and activate caspases, which are effectors of apoptosis. Opening of the PT pore induces decline in mitochondrial membrane potential, $\Delta\Psi m$, which is visualized with fluorescent indicators, JC-1, rhodamine 123, and MitoTracker Orange. Various insults induce the opening of PT pore, which is regulated by anti-apoptotic oncogenes Bcl-2 and Bcl-X$_L$ that stabilize the mitochondrial membrane and by pro-apoptotic Bax analogues that disrupt the $\Delta\Psi m$ [66]. As shown Fig. 8, $NM(R)$Sal reduced $\Delta\Psi m$ in a dose-dependent way after 30–60-min incubation evidenced by the decline in red JC-1 fluorescence. At normal hyperpolarized membrane, JC-1 aggregates in J-aggregates with red fluorescence, and at depolarized membrane, potential monomeric JC-1 shows green fluorescence. By comparison of the red/green JC-1 fluorescence, $\Delta\Psi m$ can be quantitatively assayed. Only the (R) enantiomer of NMSal induced $\Delta\Psi m$ collapse in SH-SY5Y cells, whereas the (S) enantiomer did not [30]. The enantiospecificity was confirmed also in isolated mitochondria (Akao et al., under revision), suggesting that proteins on the mitochondrial outer membrane distinguish the three-dimensional structure of NMSal and initiate the apoptotic signals. The $\Delta\Psi m$ reduction by $NM(R)$Sal was completely prevented by cyclosporin A, an inhibitor of ANT, indicating that opening of the PT pore accounts for $\Delta\Psi m$ decline.

Caspases are a family of cysteine protease and are involved in the biochemical and morphological changes associated with apoptosis [64]. $NM(R)$Sal was found to activate caspase-3 during apoptosis in SH-SY5Y cells [1]. After incubation with $NM(R)$Sal, the activated form of caspase-3 with 17 kDa was detected by Western blot analysis. In control cells, only the proenzyme with the molecular weight of 32 kDa was detected. SH-SY5Y cells were preincubated with an inhibitor of caspase-3, acetyl-L-aspartyl-L-glutamyl-L-valyl-L-aspartic aldehyde, then treated with $NM(R)$Sal for further 72 h. The inhibitor blocked the process of DNA ladder formation in a dose-dependent way, whereas untreated cells showed typical DNA fragmentation after exposure to $NM(R)$Sal.

During subsequently activated apoptotic process, GAPDH is translocated from cytosol into nuclei, from 6- to 24-h incubation with $NM(R)$Sal, as shown in Fig. 9 [31]. GAPDH was confirmed to aggregate in Golgi apparatus before the nuclear translocation [32]. Nuclear GAPDH translocation was once proposed as a signal to induce nucleosomal DNA fragmentation in some types of apoptosis.

To examine the regulation of the PT by Bcl-2 protein family, we prepared SH-SY5Y cells overexpressing Bcl-2. As reported [31], a stable cell line with Bcl-2 overexpression was prepared. Bcl-2 overexpressed cells were resistant to apoptosis induced by $NM(R)$Sal and the $\Delta\Psi m$ collapse, nuclear GAPDH translocation, and activation of caspase-3 were completely prevented, suggesting that $NM(R)$Sal-induced apoptosis is mediated by mitochondria under control of Bcl-2. Our results suggest that nuclear GAPDH is regulated by Bcl-2 and located in the downstream of apoptotic signal transduction initiated in mitochondria.

6. Discussion

As summarized in Fig. 10, the involvement of $NM(R)$ in the pathogenesis of PD is suggested by preparation of the

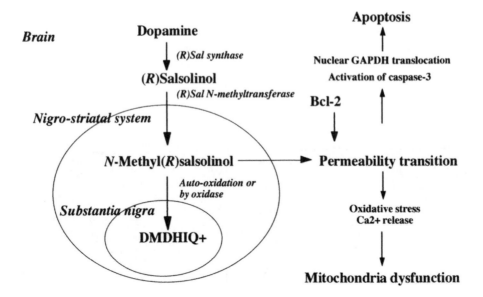

Fig. 10. Metabolism and cytotoxic mechanism of $NM(R)$Sal in the brain. (R)Sal is enantioselectively synthesized from dopamine in situ by a synthase and N-methylation by a neutral N-methyltransferase induces the accumulation in striatal dopamine neurons. By retrograde axonal flow, $NM(R)$Sal is transported to

animal and cellular models and analyses of clinical samples. The increased activity of neutral (R)Sal N-methyltransferase suggests that genetic and environmental factors may increase the activity in PD to enhance the neurotoxin level in the substantia nigra. Molecular studies on the altered characteristics of this enzyme will give us a new clue in understanding the pathogenesis of PD. Epidemiological studies on the enzyme in lymphocytes are now on the way to find out subjects with high risk factors among normal control and future studies will lead us to find preventive methods for PD using this peripheral marker.

In addition to NM(R)Sal, several endogenous alkaloids were reported to be related to the pathogenesis of PD. N-Methyl-norsalsolinol was detected in parkinsonian CSF, and the level reduced with the disease progression [39]. However, the cytotoxicity of this isoquinoline to dopamine neurons has not been confirmed [38], which is compatible with our results (Table 4). Another catechol isoquinoline, 2-methyl-4,6,7-trihydroxy-1,2,3,4-tetrahydroisoquinoline, was reported to be cytotoxic to dopamine neurons by in vivo [18] and in vitro experiments [67], but the occurrence in human brains has not been confirmed. The toxicity of THP to dopamine neurons was demonstrated by in vivo experiments [9,16]. Recently, we found that THP induced necrosis in SH-SY5Y cells, whereas papaveroline induced apoptosis [29]. It is in contradiction to the previous observation that the oxidized toxins, DMDHIQ$^+$, MPP$^+$, and 2,9-dimethylnorharmanium ion [33], induce necrosis, rather than apoptosis, by the rapid inhibition of mitochondrial Complex I and ATP synthesis. On the other hand, reduced forms of dopaminergic neurotoxins, NM(R)Sal and MPTP, mainly induced apoptosis in dopaminergic neurons. Considering that apoptosis is a common death form in parkinsonian brain, the reduced form of these alkaloids seem to be more appropriate neurotoxins to induce apoptosis in dopamine neurons in the substantia nigra after a long-term accumulation.

Acknowledgments

This work was supported by grants from the following programs: Grants-In-Aids on Priority Areas—Advanced Brain Project-(C) (W.M.) and for Scientific Research (C) (W.M.) from the Ministry of Education, Culture, Sports, and Science and Technology, Japan; Grant-In-Aid for Scientific Research on Priority Areas (C) from Japan Society for the Promotion of Science (W.M.); Grants for Longevity Sciences (W.M.), Grants for Comprehensive Research on Aging and Health (W.M. and M.N.), and Medical Frontier Strategy Research (W.M., Y.A, and M.N.) from the Ministry of Health, Labour, and Welfare, Japan.

References

[1] Y. Akao, Y. Nakagawa, W. Maruyama, T. Takahashi, M. Naoi, Apoptosis induced by an endogenous neurotoxin, N-methyl(R)salsolinol, is mediated by activation of caspase 3, Neurosci. Lett. 267 (1999) 153–156.

[2] P. Anglade, S. Vyas, F. Javoy-Agid, M.T. Herrero, P.P. Michel, J. Marquez, A. Mouatt-Priget, M. Ruberg, E.C.Y. Hirsch, Y. Agid, Apoptosis and autophagy in nigral neurons of patients with Parkinson's disease, Histol. Histopathol. 12 (1997) 25–31.

[3] M.A. Collins, J.J. Hannigan, T. Origitano, D. Moura, W. Osswald, On the occurrence, assay and metabolism of simple tetrahydroisoquinolines in mammalian tissues, Prog. Clin. Biol. Res. 90 (1982) 155–162.

[4] M.A. Collins, E.J. Neafsey, β-Carboline analogs of N-methyl-4-phenyl-1,2,3,6-tetrahydropyridine (MPTP); endogenous factors underlying idiopathic parkinsonism? Neurosci. Lett. 55 (1985) 179–184.

[5] G.C. Davis, A.C. Williams, S.P. Markey, M.H. Ebert, E.D. Caine, C.W. Reinchert, I.J. Kopin, Chronic parkinsonism secondary to intravenous injection of meperidine analogues, Psychiatry Res. 1 (1979) 249–254.

[6] B. Dipasquale, A.M. Marini, R.J. Youle, Apoptosis and DNA degradation induced by 1-methyl-4-phenylpyridinium in neurons, Biochem. Biophys. Res. Commun. 181 (1991) 1442–1448.

[7] P. Dostert, M. Strolin Benedetti, V. Bellotti, C. Allievi, G. Dordain, Biosynthesis of salsolinol, a tetrahydroisoquinoline alkaloid, in healthy subjects, J. Neural Transm. 81 (1990) 215–223.

[8] J.M. Fearnley, A.J. Lees, Ageing and Parkinson's disease: Substantia nigra regional selectivity, Brain 114 (1991) 2283–2301.

[9] H. Goto, H. Mochizuki, T. Hattori, N. Nakamura, Y. Mizuno, Neurotoxic effects of papaverine, tetrahydropapaverine and dimethoxyphenylethylamine on dopaminergic neurons in ventral mesencephalic–striatal co-culture, Brain Res. 754 (1997) 260–268.

[10] D.R. Green, J.C. Reed, Mitochondrial and apoptosis, Science 281 (1998) 1309–1312.

[11] A. Hartmann, S. Hunot, P.P. Michel, M.P. Muriel, S. Vyas, B.A. Faucheux, A. Mouatt-Prigent, H. Turmel, A. Srinivasan, M. Ruberg, G.I. Evan, Y. Agid, E.C. Hirsch, Caspase-3: A vulnerability factor and final effector in apoptotic death of dopaminergic neurons in Parkinson's disease, Proc. Natl. Acad Sci. U. S. A. 97 (2000) 2875–2880.

[12] P. Holz, K. Stocj, W. Westermann, M. Georgi, M.F. Melzig, Formation of tetrahydropapaveroline from dopamine in vitro, Nature 203 (1964) 656–658.

[13] M.D. Jacobson, M. Weil, M.G. Raff, Programmed cell death in animal development, Cell 88 (1997) 347–354.

[14] S.J. Kish, K. Shannak, A. Rajput, J.H.N. Deck, O. Hornykiewicz, Aging produces a specific pattern of striatal dopamine loss: Implications for the etiology of idiopathic Parkinson's disease, J. Neurochem. 58 (1992) 642–648.

[15] T. Kitada, S. Asakawa, N. Hattori, H. Matsumine, Y. Yamamura, S. Minoshima, Y. Yokochi, Y. Mizuno, N. Shimizu, Mutations in the parkin gene cause autosomal recessive juvenile parkinsonism, Nature 392 (1998) 605–608.

[16] I. Koshimura, H. Imai, T. Hidano, K. Endo, H. Mochizuki, T. Kondo, Y. Mizuno, Dimethoxyphenylethanolamine and tetrahydropapaverine are toxic to the nigrostriatal system, Brain Res. 773 (1997) 108–116.

[17] G. Kroemer, B. Dallaporta, M. Resche-Rigon, The mitochondrial death/life regulator in apoptosis and necrosis, Annu. Rev. Physiol. 60 (1998) 619–642.

[18] J. Liptrot, D. Holdup, O. Phillipson, 1,2,3,4-Tetrahydro-2-methyl-4,6,7-isoquinolinetriol depletes catecholamines in rat brain, J. Neurochem. 61 (1993) 2199–2206.

[19] C.D. Marsden, Parkinson's disease, Lancet 335 (1990) 948–952.

[20] W. Maruyama, D. Nakahara, M. Ota, T. Takahashi, A. Takahashi, T. Nagatsu, M. Naoi, N-Methylation of dopamine-derived 6,7-dihydroxy-1,2,3,4-tetrahydroisoquinoline, (R)-salsolinol in rat brains: In vivo microdialysis study, J. Neurochem. 59 (1992) 395–400.

[21] W. Maruyama, M. Naoi, Inhibition of tyrosine hydroxylase by a dopamine neurotoxin, 1-methyl-4-phenylpyridinium ion: Depletion of allostery to the biopterin cofactor, Life Sci. 55 (1994) 207–212.

[22] W. Maruyama, P. Dostert, M. Naoi, Dopamine-derived 1-methyl-6,7-dihydroxyisoquinolines as hydroxyl radical promoters and scavengers

in the rat brain: In vivo and in vitro studies, J. Neurochem. 64 (1995) 2635–2643.

[23] W. Maruyama, P. Dostert, K. Matsubara, M. Naoi, N-Methyl(R)salsolinol produces hydroxyl radicals : Involvement to neurotoxicity, Free Radic. Med. Biol. 19 (1995) 67–75.

[24] W. Maruyama, T. Abe, H. Tohgi, P. Dostert, M. Naoi, A dopaminergic neurotoxin, (R)-N-methylsalsolinol, increases in parkinsonian cerebrospinal fluid, Ann. Neurol. 40 (1996) 119–122.

[25] W. Maruyama, H. Narabayashi, P. Dostert, M. Naoi, Stereospecific occurrence of a parkinsonian-inducing catechol isoquinoline, N-methyl(R)-salsolinol, in the human intraventricular fluid, J. Neural Transm. 103 (1996) 1069–1076.

[26] W. Maruyama, M. Naoi, T. Kasamatsu, Y. Hashizume, T. Takahashi, K. Kohda, P. Dostert, An endogenous dopaminergic neurotoxin, N-methyl(R)salsolinol, induces DNA damage in human dopaminergic neuroblastoma SH-SY5Y cells, J. Neurochem. 69 (1997) 322–329.

[27] W. Maruyama, G. Sobue, K. Matsubara, Y. Hashizume, P. Dostert, M. Naoi, A dopaminergic neurotoxin, 1(R),2(N)-dimethyl-6,7-dihydroxy-1,2,3,4-tetra-hydroisoquinoline, N-methyl(R)salsolinol, and its oxidation product, 1,2(N)-dimethyl-6,7-dihydroxyisoquinolinium ion, accumulate in the nigrostriatal system of the human brain, Neurosci. Lett. 223 (1997) 61–64.

[28] W. Maruyama, M. Strolin Benedetti, T. Takahashi, M. Naoi, A neurotoxin N-methyl(R)salsolinol induces apoptotic cell death in differentiated human dopaminergic neuroblastoma SH-SY5Y cells, Neurosci. Lett. 232 (1997) 147–150.

[29] W. Maruyama, K. Sango, K. Iwasa, C. Minami, P. Dostert, M. Kawai, M. Moriyasu, M. Naoi, Dopaminergic neurotoxins, 6,7-dihydroxy-1-(3′,4′-dihydroxybenzyl)-isoquinolines, cause different types of cell death in SH-SY5Y cells: Apoptosis was induced by oxidized papaverolines and necrosis by reduced tetrahydropapaverolines, Neurosci. Lett. 291 (2000) 89–92.

[30] W. Maruyama, A.A. Boulton, B.A. Davis, P. Dostert, M. Naoi, Enantio-specific induction of apoptosis by an endogenous neurotoxin, N-methyl(R)salsolinol, in dopaminergic SH-SY5Y cells: Suppression of apoptosis by N-(2-heptyl)-N-methylpropargylamine, J. Neural Transm. 108 (2001) 11–24.

[31] W. Maruyama, Y. Akao, M.B.H. Youdim, G.A. Davis, M. Naoi, Transfection-enforced Bcl-2 overexpression and an antiparkinson drug, rasagiline, prevent nuclear accumulation of glyceraldehydes-3-phosphate dehydrogenase induced by an endogenous dopaminergic neurotoxin, N-methyl(R)salsolinol, J. Neurochem. 78 (2001) 727–735.

[32] W. Maruyama, T. Oya-Ito, M. Shamoto-Nagai, T. Osawa, M. Naoi, Glyceraldehyde-3-phosphate dehydrogenase is translocated into nuclei through Golgi apparatus during apoptosis induced by 6-hydroxydopamine in human dopaminergic SH-SY5Y cells, Neurosci. Lett. 321 (2002) 29–32.

[33] K. Matsubara, T. Gonda, H. Sawada, T. Uezono, Y. Kobayashi, T. Kawamura, K. Ohtaki, K. Kimura, A. Akaike, Endogenously occurring β-carboline induces parkinsonism in nonprimate animals: A possible causative protoxin in idiopathic Parkinson's disease, J. Neurochem. 70 (1998) 727–735.

[34] P.L. McGeer, S. Itagaki, H. Akiyama, E.G. McGeer, Rate of cell death in Parkinson's disease indicates active neuropathological process, Ann. Neurol. 24 (1988) 574–576.

[35] H. Mochizuki, N. Nakamura, K. Nishi, Y. Mizuno, Apoptosis induced by 1-methyl-4-phenylpyridinium ion (MPP$^+$) in ventral mesencephalic-striatal co-culture in rat, Neurosci. Lett. 170 (1994) 191–194.

[36] H. Mochizuki, G. Goto, H. Mori, Y. Mizuno, Histochemical detection of apoptosis in Parkinson's disease, J. Neurol. Sci. 137 (1996) 120–123.

[37] M. Mogi, A. Togari, T. Kondo, Y. Mizuno, O. Komure, S. Kuno, H. Ichinose, T. Nagatsu, Caspase activities and tumor necrosis factor receptor R1 (p55) are elevated in the substantia nigra from parkinsonian brain, J. Neural Transm. 107 (2000) 335–341.

[38] N. Morikawa, M. Naoi, W. Maruyama, S. Ohta, Y. Kotake, H. Kawai,

[39] T. Niwa, P. Dostert, Y. Mizuno, Effects of various tetrahydroisoquinoline derivatives on mitochondrial respiration and the electron transfer complexes, J. Neural Transm. 105 (1998) 677–688.

[39] A. Moser, J. Scholz, F. Nobbe, P. Vieregge, V. Böhme, H. Bamberg, Presence of N-methyl-norsalsolinol in the CSF: Correlation with dopamine metabolites of patients with Parkinson's disease, J. Neurol. Sci. 131 (1995) 183–189.

[40] T. Muller, S. Sallstrom Baum, P. Haussermann, D. Woitalla, H. Rommelspracher, H. Przuntek, W. Kuhn, Plasma levels of R- and S-salsolinol are not increased in "de-novo" parkinsonian patients, J. Neural Transm. 105 (1998) 239–246.

[41] T. Muller, S. Sallstrom Baum, P. Haussermann, H. Przuntek, H. Rommelspracher, W. Kuhn, R- and S-salsolinol are not increased in cerebrospinal fluid of parkinsonian patients, J. Neurol. Sci. 164 (1999) 158–162.

[42] T. Mutoh, A. Tokuda, A.M. Marini, N. Fujiki, 1-Methyl-4-phenylpyridinium kills differentiated PC12 cells with a concomitant change in protein phosphorylation, Brain Res. 661 (1994) 51–55.

[43] T. Nagatsu, M. Yoshida, An endogenous substance of the brain, tetrahydroisoquinoline, produces parkinsonism in primates with decreased dopamine, tyrosine hydroxylase and biopterin in the nigrostriatal regions, Neurosci. Lett. 87 (1988) 178–182.

[44] M. Naoi, W. Maruyama, I.N. Acworth, D. Nakahara, H. Parvez, Multi-electrode detection system for determination of neurotransmitters. In: S.H. Parvez, M. Naoi, T. Nagatsu, S. Parvez (Eds.), Methods in Neurotransmitter and Neuropeptide Research: Part 1. Techniques in the Behavioral and Neural Sciences vol. 11, Elsevier, Amsterdam, 1993, pp. 1–39.

[45] M. Naoi, W. Maruyama, P. Dostert, Binding of 1,2(N)-dimethyl-6,7-dihydroxyisoquinolinium ion to melanin; effects of ferrous and ferric ion on the binding, Neurosci. Lett. 171 (1994) 9–12.

[46] M. Naoi, W. Maruyama, J.H. Zhang, T. Takahashi, Y. Deng, P. Dostert, Enzymatic oxidation of the dopaminergic neurotoxin, 1(R),2(N)-dimethyl-6,7-dihydroxy-1,2,3,4-tetrahydroisoquinoline, into 1,2-dimethyl-6,7-dihydroxyiso-quinolinium ion, Life Sci. 57 (1995) 1061–1066.

[47] M. Naoi, W. Maruyama, P. Dostert, Y. Hashizume, D. Nakahara, T. Takahashi, M. Ota, Dopamine-derived endogenous 1(R),2(N)-dimethyl-6,7-dihydroxy-1,2,3,4-tetrahydroisoquinoline, N-methyl-(R)-salsolinol, induced parkinsonism in rat; biochemical, pathological and behavioral studies, Brain Res. 709 (1996) 285–295.

[48] M. Naoi, W. Maruyama, P. Dostert, K. Kohda, T. Kaiya, A novel enzyme enantio-selectively synthesizes (R)salsolinol, a precursor of a dopaminergic neurotoxin, N-methyl(R)salsolinol, Neurosci. Lett. 212 (1996) 183–186.

[49] M. Naoi, W. Maruyama, K. Matsubara, Y. Hashizume, A neutral N-methyltransferase activity in the striatum determines the level of an endogenous MPP$^+$-like neurotoxin, 1,2-dimethyl-6,7-dihydroxyisoquinolinium ion, in the substantia nigra of human brains, Neurosci. Lett. 235 (1997) 81–84.

[50] M. Naoi, W. Maruyama, N. Nakao, T. Ibi, K. Sahashi, M. Strolin Benedetti, (R)Salsolinol N-methyltransferase activity increases in parkinsonian lymphocytes, Ann. Neurol. 43 (1998) 212–216.

[51] M. Naoi, W. Maruyama, P. Dostert, An animal model of Parkinson's disease prepared by endogenous N-methyl(R)salsolinol, in: A. Moser (Ed.), Pharmacology of Endogenous Neurotoxins, A Handbook, Birkhäuser, Boston, 1998, pp. 41–61.

[52] M. Naoi, W. Maruyama, N-methyl(R)salsolinol and (R)salsolinol N-methyltransferase as possible pathogenic factors in Parkinson's disease, in: A. Fisher, I. Hanin, M. Yoshida (Eds.), Progress in Alzheimer's and Parkinson's Diseases, Plenum, New York, 1998, pp. 413–420.

[53] M. Naoi, W. Maruyama, K. Matsubara, K. Tipton, M. Strolin Benedetti, H. Parvez, Analysis of salsolinols, endogenous neurotoxins, in human materials, in: G.A. Quareshi, H. Parvez, P. Caudy, S. Parvez (Eds.), Progress in HPLC–HPCE: Vol. 7. Neurochemical Markers of Degenerative Nervous Diseases and Drug Addiction, VSP, Utrecht, 1998, pp. 423–460.

[54] M. Naoi, W. Maruyama, A. Akao, J. Zhang, H. Parvez, Apoptosis

induced by an endogenous neurotoxin, *N*-methyl(*R*)salsolinol, in dopamine neurons, Toxicology 153 (2000) 123–141.

[55] M.H. Polymeropoulos, J.J. Higgins, L.I. Golbe, W.G. Johnson, S.E. Ide, G.D. Iorio, G. Sanges, E.S. Stenroos, L.T. Pho, A.A. Schaffer, A.M. Lazzarini, R.L. Nussbaum, R.C. Duvoisin, Mapping of a gene for Parkinson's disease to chromosome 4q21–q23, Science 274 (1996) 1197–1199.

[56] J.H. Robbins, Alkaloid formation by condensation of biogenic amines with acetaldehyde, Clin. Res. 16 (1968) 350.

[57] M. Sandler, S. Bonham Carter, K.R. Hunter, G.M. Stern, Tetrahydroisoquinoline alkaloids: In vivo metabolites of L-DOPA in man, Nature 241 (1973) 439–443.

[58] K. Sango, W. Maruyama, K. Matsubara, P. Dostert, C. Minami, M. Kawai, M. Naoi, Enantio-selective occurrence of (*S*)-tetrahydropapaveroline in human brain, Neurosci. Lett. 283 (2000) 224–226.

[59] M. Strolin Benedetti, V. Bellotti, E. Pianezzola, P. Moro Carminati, P. Dostert, Ratio of the *R* and *S* enantiomers of salsolinol in food and human urine, J. Neural Transm. 77 (1989) 47–53.

[60] T. Takahashi, Y. Deng, W. Maruyama, P. Dostert, M. Kawai, M. Naoi, Uptake of a neurotoxin-candidate, (*R*)-1,2-dimethyl-6,7-dihydroxy-1,2,3,4-tetrahydroisoquinoline into human dopaminergic neuroblastoma SH-SY5Y cells by dopamine transport system, J. Neural Transm.: Gen. Sect. 98 (1994) 107–118.

[61] T. Takahashi, W. Maruyama, Y. Deng, P. Dostert, D. Nakahara, T. Niwa, S. Ohta, M. Naoi, Cytotoxicity of endogenous isoquinolines to human dopaminergic neuroblastoma SH-SY5Y cells, J. Neural Transm. 104 (1997) 59–66.

[62] N.A. Tatton, Increased caspase 3 and Bax immunoreactivity accompany nuclear GAPDH translocation and neuronal apoptosis in Parkinson's disease, Exp. Neurol. 166 (2000) 29–43.

[63] C.B. Thompson, Apoptosis in the pathogenesis and treatment of diseases, Science 267 (1995) 1456–1462.

[64] N.A. Thornberry, Y. Lazebnik, Caspases: Enemies within, Science 281 (1998) 1312–1316.

[65] Y. Tsujimoto, S. Shimizu, Bcl-2 family: Life-or-death switch, FEBS Lett. 336 (2000) 6–10.

[66] Y. Tsujimoto, S. Shimizu, VDAC regulation by the Bcl-2 family of proteins, Cell Death Differ. 7 (2001) 1174–1181.

[67] J.M. Willets, D.G. Lambert, J. Lunec, H.R. Griffiths, O. Phillipson, Neurotoxicity of 1,2,3,4-tetrahydro-2-methyl-4,6,7-isoquinolinetriol (TMIQ) and effects on catecholamine homeostasis in SH-SY5Y cells, Environ. Toxicol. Pharmacol. 2 (1996) 59–68.

NEUROTOXICOLOGY
AND
TERATOLOGY

Neurotoxicology and Teratology 24 (2002) 593–598

www.elsevier.com/locate/neutera

Review article

N-methylation underlying Parkinson's disease

Kazuo Matsubara[a,*], Koji Aoyama[b], Manabu Suno[a], Toshio Awaya[a]

[a]*Department of Hospital Pharmacy and Pharmacology, Asahikawa Medical College, Asahikawa 078-8510, Japan*
[b]*Department of Internal Medicine III, Shimane Medical University, Izumo 693-8501, Japan*

Received 14 December 2001; accepted 24 January 2002

Abstract

The discovery of 1-methyl-4-phenyl-1,2,2,6-tetrahydropyridine (MPTP) leads to the hypothesis that Parkinson's disease (PD) is may be initiated or precipitated by environmental or endogenous toxins by the mechanism similar to that of MPTP in genetically-predisposed individuals. Endogenous analogs of MPTP, such as β-carbolines (βCs) and tetrahydroisoquinolines, have been proposed as possible causative candidates causing PD and are bioactivated into potential neurotoxins by *N*-methylation enzyme(s). These *N*-methylated βCs and tetrahydroisoquinoline have been higher cerebrospinal levels in parkinsonian patients than age-matched controls. Thus, there is a hypotheses to influence the pathogenesis of PD, that is, the excess enzyme activity to activate neurotoxins, such as *N*-methyltransferase, might be higher in PDs. Indeed, simple βCs, via *N*-methylation steps, induced bradykinesia with the decreased dopamine contents in the striatum and midbrain in C57/BL mice. In younger (65 years old) PD patients, the excretion amount of N^1-methyl-nicotinamaide was significantly higher than that in younger controls. The protein amount of nicotinamide *N*-methyltransferase (NNMT) was also significantly higher in younger PD patients than that in younger controls. These findings described here would indicate that the excess *N*-methylation ability for azaheterocyclic amines, such as βCs, before the onset had been implicated in PD pathogenesis. On the other hand, the contribution of aberrant cytochrome P450 or aldehyde oxidase activity acting on the pyridine ring, that could act as detoxification routes of endogenous neurotoxins, would be small in the etiology of PD. © 2002 Elsevier Science Inc. All rights reserved.

Keywords: β-Carboline; *N*-methylation; Cerebrospinal fluid; Parkinson's disease; Neurotoxin

1. Introduction

Epidemiological studies indicate that idiopathic Parkinson's disease (PD) is associated with certain environmental factors, such as early rural life [10,22,28], drinking water from rural wells [22,28], exposure to certain pesticides [29], wood preservatives [29] and industrial toxicants [31,32]. The discovery of 1-methyl-4-phenyl-1,2,3,6-tetrahydropyridine (MPTP) leads to the hypothesis that PD may be initiated or precipitated by environmental or endogenous toxins by the mechanism similar to that of MPTP in genetically predisposed individuals [9,31]. Endogenous analogs of MPTP, such as β-carbolines (βCs) and tetrahydroisoquinolines, have been proposed as possible causative candidates causing PD [4,11,23,24] and are bioactivated into potential neurotoxins by *N*-methylation enzyme(s)

[19,26,27]. These *N*-methylated βC and tetrahydroisoquinoline have been higher in cerebrospinal levels in parkinsonian patients than age-matched controls [12,18]. Thus, the involvement of aberrant *N*-methylation of azaheterocyclic amines in the etiology of PD has been argumentative.

2. βC *N*-methyltransferase

Indoleamine-related βCs have been proposed as promising endogenous MPTP-like protoxins [4]. We have previously reported that enzymes in mammalian brain methylate βCs, sequentially forming 2-mono-[*N*]-methylated and neurotoxic 2,9-di-[*N*,*N'*]-methylated β-carbolinium cations (2,9-Me₂βC⁺s) (Fig. 1) [14,19]. These βC⁺s are structural analogs of MPP⁺ with a nitrogen bridge. βC⁺s not only inhibit DA reuptake [5,16,21] and tyrosine hydroxylase [17], but also function as NADH-linked respiratory inhibitors in isolated mitochondria [1,4]. Among βC species, norharman (NH) could be an underlying factor in idiopathic PD that is sequentially methylated to form 2,9-MeNH⁺ in

* Corresponding author. Tel.: +81-166-69-3240; fax: +81-166-65-1392.

E-mail address: kmatsuba@asahikawa-med.ac.jp (K. Matsubara).

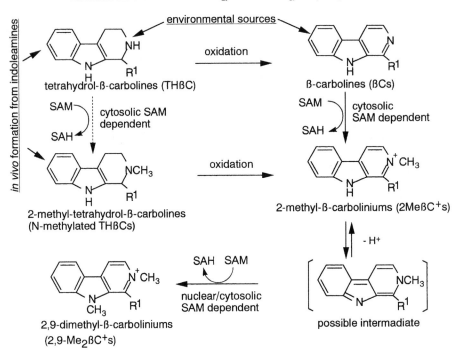

Fig. 1. Metabolic bioactivation routes to form neurotoxic βC$^+$s in mammalian brain [14,19]. R^1=H: NH; R^1=CH^3: harman.

vitro and in vivo, since a higher cerebrospinal fluid (CSF) level of 2,9-MeNH$^+$ has been found in the CSF of patients with PD [18]. The striatal NH$^+$ is attenuated by the treatment with L-deprenyl (Fig. 2) [20]. Systematic administration of simple βC, NH, induces bradykinesia (Fig. 3) with the decreased dopamine contents in the striatum and midbrain of C57/BL mouse [15]. In this animal model, tyrosine hydroxylase-positive cells in the substantia nigra pars compacta have been diminished (Fig. 4 and Table 1). The formation of a toxic metabolite in the central nervous system, 2,9-Me$_2$NH$^+$ (Fig. 5), has been suggested to be responsible for the NH toxicity. It has been suggested that 9[indole]-nitrogen methylation is the limiting step in the development of simple βC toxicity.

A higher concentration of 2,9-Me$_2$NH$^+$ localized in the nigra than in the cortex, and we observed the S-adenosyl-L-

methionine (SAM)-dependent methylation of 2[β]- and 9[indole]-nitrogens of βCs in non-PD human brains [13] and PD cases [7]. The activity of βC-2[β]-N-methyltrans-

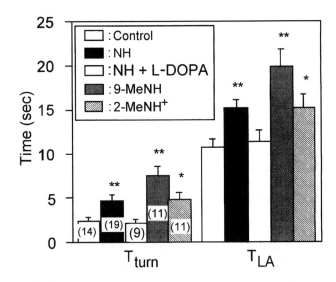

Fig. 3. Subchronic treatments with βCs induced behavioral bradykinesia in C57BL/6 mice [15]. The test was done 2 days after the last injection. NH (0.5 mmol/kg), 2-MeNH (0.35 mmol/kg), 9-MeNH (0.25 mmol/kg) or vehicle was given intraperitoneally twice per day, 12 h apart, for 7 days. An intraperitoneal injection of L-DOPA (100 mg/kg) with carbidopa (10 mg/kg) was performed 1 h before the pole test. T_{turn} was defined as the time from which the mouse released either forefoot from the pole to turn downward completely. T_{LA} was defined as the time until all four feet reached the floor after turning on the top of the pole. Data are expressed as mean±S.E.M. The number of experimental animals used is in parentheses. *$P<.05$, **$P<.01$ compared with controls.

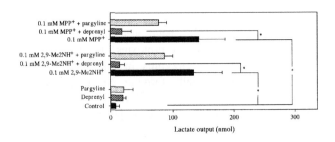

Fig. 2. Effects of intrastriatal applications of L-deprenyl (10 μM) and pargyline (20 μM) on the efflux of lactate induced by either MPP$^+$ or 2,9-Me$_2$NH$^+$ [20]. The total lactate efflux during and after (240 min) the administration of the cation was obtained by subtraction of basal output of lactate. Number of animals used in each group was seven or eight. *$P<.01$.

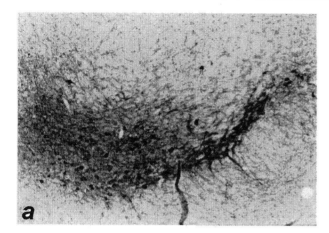

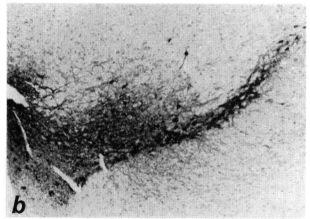

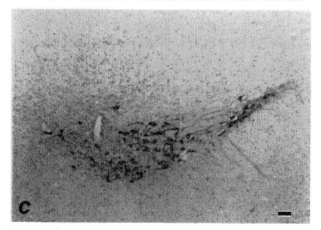

Fig. 4. Immunohistochemical examination on the nigra 2 days after the last injection in mice treated with βCs [15]. NH (0.5 mmol/kg), 9-MeNH (0.25 mmol/kg) or vehicle was given intraperitoneally twice per day, 12 h apart, for 7 days. Neurons were stained with TH antibody. (A) Control, (B) NH-treated and (C) 9-MeNH-treated mice. Note that loss of TH-immunoreactive nerve cells in the SNc was significant in the mice treated with NH (B) and 9-MeNH (C). Scale bars: 50 μm. *Lateral ventricle.

Table 1
TH immunohistochemistry of βC-treated mice in the substantia nigra

Drug treatment	Control	NH	9-MeNH
SNc	71.8±5.1	54.8±3.2*	47.2±4.1*
VTA	113.8±8.8	110.3±8.9	77.4±6.6*

Number of cell bodies (mean±S.E.M. of three in each group) that stained with TH antibody was counted 2 days after the last injection. The mice received NH (14×0.5 mmol/kg), 9[indole]N-methylated NH (14×0.25 mmol/kg) or vehicle at 12-h intervals. Cell numbers are expressed as the mean number per section.
* $P<.01$ vs. control.

be increased fourfold in PD frontal cortex [6]. On the other hand, no significant difference has been observed in any brain region in particulate and supernatant fraction βC 2[β]-N-methyltransferase activity or particulate fraction βC 9N-methyltransferase activity. This aberrant 9[indole]-N-methylation might be involved in idiopathic PD. These data also supported the proposal that endogenous MPP⁺ analogs such as 2,9-Me$_2$βC⁺s are plausible factors in the etiology of idiopathic PD.

3. N-methylation ability for azaheterocyclic amines in Parkinson's patients

There are two metabolic hypotheses to influence the pathogenesis of PD: (i) the excess enzyme activity to activate neurotoxins, such as N-methyltransferase, might be higher in PDs. Indeed, in patients with PD, the activity of neutral N-

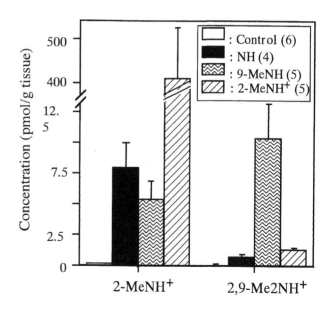

Fig. 5. Formation of active metabolite, 2,9-Me$_2$NH⁺, in the cerebrum of mice 8 h after three consecutive intraperitoneal injections of βCs with 12-h intervals [15]. The doses of NH, 2-MeNH⁺ and 9-MeNH were 0.5, 0.35 and 0.25 mmol/kg, respectively. 2,9-Me$_2$NH⁺ formation after βCs was statistically significant [$F(3,15)=13.922$, $P<.01$]. Data are expressed as mean±-S.E.M. The number of experimental animals used is in parentheses.

ferase is primarily localized in the cytosol. This 2[β]-N-methylation enzyme for βC has been recently identified, at least in part, as phenylethanolamine N-methyltransferase [8]. The enzyme catalyzing 9[indole]-N-methylation of 2-MeβC⁺ has not been characterized yet. However, βC 9[indole]-N-methyltransferase activity has been found to

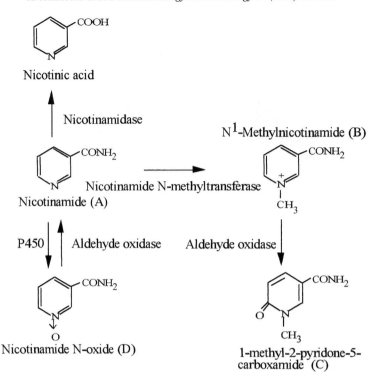

Fig. 6. Metabolic pathways of nicotinamide. The major metabolic pathway in nicotinamide (A) metabolism is N-methylation step to form N^1-methylnicotinamide (B), which is further catalyzed by aldehyde oxidase to 1-methyl-2-pyridone-5-carboxyamide (C). The conversion of nicotinamide to its N-oxide (D) is primary catalyzed by cytochrome P450. Nicotinamide N-oxide is reconverted to nicotinamide by aldehyde oxidase.

methyltransferase, which can methylate catechol isoquinolines, is higher in comparison with that in controls [25]; (ii)

the activity of cytochrome P450 mono-oxygenase, which could detoxify endogenous neurotoxins, might be lower in the patients [30]. Thus, the determination of aberrant metabolic state in PD patients is an intriguing issue.

Nicotinamide has a pyridine ring in its structure and is metabolized through the pathways similar to those for the

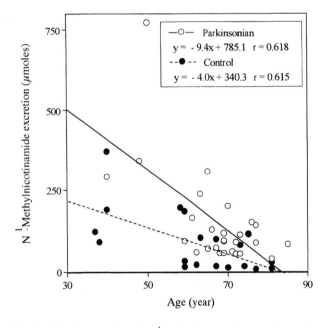

Fig. 7. Age-related decline in N^1-methylnicotinamide excretion. After dosing with 100 mg of nicotinamide and 200 ml of water, the urine was collected for 4 h [3]. The excretion of N^1-methylnicotinamide decreased along with aging both in PD patients and controls ($P<.01$ in both groups). The decline rate of N^1-methylnicotinamide excretion in PD patients was significantly greater than that in controls; the decline rate is more than twofold higher in PD patients ($P<.05$).

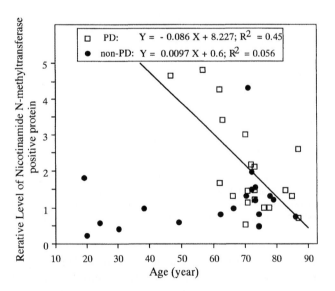

Fig. 8. The relationship between the relative amount of NNMT in CSF and age [2]. The age-associated decline in the amount of NNMT-positive protein was observed only in PD patients ($r=.543$, $P<.02$), but not in control patients.

endogenous neurotoxins (Fig. 6). To estimate the *N*-methylation ability for azaheterocyclic amines in PD patients, nicotinamide was dosed in 100 mg to 26 PDs, and 20 controls consisted of 16 other neurogenic disease patients and 4 healthy volunteers [3]. The urine was collected for 4 h, and then urinary metabolites analyzed by an improved HPLC method. The urinary excretions of nicotinamide metabolites were significantly affected by aging. The excretion of N^1-methylnicotinamide decreased along with aging both in PD patients and controls. In younger (65 years old or younger) PD patients, the excretion amount of N^1-methylnicotinamide was significantly higher than that in younger controls (Fig. 7). The decline rate of N^1-methylnicotinamide excretion in PDs was significantly greater than that in controls; the rate is more than twofold higher in PD patients. The age-associated decrease in 1-methyl-2-pyridone-5-carboxyamide excretion was observed only in PD patients, but not in controls. The total excreted amount of *N*-methylated metabolites (N^1-methylnicotinamide plus 1-methyl-2-pyridone-5-carboxyamide) was also observed in the age-related decline in both groups. Aging did not influence the urinary excretions of nicotinamide and nicotinamide-*N*-oxide.

4. Nicotinamide *N*-methyltransferase in the CSF of PD patients

Among several *N*-methyltransferases that could be involved in *N*-methylation of azaheterocyclic amines to form neurotoxic products in mammals, nicotinamide *N*-methyltransferase (NNMT) is the sole enzyme that has been well characterized. NNMT is a SAM-dependent cytosolic enzyme that catalyzes the *N*-methylation of nicotinamide to form *N*-methylnicotinamide and exists in many tissues. NNMT catalyzes *N*-methylation of nicotinamide and other pyridines to form pyridinium ions. The protein amount of NNMT was measured in the lumber CSF of PD patients by immunoblot analysis using antihuman NNMT antibody [2]. In younger (65 years old or younger) PD patients, the relative level of NNMT protein was significantly higher than that in younger controls. The NNMT protein was significantly affected by aging: the amount decreased along with aging in PD patients (Fig. 8) [2]. These results suggested that the excess NNMT in the central nervous system might be implicated in the PD pathogenesis.

5. Conclusion

These findings described here would indicate that the excess *N*-methylation ability for azaheterocyclic amines, such as βCs, before the onset had been implicated in PD pathogenesis. On the other hand, the contribution of aberrant cytochrome P450 or aldehyde oxidase activity acting on the pyridine ring, which could act as detoxification routes of endogenous neurotoxins, would be small in the etiology of PD.

References

[1] R. Albores, E.J. Neafsey, G. Drucker, J.Z. Fields, M.A. Collins, Mitochondrial respiratory inhibition by *N*-methylated-β-carboline derivatives structurally resembling *N*-methyl-4-phenylpyridine, Proc. Natl. Acad. Sci. USA 87 (1990) 9368–9372.

[2] K. Aoyama, K. Matsubara, M. Kondo, Y. Murakawa, M. Suno, K. Yamashita, S. Yamaguchi, S. Kobayashi, Nicotinamide-*N*-methyltransferase is higher in the lumbar cerebrospinal fluid of patients with Parkinson's disease, Neurosci. Lett. 298 (2001) 78–80.

[3] K. Aoyama, K. Matsubara, K. Okada, S. Fukushima, K. Shimizu, S. Yamaguchi, T. Uezono, M. Satomi, N. Hayase, S. Ohta, H. Shiono, S. Kobayashi, *N*-methylation ability for azaheterocyclic amines is higher in Parkinson's disease: Nicotinamide loading test, J. Neural Transm. 107 (2000) 985–995.

[4] M.A. Collins, E.J. Neafsey, K. Matsubara, R.J. Cobuzzi Jr., H. Rollema, Indole-*N*-methylated β-carbolinium ions as potential brain-bioactivated neurotoxins, Brain Res. 570 (1992) 154–160.

[5] G. Drucker, K. Raikoff, E.J. Neafsey, M.A. Collins, Dopamine uptake inhibitory capacities of β-carboline and 3,4-dihydro-β-carboline analogs of *N*-methyl-4-phenyl-1,2,3,6-tetrahydropyridine (MPTP) oxidation products, Brain Res. 509 (1990) 125–133.

[6] D.A. Gearhart, M.A. Collins, J.M. Lee, E.J. Neafsey, Increased beta-carboline 9*N*-methyltransferase activity in the frontal cortex in Parkinson's disease, Neurobiol. Dis. 7 (2000) 201–211.

[7] D.A. Gearhart, E.J. Neafsey, M.A. Collins, Characterization of brain beta-carboline-2-*N*-methyltransferase, an enzyme that may play a role in idiopathic Parkinson's disease, Neurochem. Res. 22 (1997) 113–121.

[8] D.A. Gearhart, E.J. Neafsey, M.A. Collins, Phenylethanolamine *N*-methyltransferase has β-carboline 2*N*-methyltransferase activity; hypothetical relevance to Parkinson's disease, Neurochem. Int. 40 (2002) 611–620.

[9] P. Jenner, A.H.V. Shapira, C.D. Marsden, New insights into the cause of Parkinson's disease, Neurology 42 (1992) 2241–2250.

[10] W. Koller, B. Vetere-Overfield, C. Gray, C. Alexander, T. Chin, J. Dolezal, R. Hassanein, C. Tanner, Environmental risk factors in Parkinson's disease, Neurology 40 (1990) 1218–1221.

[11] Y. Kotake, Y. Tasaki, Y. Makino, S. Ohta, M. Hirobe, 1-Benzyl-1,2,3,4-tetrahydroisoquinoline as a parkinsonism-inducing agent: A novel endogenous amine in mouse brain and parkinsonian CSF, J. Neurochem. 65 (1995) 2633–2638.

[12] W. Maruyama, T. Abe, H. Tohgi, P. Dostert, M. Naoi, A dopaminergic neurotoxin, (*R*)-*N*-methylsalsolinol, increases in parkinsonian cerebrospinal fluid, Ann. Neurol. 40 (1996) 119–122.

[13] K. Matsubara, M.A. Collins, A. Akane, J. Ikebuchi, E.J. Neafsey, M. Kagawa, H. Shiono, Potential bioactivated neurotoxicants, *N*-methylated β-carbolinium ions, are present in human brain, Brain Res. 610 (1993) 90–96.

[14] K. Matsubara, M.A. Collins, E.J. Neafsey, Mono-*N*-methylation of 1,2,3,4-tetrahydro-β-carbolines in brain cytosol: Absence of indole methylation, J. Neurochem. 59 (1992) 505–510.

[15] K. Matsubara, T. Gonda, H. Sawada, T. Uezono, Y. Kobayashi, T. Kawamura, K. Ohtaki, K. Kimura, A. Akaike, Endogenously occurring β-carboline induces parkinsonism in nonprimate animals: A possible causative protoxin in idiopathic Parkinson's disease, J. Neurochem. 70 (1998) 727–735.

[16] K. Matsubara, T. Idzu, Y. Kobayashi, T. Gonda, H. Okunishi, K. Kiumura, Differences in dopamine efflux induced by MPP$^+$ and β-carbolinium in the striatum of conscious rats, Eur. J. Pharmacol. 315 (1996) 145–151.

[17] K. Matsubara, T. Idzu, Y. Kobayashi, D. Nakahara, W. Maruyama, S. Kobayashi, K. Kimura, M. Naoi, *N*-methyl-4-phenylpyridinium and an endogenously formed analog, *N*-methylated β-carbolinium, inhibit

striatal tyrosine hydroxylation in freely moving rats, Neurosci. Lett. 199 (1995) 199–202.

[18] K. Matsubara, S. Kobayashi, Y. Kobayashi, K. Yamashita, H. Koide, M. Hatta, K. Iwamoto, O. Tanaka, K. Kimura, β-Carbolinium cations, endogenous MPP$^+$ analogs, in the lumbar cerebrospinal fluid of parkinsonian patients, Neurology 45 (1995) 2240–2245.

[19] K. Matsubara, E.J. Neafsey, M.A. Collins, Novel S-adenosylmethionine-dependent indole-N-methylation of β-carbolines in brain particulate fractions, J. Neurochem. 59 (1992) 511–518.

[20] K. Matsubara, T. Senda, T. Uezono, T. Awaya, S. Ogawa, K. Chiba, K. Shimizu, N. Hayase, K. Kimura, L-Deprenyl prevents the cell hypoxia induced by dopaminergic neurotoxins, MPP$^+$ and β-carbolinium: A microdialysis study in rats, Neurosci. Lett. 302 (2001) 65–68.

[21] K. Matsubara, T. Senda, T. Uezono, S. Fukushima, S. Ohta, K. Igarashi, M. Naoi, Y. Yamashita, K. Ohtaki, N. Hayase, S. Akutsu, K. Kimura, Structural significance of azaheterocyclic amines related to Parkinson's disease for dopamine transporter, Eur. J. Pharmacol. 348 (1998) 77–84.

[22] A. Morano, F.J. Jimenez-Jimenez, J.A. Molina, M.A. Antolin, Risk-factors for Parkinson's disease: Case–control study in the province of Caceres, Spain, Acta Neurol. Scand. 89 (1994) 164–170.

[23] T. Nagatsu, M. Ysohida, An endogenous substrate of brain, tetrahydroisoquinoline, produces Parkinsonism in primates with decreased dopamine, tyrosine hydrozylase and biopterin in the nigrostriatal regions, Neurosci. Lett. 87 (1988) 178–182.

[24] M. Naoi, P. Dostert, M. Yoshida, T. Nagatsu, N-methylated tetrahydroisoquinolines as dopaminergic neurotoxins, Adv. Neurol. 60 (1993) 212–217.

[25] M. Naoi, W. Maruyama, N. Nakao, T. Ibi, K. Sahashi, M.S. Benedetti, (R)Salsolinol N-methyltransferase activity increases in parkinsonian lymphocytes, Ann. Neurol. 43 (1998) 212–216.

[26] M. Naoi, W. Maruyama, T. Niwa, T. Nagatsu, Novel toxins and Parkinson's disease: N-methylation and oxidation as metabolic bioactivation of neurotoxin, J. Neural Transm. 41 (1994) 197–205 (Supplement).

[27] M. Naoi, S. Matsuura, T. Takahashi, T. Nagatsu, A N-methyl-transferase in human brain catalyses N-methylation of 1,2,3,4-tetrahydroisoquinoline, a precursor of a dopaminergic neurotoxin, N-methylisoquinolinium ion, Biochem. Biophys. Res. Commun. 161 (1989) 1213–1219.

[28] A.H. Rajput, R.J. Uitti, W. Stern, W. Laverty, K. O'Donnell, D. O'Donnell, W.K. Yuen, A. Dua, Geography, drinking water chemistry, pesticides and herbicides and the etiology of Parkinson's disease, Can. J. Neurol. Sci. 14 (1987) 414–418.

[29] A. Seidler, W. Hellenbrand, B.P. Robra, P. Vieregge, P. Nischan, J. Joerg, W.H. Oertel, G. Ulm, E. Schneider, Possible environmental, occupational, and other etiologic factors for Parkinson's disease: A case–control study in Germany, Neurology 46 (1996) 1275–1284.

[30] C.A.D. Smith, A.C. Gough, P.N. Leigh, E.A. Summers, A.E. Harding, D.M. Maranganore, S. Sturman, A.H.V. Schapira, A. Williams, N.K. Spurr, C.R. Wolf, Debrisoquine hydroxylase gene polymorphism and susceptibility to Parkinson's disease, Lancet 339 (1992) 1375–1377.

[31] C.M. Tanner, J.W. Langston, Do environmental toxins cause Parkinson's disease? A critical review, Neurology 40 (1990) 17–30.

[32] C.M. Tanner, R. Ottman, S.M. Goldman, J. Ellenberg, P. Chan, R. Mayeux, J.W. Langston, Parkinson disease in twins: An etiologic study, JAMA, J. Am. Med. Assoc. 281 (1999) 341–346 (see comments).

NEUROTOXICOLOGY

AND

TERATOLOGY

Neurotoxicology and Teratology 24 (2002) 599–605

www.elsevier.com/locate/neutera

Review article

Apoptotic molecules and MPTP-induced cell death

A. Nicotra[a,*], S.H. Parvez[b]

[a]Dipartimento di Biologia Animale e dell'Uomo, Università di Roma La Sapienza, Viale dell'Università 32, 00185 Rome, Italy
[b]Institut Alfred Fessard of Neurosciences CNRS, UPR 2212-Bât 5, Chateau CNRS, 91190 Gif Sur Yvette, France

Received 10 May 2001; accepted 24 June 2001

Abstract

MPTP-induced neurotoxicity is one of the experimental models most commonly used to study the pathogenesis of Parkinson's disease (PD). MPTP administered in vivo to mice causes selective loss of dopaminergic neurons in the substantia nigra (SN), as in this disease. Cell death may be induced in vitro by MPP$^+$, the active metabolite of MPTP, when neuronal cell cultures are used. Biochemical mechanisms underlying cell death induced by MPTP/MPP$^+$ still remain to be clarified completely. This article reviews some recent findings linking the effects of MPTP/MPP$^+$ with molecules typically involved in apoptotic pathways. This type of research has made extensive use of genetically manipulated systems such as transgenic mice and transfected cell lines. Evidence has emerged to suggest that Bcl-2, Bax, JNK, and caspases are implicated in neurotoxic effects due to in vivo MPTP administration to mice. Different neuronal cell lines such as MN9D cells, SH-SY5Y cells, cerebellar granule neurons, cortical neurons, and GH3 cells were also tested to investigate the possible involvement of Bcl-2, Bax, and caspases in in vitro MPP$^+$-induced neurotoxicity. © 2002 Elsevier Science Inc. All rights reserved.

Keywords: MPTP; Apoptosis; Necrosis

1. Introduction

Parkinson's disease (PD) is characterized by the selective loss of dopaminergic neurons in the substantia nigra (SN). Even after numerous studies, the cause of dopaminergic cell degeneration in SN of PD patients still has not been identified with certainty [4].

In order to overcome the obvious difficulties involved in the post mortem study of PD patients, several experimental models have been developed for PD [14]. The most frequently used of these is the one produced by MPTP administration on mice. MPTP is a neurotoxin that can induce symptoms similar to those observed in PD in humans, nonhuman primates, and mice [38]. In addition, MPTP induces selective loss of dopaminergic neurons in mice SN. In order to be active, MPTP requires the activity of monoamine oxidase B and dopamine transporter [39,45]. Indeed, MPTP must be converted into MPP$^+$, the true toxic agent, by the monoamine oxidase B of the inner mitochondrial membrane. At the level of the central nervous system, this process takes place mainly in the glial cells. MPP$^+$ is thus selectively taken up by dopaminergic neurons via the dopamine transporter of the plasmatic membrane. In simplified in vitro systems, MPP$^+$ is used directly. MPP$^+$ is concentrated inside the mitochondria where it potently inhibits complex I of the electron transport system. This results in ATP depletion, loss of mitochondrial membrane potential, and the formation of reactive oxygen species (ROS). Bioenergetic and oxidative stresses create a situation that ultimately leads to cell death. A similar scenario has been proposed also for PD [18]. Further contributions to neurotoxicity due to MPTP apparently come from the breakdown of calcium homeostasis, the induction of glutamate-mediated excitotoxicity, and the stimulation of nitric oxide synthase to form nitric oxide [40].

However, it has also been suggested that MPP$^+$ may act independently of inhibition of the electron transport chain (ETC). This seems to be the case for the RhoO cells, which are mutant cells devoid of ETC activity [24]. MPP$^+$ toxicity in Rho and RhoO cells is comparable and causes a similar proportion of necrotic and apoptotic cell deaths [22].

Many doubts still persist concerning the action mechanism of MPTP/MPP$^+$. One controversial aspect is whether MPTP/MPP$^+$ induces structural changes related to cell death via apoptosis or necrosis. Investigations aimed at

---

* Corresponding author. Fax: +39-6-495-8259.
E-mail address: antonietta.nicotra@uniroma1.it (A. Nicotra).

clarifying this aspect were carried out both after in vivo administration of MPTP to mice and by treating neuronal cultures with MPP$^+$ [35]. Methods allowing a distinction between apoptosis and necrosis were then applied. However, the results gave no univocal explanation, although it is generally accepted that low MPTP/MPP$^+$ doses lead to apoptosis while high MPTP/MPP$^+$ doses, causing acute toxicity, lead to necrotic degeneration. Nevertheless, in the MN9D cell line similar biochemical and structural changes occur after treatment with low and high MPP$^+$ concentrations (range 5–100 µM). It may be inferred that this represents a case of a distinct form of cell death resembling both apoptosis and necrosis [6].

This article reviews the most recent results of research aimed at relating the effects of MPTP/MPP$^+$ to typical apoptotic molecules. The apoptotic signaling pathway involves numerous molecules, which act as regulators or effectors. A brief survey will first be made of the biochemical mechanisms of cell death, accompanied by a description of several properties of apoptotic molecules that have already been taken into consideration in research on the effects of MPTP/MPP$^+$. In addition, the papers aimed at detecting apoptotic molecules in the SN neurons of PD patients will be discussed.

2. Apoptotic mechanisms: an overview

The biochemical mechanisms regulating the process of cell death have been extensively investigated over the past decade and this has led to considerable progress being achieved in this field. It thus emerged that the mechanisms involved are varied and complex, although a basic pattern emerges.

It is now clear that the mitochondria lie at the centre of the cell death regulation process, at least in mammals [13,15,27,44]. The stimuli leading to cell death (oxidative stress, hypoxia, cytokines, hormones, radiation, etc.) are transmitted through signal transduction to the mitochondrion. The latter organelle then translates the signals into a cell death response. The entire cell death process may be subdivided into three distinct phases [27,44]. The first phase of induction or initiation takes place in the cytosol where different stimuli lead to definite concentrations of apoptogenic or pro-apoptotic molecules that act on the mitochondria in different ways, causing mitochondrial membrane permeabilization. When the permeability of the mitochondrial membrane is affected, the cell is forced to commit suicide, thus opening the effector or decision phase, during which components capable of activating hydrolases, such as protease and nuclease, leave the mitochondrial intermembrane space and enter the cytosol. This marks the beginning of the degradation phase, which is characterized by the digestion of several proteins, fragmentation nuclear DNA, and the onset of the major morphological changes typical of cell death. Many molecules are involved in the apoptotic signal, which reflect the variety of the stimuli that can cause cell death. They include calcium ions, nitric oxide, ROS, ceramide, and several proteins such as transcription regulators, protein kinases (JNK), and Bcl-2 and its homologs [20,48].

Bcl-2 is a 26-kDa protein preferentially located at contact sites between the inner and outer mitochondrial membranes [41]. Insertion into the mitochondrial membrane occurs via its hydrophobic C-terminus. Bcl-2 belongs to a very interesting family of proteins including members that are capable both of acting as agonists in cell death, such as Bax and Bak, and others capable of inhibiting it, such as Bcl-2 and Bcl-X$_L$ [41]. The various members possess variable regions of homology that determine the capacity to interact with other members of the same family as well as with unrelated proteins.

The following model illustrates the action mechanism of Bcl-2 and its homologs in mediating mitochondrial membrane permeabilization. Bax is able to insert into the outer mitochondrial membrane and after forming heterodimers with other members of the family may form a specific channel for cytochrome-c transport [9] and thus leads to an increase in mitochondrial permeability and cell death. Bcl-2 might function as an inhibitor of Bax, thus inhibiting cell death. The ratio between the various members of the family could thus be instrumental in deciding between cell survival and cell death.

The soluble factors released into the cytosol during the decision phase include the already mentioned cytochrome-c [12]. The latter forms a complex called apoptosome with other components like Apaf-1 and a procaspase-9 [3]. This leads to activation of the caspase-9, which then activates a procaspase-3. Numerous cellular proteins are then hydrolyzed in a specific way and endonucleases activated. The major structural changes of apoptosis start to become visible. The caspases, several of which are known as ICE (interleukin-converting enzymes), form a family with at least 12 members that cleave after aspartate residues [23]. All are synthesized as inactive enzymes that, in order to be activated, require proteolytic digestion. The caspases are capable of inducing cell neuronal death when overexpressed. Conversely, the absence of caspases leads to anomalous neuronal development [28]. Several caspase inhibitors are known. For example, Ac-YVAD-CHO and Ac-DEVD-CHO are specific inhibitors for caspase-1 and -3, respectively. Others, such as ZVAD-fmk and BAF are instead broad-spectrum inhibitors. By using a suitable inhibitor, it is possible to reduce apoptotic cell death, but not necrotic cell death [2].

Cytochrome-c is not the only agent released from mitochondria. For example, the mitochondria also release a factor called AIF (apoptosis-inducing factor), a flavoprotein that enters the nuclei, where it induces chromatin condensation and DNA fragmentation [8]

The complexity of the process is shown by the fact that the same factor can act both during the induction phase and in the decision phase but with different functions. For example, caspases can also act by inactivating Bcl-2. Bax can promote cell survival in some cases [31].

According to another alternative model, different stimuli may cause a transient opening of the mitochondrial membrane transition pore. After the opening of this channel, the mitochondria swell and the rupture of the outer membrane occurs. In this way, cytochrome-*c* may be released into the cytosol [15,54]. The opening of the channel is believed to initiate apoptotic or necrotic death cascade depending on the degree of the permeability transition. When it is large, rapid, and persistent, necrosis occurs before the caspases can begin acting. Whenever the transition is more moderate and it is possible to regenerate ATP levels, the caspases are activated and the structural changes of apoptosis appear.

3. Apoptotic molecules and MPTP/MPP$^+$ effects

3.1. JNK

The molecules identified as possible pro-apoptotic signals include JNK (c-*jun* N-terminal kinase), a component of the stress-activated protein kinase (SAPK) system that seems to be involved in some cases of neurodegeneration. JNK is itself activated by other apoptotic signals and then it phosphorylates and inactivates the anti-apoptotic protein Bcl-X$_L$ [47].

An interesting observation is that CEP-1347/KT-7515, an inhibitor of JNK activation [29], reduces the neurotoxic effects of MPTP administered in vivo (a single 40-mg/kg dose) to mice [42]. In particular, there is a decrease in the death of nigrostriatal dopaminergic neurons. In later research [43], the involvement of JNK was confirmed, demonstrating that treatment with MPTP causes a rapid increase in phosphorylated JNK (2.5-fold) and JNK kinase (5-fold) levels in the nigrostriatal system. JNK kinase, also called MKK4, is an activator of JNK. The JNK signaling cascade might thus be part of the death machine set in motion by MPTP.

3.2. Bcl-2

Several research groups have attempted to demonstrate that the negative regulator Bcl-2 is implicated in MPTP-induced cell-death. Using a human neuroblastoma cell line (SH-SY5Y cells) that natively expresses Bcl-2, it was shown that treatment with 1 mM MPP$^+$ for several days causes an increase of Bcl-2 protein [19]. However, staurosporine, a protein kinase inhibitor, inhibits Bcl-2 increase without reducing the number of cell deaths. This seems to indicate that in these cells the observed increase in Bcl-2 is actually not implicated in the cell death pathway initiated by MPP$^+$. Different conclusions were reached by other researchers using the same cell line [49]. In SH-SY5Y cells treated with 5 mM MPP$^+$ for 18–24 h, Western blotting analysis revealed a substantial increase in Bcl-2 and Bcl-X$_1$.

The results of semiquantitative RT-PCR showed that the treatment also causes a parallel increase in bcl-2 mRNA synthesis (40% after 8 h). One further interesting observation emerges from confocal microscopic examination. MPP$^+$ treatment stimulates an increased mitochondrial colocalization of Bcl-2 and thus of its anti-apoptotic function. The increase in Bcl-2 and Bcl-X$_1$ could be a adaptive survival response to the product of ROS and to oxidative stress caused by MPP$^+$. This is believed to be demonstrated by the finding that MPP$^+$ is capable of stimulating an increase in the two anti-apoptotic proteins only if the mitochondrial ETC is functioning in the cells. Indeed, SH-SY5Y cells devoid of ECT and mitochondrial DNA (SH-SY5Yρ°) show no increase in Bcl-2 or Bcl-X$_1$ after treatment with MPP$^+$.

Other laboratories have used genetic manipulation to demonstrate the role of Bcl-2 in MPTP-induced cell death. As reported above, after treatment with MPP$^+$ (5–100 μM), the dopaminergic cell line MN9D [5] undergoes a specific kind of cell death that displays features of both apoptosis and necrosis [6]. When this cellular line was stably transfected with human Bcl-2 (MN9D/Bcl-2), MPP$^+$ was less effective in causing cell death [6,37]. When MN9D cells were transfected with the C-terminal deletion mutant of human Bcl-2, the percentage of survival of the cells after MPP$^+$ treatment was intermediate between that observed in normal MN9D cells and MN9D/Bcl-2 cells. This finding suggested that the C-terminal membrane anchorage domain of Bcl-2 was required for the attenuation of MPP$^+$ effects [6].

In other experiments [36], cortical neuronal cultures were tested. Cells were prepared from transgenic mice overexpressing human Bcl-2 under the control of neuron-specific enolase promoter (NSE-*hbcl*-2). After 50-μM MPP$^+$ treatment for 2–4 h, the dead cells were considerably fewer in transgenic cultures than in wild-type cultures. In addition, in vivo experiments were performed. In this case, 10 mg/kg MPTP (four times at 24-h intervals) was injected into both wild-type and transgenic mice. It was observed that after 10 days, striatal dopamine level decreased by 32% in wild-type mice, while it was not affected in the heterozygous NSE-*hbcl*-2 mice. The effects of MPTP on the loss of dopaminergic neurons are thus reduced by overexpression of Bcl-2.

3.3. Bax

As we have seen, Bax is a member of the Bcl-2 family that acts as a promoter of cell death. MPTP administered in vivo to mice induces an increase in the amount of Bax in the SN [17]. Adult mice received four intraperitoneal injections of 20 mg/kg MPTP at 4-h intervals and were killed 3 and 6 days after the first injection. The levels of *bax* mRNA in the SN and Bax immunoreactivity in dying neurons were both measured. After treatment with MPTP, a strong increase was observed in both parameters. Different results were obtained using MN9D cells [6]. Treatment with 100 μM MPP$^+$ for up to 24 h did not cause any increase in Bax expression. MN9D cells overexpressing murine Bax (MN9D/ Bax) were also used [7]. When normal MN9D cells and MN9D/Bax cells were treated with MPP$^+$, no differences were found in the rate of cell death.

3.4. Caspases

The onset of caspase-3-like activity following in vitro treatment with MPP$^+$ was demonstrated after the exposure of cerebellar granule neurons to 60 μM MPP$^+$ [11]. Treatment resulted in a concentration- and time-dependent activation of caspase-3-like activity in cytosolic extracts. The increase was fivefold greater than in untreated cells. Moreover, treatment with 100 μM MPP$^+$ causes a significant increase in cytochrome-c in cytosolic extracts. This increase precedes structural changes associated with apoptosis. In other experiments, it was endeavored to determine whether caspase inhibitors Ac-YVAD-CHO (caspase-1-like inhibitor) and Ac-DEVD-CHO (caspase-3-like inhibitor) were capable of attenuating the neurotoxic effects of 60 μM MPP$^+$ administered for 72 h to cerebellar granule neurons [11]. It was thus ascertained that only Ac-DEVD-CHO is capable of reducing the effects of the neurotoxin as determined by DNA laddering analysis and Hoechst staining of the nuclei. In the same laboratory [10], the effect of caspase inhibitors was analyzed also on rat ventral mesencephalon cultures containing dopamine neurons. The results show that in this cell line a caspase-3-like protease is also involved in the apoptotic cell death caused by low concentrations (1–100 M for 72 h) of MPP$^+$.

A caspase inhibitor was used also in GH3 cells (a clonal strain from the rat anterior pituitary) treated with 200 μM MPP$^+$ for 24–48 h [53]. The broad-spectrum caspase inhibitor ZVAD-fmk (50–100μM) inhibited DNA laddering induced by MPP$^+$. It is interesting to note that MPP$^+$ treatment alone did not cause any increase in Bcl-2.

Further evidence of caspase involvement in MPTP effects was obtained in vivo using a transgenic mouse strain expressing a dominant negative mutant ICE. Wild-type and mutant mice were treated with 15 mg/kg MPTP every 2 h in four doses [25]. Animals were sacrificed at 1 week and the striatal levels of dopamine and its metabolites such as DOPAC and HVA were measured by HPCL. While a significant reduction in the above compounds was observed in wild-type mice, as well as increased neuronal loss in the SN, these variations were not found in transgenic mice. Mutant mice are thus resistant to the neurotoxic effects of MPTP.

Interesting results were also obtained using transgenic mice overexpressing Bcl-2 [50]. Striatal injections of MPP$^+$ (15 mM, 0.75 μl) in wild-type mice induced an increase in caspase-2. However, this increase was much less marked if the injections were made to mice overexpressing Bcl-2 [52].

It is likely that MPP$^+$ effects do not always require caspase action. Indeed, in MN9D cells, the percentage of cell deaths induced by MPP$^+$ (100 μM for 40 h) was not reduced by cotreatment with ZVAD-fmk (50–400 μM) or

BAF (up to 200 μM), two broad-spectrum caspase inhibitors [6,7].

The presence of caspase-3 activity in a tissue may be detected using an antibody (CM1) that specifically recognizes the activated caspase-3. This experimental approach was used in primary cultures of rat mesencephalon to investigate whether MPP$^+$ treatment triggers caspase-3 activation [16]. It was also attempted to investigate the relation between the period of caspase-3 activation and the onset of cell death. Cells were treated per 12–72 h with 1 μM MPP$^+$ in such a way as to induce apoptosis. After 12-h treatment, a 3.4% cell loss was observed in dopaminergic cells (TH-positive), rising to 78.5% after 72 h. The number of dopaminergic cells binding the antibody CM1 after 12-h treatment is more than double of that measured in untreated cells. However, after 72 h of treatment, the same value is found to have fallen by about half versus the controls. These results show that caspase-3 activation is required during the effector phase of cell death and is not a consequence of cell death. It might thus be possible to block this activity and thus prevent cell death from progressing.

4. Apoptotic molecules evidenced in the SN of PD patients

Only a few studies have so far been carried out on the presence of apoptotic molecules in the SN of PD patients. Bcl-2 protein levels were measured both using a highly sensitive immunochemical method [33] and by Western blotting [30]. The results of both methods pointed to a significant increase in Bcl-2 in PD patients versus controls. Using Western blotting, as much as a threefold increase was obtained. The increase in Bcl-2 does not occur in the whole brain. For example, no differences were found in the cerebral cortex. The up-regulation of this protein could have a compensatory function aimed at maintaining the viability of neurons subjected to oxidative stress such as the dopaminergic neurons of the NS. The results obtained by measuring the bcl-2 mRNA levels by means of in situ hybridization are not in full agreement with previous observations. Indeed, the bcl-2 mRNA levels were found to be comparable in the NS of both PD patients and controls [50]. However, the authors point out that the NS neurons from PD patients are in a distressed state and several of their cell functions are reduced. The fact that bcl-2 mRNA is maintained at a level comparable to that found in controls is probably an indication of its importance in the protection of neurons from degeneration. However, another article [51] is in complete disagreement with the results described in Refs. [30,33]. In the NS of PD patients, the immunoreactivity towards Bcl-2 does not differ from that in controls. This is true also for Bax and Bcl-X$_1$. Furthermore, no difference in immunoreactivity is found between neurons containing Lewy bodies and the remaining melanin-containing neurons. The latter observation, referring only to Bcl-2 and Bax, is reported also in Ref. [46].

The other apoptotic molecules examined in PD patients are the caspases. The activity of caspase-1 and -3 was measured using fluoropeptides as substrate [34]. In the nigrostriatal regions from PD patients, the activity of the two caspases increased 1.5- and 2-fold, respectively, versus controls. Using the antibody CM1 specific to activated caspase-3, it was found that the number of dopaminergic neurons of the NS of PD patients positive for the antibody is higher than in controls [16]. Activated caspase-3 is, however, present in neurons in which the chromatin condensation typical of the final stage of apoptosis has not yet occurred. Caspase-3 activity is thus apparently required at the beginning of the effector phase of apoptosis. A pro-apoptotic environment thus exists in the NS of PD patients.

5. Conclusions

The data illustrated in this short review indicate that several apoptotic molecules may be involved in MPTP/MPP$^+$-induced neurodegeneration. Among these, special attention has been dedicated by the various laboratories to death-inhibiting Bcl-2 and the caspases (Tables 1–3). The neuroprotective effects of Bcl-2 overexpression after treatment with MPTP/MPP$^+$ are often reported. In addition, caspase participation seems to be an event associated with MPTP/MPP$^+$-induced cell death. Nevertheless, it must be emphasized that the involvement of apoptotic molecules in cell death due to MPTP/MPP$^+$ is not in itself conclusive evidence that an apoptotic type pathway was actually followed. It has in fact been found that Bcl-2 may also have inhibiting effects on necrotic cell death [21]. Evidence linking MPTP/MPP$^+$ with apoptotic molecules could, however, contribute to increasing our knowledge of PD

pathogenesis. Very few studies exist on the involvement of apoptotic molecules in the NS of PD patients and the results obtained are not completely in agreement. On the other hand, also in the case of PD, it remains to be definitively clarified whether NS neurons degenerate by apoptosis or necrosis [1,26,32]. Studies on PD patients are rendered difficult by the fact of having to use post mortem tissues and that a statistically valid number of cases is not always available. New treatments designed to slow down PD progress could be developed in view of the fact that pro-apoptotic molecules may be involved in the degenerative process.

Table 1
Evidence of involvement of apoptotic molecules in cell death induced in SN by in vivo administration of MPTP or MPP$^+$ to mice

Experimental models	Treatments	Effects
Normal mice	20 mg/kg MPTP, four doses	increase of bax mRNA and Bax protein [17]
Normal mice	40 mg/kg MPTP, single dose + inhibitor of JNK	reduced cell loss [42]
Normal mice	40 mg/kg MPTP, single dose	increase of phosphorylated JNK and JNK kinase [43]
Transgenic mice overexpressing Bcl-2	10 mg/kg MPTP, four doses	reduced cell loss [36]
Transgenic mice overexpressing Bcl-2	striatal injections of 15 mM MPP$^+$ (0.75 μl)	reduced caspase-2 [48]
Transgenic mice ICE deficient	15 mg/kg MPTP, four doses	reduced cell loss and increase of DA, DOPAC, and HVA [25]

The effects reported for normal mice are compared to untreated mice. Transgenic mice are compared to treated wild-type mice.

Table 2
Evidence of involvement of Bcl-2/Bax in cell death induced by in vitro treatment of different cell lines with MPP$^+$

Cell lines	Treatments	Effects
Cortical neurons overexpressing Bcl-2	50 μM MPP$^+$, 2–4 h	reduced cell death [36]
GH3 cells	200 μM MPP$^+$, 24–48 h	no increase in Bcl-2 [53]
MN9D	100 μM MPP$^+$, 24 h	no increase in Bax [6]
MN9D overexpressing Bax	100 μM MPP$^+$, 24 h	unaffected cell death [7]
MN9D overexpressing Bcl-2	10–50 μM MPP$^+$, 40–48 h	reduced cell death [37]
SH-SY5Y	1 mM MPP$^+$, several days	increase of Bcl-2 not related to cell death [19]
SH-SY5Y	5 mM MPP$^+$, 18–24 h	increase of Bcl-2, Bcl-X$_1$, and bcl-2 mRNA [49]
SH-SY5Yρ°	5 mM MPP$^+$, 18–24 h	no increase of Bcl-2, Bcl-X$_1$, and bcl-2 mRNA [49]

The effects reported in the case of normal cell lines are compared to untreated cells. Genetically manipulated cells are compared to treated normal cell lines.

Table 3
Evidence of involvement of caspases in cell death induced by in vitro treatment of different cell lines with MPP$^+$

Cell lines	Treatments	Effects
Cerebellar granule neurons	60 μM MPP$^+$, 48 h	activation of caspase-3 [11]
Cerebellar granule neurons	60 μM MPP$^+$ + Ac-YVAD-CHO, Ac-DEVD-CHO	cell death reduced by Ac-DEVD-CHO [11]
Dopaminergic neurons (rat ventral mesencephalon)	1-100 μM MPP$^+$ (72 h) + Ac-YVAD-CHO, Ac-DEVD-CHO, and ZVAD-fmk	cell death reduced by Ac-DEVD-CHO and ZVAD-fmk [10]
Dopaminergic neurons (rat ventral mesencephalon)	1 μM MPP$^+$, 12–72 h	caspase-3 activation [16]
GH3 cells	200 μM MPP$^+$, 24–48 h + ZVAD-fmk	reduced cell death [53]
MN9D	100 μM MPP$^+$, 40 h + ZVAD-fmk, BAF	unaffected cell death [6,7]

References

[1] P. Anglade, S. Vyas, F. Javoy-Agid, M.T. Herrero, P.P. Michel, J. Marquez, A. Mouatt-Prigent, M. Ruberg, E.C. Hirsch, Y. Agid, Apoptosis and autophagy in nigral neurons of patients with Parkinson's disease, Histol. Histopathol. 12 (1997) 25–31.

[2] R.C. Armstrong, T.J. Aja, K.D. Hoang, S. Gaur, X. Bai, E.S. Alnemri, G. Litwack, D.S. Karanewsky, L.C. Fritz, K.J. Tomaselli, Activation of the CED3/ICE-related protease CPP32 in cerebellar granule neurons undergoing apoptosis but not necrosis, J. Neurosci. 17 (1997) 553–562.

[3] I. Budijardjo, H. Oliver, M. Lutter, X. Luo, X. Wang, Biochemical pathways of caspase activation during apoptosis, Annu. Rev. Cell Biol. 15 (1999) 269–290.

[4] R.E. Burke, Programmed cell death and Parkinson's disease, Mov. Disord. 13 (1998) 17–23.

[5] H.K. Choi, L.A. Won, P.J. Kontur, D.N. Hammond, A.P. Fox, B.H. Wainer, P.C. Hoffmann, A. Heller, Immortalization of embryonic mesencephalic dopaminergic neurons by somatic cell fusion, Brain Res. 552 (1991) 67–76.

[6] W.S. Choi, L.M.T. Canzoniero, S.L. Sensi, K.L. O'Malley, B.J. Gwag, S. Sohn, J.-E. Kim, T.H. Oh, E.B. Lee, Y.J. Oh, Characterization of MPP$^+$-induced cell death in a dopaminergic neuronal cell line: Role of macromolecule synthesis, cytosolic calcium, caspase, and Bcl-2 related proteins, Exp. Neurol. 159 (1999) 274–282.

[7] W.S. Choi, S.Y. Yoon, T.H. Oh, E.J. Choi, K.L. O'Malley, Y.J. Oh, Two distinct mechanisms are involved in 6-hydroxydopamine- and MPP$^+$-induced dopaminergic neuronal cell death: Role of caspases, ROS, and JNK, J. Neurosci. Res. 57 (1999) 86–94.

[8] E. Daugas, S.A. Susin, N. Zamzami, K. Ferri, T. Irinopoulos, N. Larochette, M.C. Prevost, B. Leber, D. Andrews, J. Penninger, G. Kroemer, Mitochondrio-nuclear redistribution of AIF in apoptosis and necrosis, FASEB J. 14 (2000) 729–739.

[9] S. Desagher, A. Osen-Sand, A. Nichols, R. Eskes, S. Montessuit, S. Lauper, K. Maundrell, B. Antonsson, J.-C. Martinou, Bid-induced conformational changes of Bax is responsible for mitochondrial cytochrome c release during apoptosis, J. Cell. Biol. 144 (1999) 891–901.

[10] R.C. Dodel, Y. Du, K.R. Bales, Z.D. Ling, P.M. Carvey, S.M. Paul, Peptide inhibitors of caspase-3-like proteases attenuate 1-methyl-4-phenylpyridinium-induced toxicity of cultured fetal rat mesencephalic dopamine neurons, Neuroscience 86 (1998) 701–707.

[11] Y. Du, R.C. Dodel, K.R. Bales, R. Jemmersom, E. Hamilton-Byrd, S.M. Paul, Involvement of a caspase-3-like cysteine protease in 1-methyl-4-phenylpyridinium-mediated apoptosis of cultured cerebellar granule neurons, J. Neurochem. 69 (1997) 1382–1388.

[12] R. Eskes, B. Antonsson, A. Osen-Sand, S. Montessuit, C. Richter, R. Sadoul, G. Mazzei, A. Nichols, J.-C. Martinou, Bax-induced cytochrome c release from mitochondria is independent of the permeability transition pore but highly dependent on Mg^{2+} ions, J. Cell. Biol. 143 (1998) 217–224.

[13] K.F. Ferri, G. Kroemer, Mitochondria—the suicide organelles, BioEssays 23 (2001) 111–115.

[14] M. Gerlach, P. Riederer, Animal models of Parkinson's disease: An empirical comparison with the phenomenology of the disease in man, J. Neural. Transm. 103 (1996) 987–1041.

[15] A.P. Halestrap, E. Doran, G.P. Gillespie, A. O'Toole, Mitochondria and cell death, Biochem. Soc. Trans. 28 (2000) 170–177.

[16] A. Hartmann, S. Hunot, P.P. Michel, M.-P. Muriel, S. Vyas, B.A. Faucheux, A. Mouatt-Prigent, H. Turmel, A. Srinivasan, M. Ruberg, G.I. Evan, Y. Agid, E.C. Hirsch, Caspase-3: A vulnerability factor and final effector in apoptotic death of dopaminergic neurons in Parkinson's disease, Proc. Natl. Acad. Sci. U. S. A. 97 (2000) 2875–2880.

[17] L. Hassouna, H. Wickert, M. Zimmermann, F. Gillardon, Increase in bax expression in substantia nigra following 1-methyl-4-phenyl-1,2,3,6-tetrahydropyridine (MPTP) treatment of mice, Neurosci. Lett. 204 (1996) 85–88.

[18] E.C. Hirsch, Mechanism and consequences of nerve cell death in Parkinson's disease, J. Neural. Transm., (Suppl. 56) (1999) 127–137.

[19] Y. Itano, Y. Nomura, 1-Methyl-4-phenylpyridinium ion (MPP$^+$) causes DNA fragmentation and increases the Bcl-2 expression in human neuroblastoma, SH-SY5Y cells through different mechanisms, Brain Res. 704 (1995) 240–245.

[20] S. Johnson Webb, D.J. Harrison, A.H. Wyllie, Apoptosis: An overview of the process and its relevance in disease, Adv. Pharmacol. 41 (1997) 1–34.

[21] D.J. Kane, T. Ord, R. Anton, D.E. Bredesen, Expression of bcl-2 inhibits necrotic neural cell death, J. Neurosci. Res. 40 (1995) 269–275.

[22] U. Khan, B. Filiano, M.P. King, S. Przedborski, Is Parkinson's disease (PD) an extra-mitochondrial disorder? Neurology 48 (1997) A201.

[23] V.J. Kidd, Proteolytic activities that mediates apoptosis, Annu. Rev. Physiol. 60 (1998) 533–573.

[24] M.P. King, G. Attardi, Human cells lacking mtDNA: Repopulation with exogenous mitochondria by complementation, Science 246 (1989) 500–503.

[25] P. Klevenyi, O. Andreassen, R.J. Ferrante, J.R. Schleicher Jr., R.M. Friedlander, M.F. Beal, Transgenic mice expressing a dominant negative mutant interleukin-1β converting enzyme show resistance to MPTP neurotoxicity, NeuroReport 10 (1999) 635–638.

[26] S. Kösel, R. Egensperger, U. Eitzen, P. Mehraein, M.B. Graeber, On the question of apoptosis in the parkinsonian substantia nigra, Acta Neuropathol. 93 (1997) 105–108.

[27] G. Kroemer, J.C. Reed, Mitochondrial control of cell death, Nat. Med. 6 (2000) 513–519.

[28] P. Li, H. Allen, S. Banerjee, S. Franlin, L. Herzog, C. Johnston, J. McDowell, M. Paskind, L. Rodman, J. Salfeld, E. Towne, D. Tracey, S. Wardwell, F.-Y. Wei, W. Wong, R. Kamen, T. Seshadri, Mice deficient in IL-1β-converting enzyme are defective in production of mature IL-1 β and resistant to endotoxic shock, Cell 80 (1995) 401–411.

[29] A.C. Maroney, J.P. Finn, D. Bozyczko-Coyne, T.M. O'Kane, N.T. Neff, A.M. Tolkovski, D.S. Park, C.Y.I. Yan, C.M. Troy, L.A. Greene, CEP-1347 (KT7515), an inhibitor of JNK activation, rescues sympathetic neurons and neuronally differentiated PC12 cells from death evoked by three distinct insults, J. Neurochem. 73 (1999) 1901–1912.

[30] K.-A. Marshall, S.E. Daniel, N. Cairns, P. Jenner, B. Halliwell, Upregulation of the anti-apoptotic protein Bcl-2 may be an early event in neurodegeneration: Studies on Parkinson's and incidental Lewy body disease, Biochem. Biophys. Res. Commun. 240 (1997) 84–87.

[31] G. Middleton, G. Nunez, A.M. Davies, Bax promotes neuronal survival and antagonizes the survival effects of neurotrophic factors, Development 122 (1996) 695–701.

[32] H. Mochizuchi, K. Goto, H. Mori, Y. Mizuno, Histochemical detection of apoptosis in Parkinson's disease, J. Neurol. Sci. 137 (1996) 120–123.

[33] M. Mogi, M. Harada, T. Kondo, Y. Mizuno, H. Narabayashi, F. Riederer, T. Nagatsu, Bcl-2 protein is increased in the brain from parkinsonian patients, Neurosci. Lett. 215 (1996) 137–139.

[34] M. Mogi, A. Togari, T. Kondo, Y. Mizuno, O. Komure, S. Kuno, H. Ichinose, T. Nagatsu, Caspase activities and tumor necrosis factor receptor R1 (p55) level are elevated in the substantia nigra from parkinsonian brain, J. Neural. Transm. 107 (2000) 335–341.

[35] A. Nicotra, S.H. Parvez, Cell death induced by MPTP, a substrate for monoamine oxidase B, Toxicology 153 (2000) 157–166.

[36] D. Offen, P.M. Beart, N.S. Cheung, C.J. Pascoe, A. Hochman, S. Gorodin, E. Melamed, R. Bernard, O. Bernard, Transgenic mice expressing human Bcl-2 in their neurons are resistant to 6-hydroxydopamine and 1-methyl-4-phenyl-1,2,3,6-tetrahydropyridine neurotoxicity, Proc. Natl. Acad. Sci. U. S. A. 95 (1998) 5789–5794.

[37] Y.J. Oh, S.C. Wong, M. Moffat, K.L. O'Malley, Overexpression of Bcl-2 attenuates MPP$^+$ but not 6-ODHA, induced cell death in a dopaminergic neuronal cell line, Neurobiol. Dis. 2 (1995) 157–167.

[38] C.W. Olanow, W.G. Tatton, Etiology and pathogenesis of Parkinson's disease, Annu. Rev. Neurosci. 22 (1999) 123–144.

[39] S. Przedborski, V. Jackson-Lewis, Mechanisms of MPTP toxicity, Mov. Disord. 13 (Suppl. 1) (1998) 35–38.

[40] S. Przedborski, V. Jackson-Lewis, R. Yokoyama, T. Shibata, V.L. Dawson, T.M. Dawson, Role of neuronal nitric oxide in 1-methyl-4-phenyl-1,2,3,6-tetrahydropyridine (MPTP)-induced dopaminergic neurotoxicity, Proc. Natl. Acad. Sci. U. S. A. 93 (1996) 4565–4571.

[41] J.C. Reed, Double identity for proteins of the Bcl-2 family, Nature 387 (1997) 773–776.

[42] M.S. Saporito, E.M. Brown, M.S. Miller, S. Carswell, CEP-1347/KT-7515, an inhibitor of c-*jun* N-terminal kinase activation, attenuates the 1-methyl-4-phenyl tetrahydropyridine-mediated loss of nigrostriatal dopaminergic neurons in vivo, J. Pharmacol. Exp. Ther. 288 (1999) 421–427.

[43] M.S. Saporito, B.A. Thomas, R.W. Scott, MPTP activates c-Jun NH$_2$-terminal kinase (JNK) and its upstream regulatory kinase MKK4 in nigrostriatal neurons in vivo, J. Neurochem. 75 (2000) 1200–1208.

[44] S.A. Susin, N. Zamzani, G. Kroemer, Mitochondria as regulators of apoptosis: Doubt no more, Biochim. Biophys. Acta 1366 (1998) 151–165.

[45] K.F. Tipton, T.P. Singer, Advances in our understanding of the mechanisms of the neurotoxicity of MPTP and related compounds, J. Neurochem. 61 (1993) 1191–1206.

[46] A. Tortosa, E. Lopez, I. Ferrer, Bcl-2 and Bax proteins in Lewy bodies from patients with Parkinson's disease and diffuse Lewy body disease, Neurosci. Lett. 238 (1997) 78–80.

[47] C. Tournier, P. Hess, D.D. Yang, J. Xu, T.K. Turner, A. Nimnual, D. Bar-Sagi, S.N. Jones, R.A. Flaverl, R.J. Davis, Requirement of JNK for stress-induced activation of the cytochrome *c*-mediated death pathway, Science 5 (2000) 870–874.

[48] D.L. Vaux, A. Strasser, The molecular biology of apoptosis, Proc. Natl. Acad. Sci. U. S. A. 93 (1996) 2239–2244.

[49] G.A. Veech, J. Dennis, P.M. Keeney, C.P. Fall, R.H. Swerdlow, W.D. Parker Jr., J.P. Bennet Jr., Disrupted mitochondrial electron transport function increases expression of anti-apoptotic Bcl-2 and Bcl-X$_L$ protein in SH-SY5Y neuroblastoma and in Parkinson disease cybrid cells through oxidative stress, J. Neurosci. Res. 61 (2000) 693–700.

[50] S. Vyas, F. Javoy-Agid, M.T. Herrero, O. Strada, F. Boissiere, U. Hibner, Y. Agid, Expression of Bcl-2 in adult human brain regions with special reference to neurodegenerative disorders, J. Neurochem. 69 (1997) 223–231.

[51] U. Wüllner, J. Kornhuber, M. Weller, J.B. Schulz, P.A. Löschmann, P. Riederer, T. Klockgether, Cell death and apoptosis regulating protein in Parkinson's disease—a cautionary note, Acta Neuropathol. 97 (1999) 408–412.

[52] L. Yang, R.T. Matthews, J.B. Schulz, T. Klockgether, A.W. Liao, J.-C. Martinou, J.B. Penney, B.T. Hyman, M.F. Beal, 1-Methyl-4-phenyl-1,2,3,6-tetrahydropyridine neurotoxicity is attenuated in mice overexpressing Bcl-2, J. Neurosci. 18 (1998) 8145–8152.

[53] N. Yoshinaga, T. Murayama, Y. Nomura, Apoptosis induction by a dopaminergic neurotoxin, 1-methyl-4-phenylpyridinium ion (MPP+) and inhibition by epidermal growth factor in GH3 cells, Biochem. Pharmacol. 60 (2000) 111–120.

[54] S. Zamzani, S.A. Susin, P. Marchetti, T. Hirsch, I. Gomez-Monterrey, M. Castedo, G. Kroemer, Mitochondrial control of nuclear apoptosis, J. Exp. Med. 183 (1996) 1533–1544.

Neurotoxicology and Teratology 24 (2002) 607–620

NEUROTOXICOLOGY

AND

TERATOLOGY

www.elsevier.com/locate/neutera

Review article

MPTP
Insights into parkinsonian neurodegeneration

Samuel G. Speciale*

Department of Psychiatry, University of Texas Southwestern Medical School, Dallas, TX 75390, USA

Received 28 January 2002; accepted 28 January 2002

Abstract

MPTP burst upon the medical landscape two decades ago, first as a mysterious parkinsonian epidemic, triggering an unparalleled quest for the toxin's identity, and closely followed by an intense pursuit of its cellular mechanisms of action. MPTP treatment created an animal model of many features of Parkinson's disease (PD), used primarily in primates and later in mice. The critical role of oxidative stress damage to vulnerable dopamine neurons, as well as for neurodegenerative diseases in general, emerged from MPTP neurotoxicity. A remarkable cross-fertilization of basic and clinical findings, including genetic and epidemiologic studies, has greatly advanced our understanding of PD and revealed multiple factors contributing to the parkinsonian phenotypes. Brain imaging localizes sites of action and provides potential presymptomatic diagnostic testing. Epidemiologic reports linking PD with pesticide exposure were complimented by supportive evidence from biochemical studies of MPTP and structurally related compounds, especially after low-level, long-term exposure. Genetic studies on the role of risk genes, such as α-synuclein or parkin, have been validated by biochemical, anatomical and neurochemical investigations showing factors interacting to produce pathophysiology in the animal model. Focusing on the pivotal role of mitochondria, subcellular pathways participating in cell death have been clarified by unraveling similar sites of action of MPTP. Along the way, compounds antagonizing or potentiating MPTP effects indicated new PD therapies, some of the former achieving clinical trials. The future is encouraging for combating PD and will continue to benefit from the MPTP neurotoxicity model. © 2002 Elsevier Science Inc. All rights reserved.

Keywords: MPTP; Parkinsonism; Dopamine; Oxidative stress; Neurotoxicity; Neurodegeneration

1. Introduction

A complete review of the voluminous literature comprising both basic and clinical aspects of MPTP research is beyond the scope of this article (a PubMed search of "MPTP" as a keyword yielded 2713 citations at the start of 2002). Since its identification as a potent and specific neurotoxin, most books, symposia or reviews dealing with neurotoxicity and neurodegeneration generally, or Parkinson's disease (PD) specifically, have contained one or more contributions dealing with aspects of the MPTP model of neurotoxicity. An early book [114], reviews [100,166] and several multiarticle journal issues [Life Sci. 36 (3) (1985); Life Sci. 40 (8) (1987); Eur. Neurol. 26 (Suppl. 1) (1987)] summarized the initial flood of studies stimulated by the striking clinical observation of parkinsonism in a relatively young, drug-abusing population and the ensuing detective work involving laboratories around the world that identified the causative agent as MPTP. Subsequent reviews addressed additional biochemical steps in active toxin formation [177], substances synergizing or antagonizing MPTP toxicity [129], or natural compounds related to MPTP [121] that might cause PD via similar mechanisms, as well as more recent examinations of subcellular molecular mechanisms of neuronal cell death [16]. Overall, although MPTP does not exactly reproduce PD, it has been an extremely valuable tool to model many features of PD and has led to a better understanding of crucial aspects of subcellular events participating in the evolution of the PD clinical syndrome and neurotoxicity. Some are primary, acute components of neurodegeneration, while others are secondary modifiers, but which might play roles in the chronic, long-term manifestations, and suggest targets for innovative therapeutic interventions.

The history of MPTP has proceeded in a number of critical phases: description of the clinical syndrome, iden-

* 10460 Remington Lane, Dallas, TX 75229, USA. Tel.: +1-214-350-2611.

E-mail address: specialetsys@hotmail.com (S.G. Speciale).

tification of the causative agent, elucidation of the formation and distribution of the active toxin, animal models and their variable sensitivities, cellular actions producing neurotoxicity, use as a model for aspects of neurodegenerative syndromes and testing of neuroprotective agents or strategies. This review will first briefly describe PD, then trace the characterization of MPTP neurotoxicity and its cellular targets and lastly present new studies utilizing MPTP to understand factors that contribute to the parkinsonian syndrome and might indicate new therapeutic modalities.

1.1. PD

PD is a progressive neurodegenerative disease of unknown etiology, which produces resting tremor, akinesia and rigidity [135]. The major symptoms of the disease are related to the loss of the midbrain dopamine (DA) neurons within the substantia nigra (SN), which normally project to the caudate nucleus and putamen [72,84]. Genetic transmission only represents a small percentage [45,164], mostly early-onset cases, shown in an important twins study [174] and involving mutations in the α-synuclein and parkin genes [60]. Environmental factors [36], especially pesticides, have been strongly implicated [11,107], but have not yet been proven. Insights into the pathophysiology of PD from exogenous sources have been derived from biochemical, neurophysiological and anatomical studies of the MPTP animal model of the disease.

1.2. PD neurochemical pathology

Specific catecholamine (CA) nuclei within the brainstem, especially the SN and the locus coeruleus (LC; nucleus A6, the primary noradrenergic cell group extensively innervating the forebrain, as well as some posterior structures), exhibit significant cell losses in PD. The DA subnuclei, including the SN (nucleus A9), ventral tegmental areas (nucleus A10) and retrorubral area (nucleus A8), collectively contain about 50% fewer cells than in age-matched controls [13,17]. Using computer imaging techniques, the greatest reductions in midbrain DA cells occur within nuclei A8 and A9 (>60%, predominantly in the ventral tier of A9), but also within nucleus A10 (>40%) [61]. The pattern of cell loss is similar to the pattern of striatal (as opposed to cortical or limbic) innervation, suggesting that in PD, these DA neurons are preferentially lost. In contrast, hypothalamic DA neurons (e.g., arcuate and periventricular) are not reduced in PD [119]. There is considerable loss of noradrenergic neurons in the LC [111,112] and in the dorsal motor nucleus of the vagus (nucleus A3). Several other neurotransmitter systems have been shown to be decreased in PD brains [3], but the CA neuron loss appears to be the primary anatomical defect. This important aspect of selective neuronal vulnerability has been clarified with some success using in vivo and in vitro MPTP studies.

1.3. Discovery of the MPTP clinical syndrome

In the early 1980s, a number of young adults were seen at several California emergency rooms with bizarre symptoms, including rigidity, bradykinesia, some tremor and postural instability, which responded to L-DOPA treatment, as with PD [102]. The appearance of this parkinsonian syndrome at this age was surprising in that PD usually manifests itself in the late 50 or early 60 years of age, although the range can be greater. They turned out to be addicts, who were taking homemade heroin-related drugs. An earlier episode [40] in a student in the East with similar symptoms, who had injected a meperidine derivative he synthesized with some short-cuts or sloppy technique, was also L-DOPA-responsive. The student later died of a drug overdose and neuropathological brain examination revealed SN neuron loss, similar to PD. Analysis of residual drug from the California cases confirmed the presence of MPTP, also produced by seemingly minor changes in the synthesis protocol. Additional literature search found reports of chemists with puzzling parkinsonian syndromes, in retrospect due to their exposure to MPTP, which was used as an industrial synthetic intermediary [101]. This extraordinary medical detective work identified this pharmacological tool that has been utilized over the past two decades in a wide variety of animal and in vitro studies to provide valuable insights into underlying neuropathological processes in PD, as well as for neurodegenerative diseases in general.

2. Animal studies

2.1. Primate studies

Shortly after MPTP was identified as the causative agent of the human clinical syndrome, experiments were undertaken to create animal models to study its biochemical actions more thoroughly. Primates are sensitive to MPTP, which produces a severe parkinsonian syndrome [19,89] at similar doses as in humans (unlike mice, which require much higher doses for toxicity, but are more widely used in biochemical, toxicological and genetic studies; see below). MPTP is highly lipophilic and readily enters the brain, but its levels decline rapidly, replaced by its charged metabolite, 1-methyl-4-phenylpyridinium (MPP^+), which is found in the brains of MPTP-treated monkeys up to 20 days later [115]. In addition to clinical symptoms and pharmacological responsivity, postmortem examination showed SN neuron loss and striatal DA depletion similar to PD [20]. Minimal effects on the ventral tegmental DA neurons, or little depletion of norepinephrine and serotonin brain concentrations was thought to be a limitation of the MPTP animal model of PD, although it might only reflect dosage and acute vs. chronic toxin exposure. In addition, MPTP-treated animals (including humans) do not exhibit Lewy bodies (LB) [50], a hallmark neuronal inclusion of PD, but old treated primates have inclusions that are somewhat similar [51].

2.2. Rodent studies

In spite of early studies reporting that rodents were resistant to MPTP, Heikkila et al. [75] demonstrated that mice, administered multiple high MPTP doses, exhibited striatal DA reductions, SN neuron losses and behavioral impairments, paving the way for extensive biochemical investigations on the mechanisms of MPTP neurotoxicity. However, depending on the endpoint tested, MPTP effects in mice varied with dose, route, number and timing of injections, as well as gender, age [86], strain and even supplier [73,177], which created confusion, until species appropriate dosing and the various complex steps in MPTP metabolism, distribution and toxic effects were clarified (see below). However, even at high dosages, rats are only mildly and transiently affected, which itself is interesting and perhaps reflects an evolutionary advantage of neuroprotective processes [169,193].

3. Mechanisms of MPTP neurotoxicity

3.1. Metabolism to active toxin

Once a suitable laboratory animal model for MPTP effects was demonstrated in mice with sufficient dosing, systematic studies on its mechanism of action were undertaken. Unlike primates, in mice, MPTP is rapidly metabolized to MPP^+ and neither chemical species is detectable a few hours after a single dose [115], partly explaining the relative murine resistance and the necessity for multiple high doses to achieve neurotoxicity. Depending on the dosing regimen and mouse strain utilized, 80–90% depletion of striatal DA (a convenient and widely used measure of MPTP effect) could be achieved, without affecting other brain regional monoamine neurotransmitters (e.g., serotonin or amino acids) [75,78].

3.1.1. Blockade of toxin formation

Pretreatment with monoamine oxidase (MAO) B inhibitors (e.g., deprenyl), but not MAO A inhibitors (e.g., clorgyline), blocked the MPTP-induced depletion of striatal DA concentrations, an important early clue to its metabolism and neuronal specificity [76]. In vitro brain metabolic studies with these inhibitors also demonstrated that MAO B is an essential component, forming MPP^+ from MPTP via a short-lived intermediary, 1-methyl-4-phenyl-2,3-dihydropyridinium ($MPDP^+$) [31]. Although DA neurons do not contain MAO B, it is present in proximate glia, suggesting the participation of these cells in MPTP metabolism and toxicity to DA neurons. Direct injections of MPP^+ into the striatum or median forebrain bundle (containing SN axons projecting to the striatum) produced

similar neurotoxicity as systemic MPTP treatment but were not antagonized by MAO B inhibitors [78]. They further demonstrated that pretreatment with specific DA neuronal (plasma) reuptake or transporter blockers (e.g., mazindol), but not those specific for other brain neuronal systems, protected animals from MPTP neurotoxicity, giving additional neurochemical specificity to MPTP's primary actions on DA neurons. The importance of the DA transporter is further illustrated in transgenic mice lacking it, which are resistant to MPTP neurotoxicity [15]. The degree of MPTP-induced striatal damage was shown to be proportional to striatal MPP^+ concentrations [63]. In vitro uptake of MPP^+ via the DA transport system in striatal synaptosomes follows kinetics very similar to those for DA itself [87]. Thus, glial MAO B forms $MPDP^+$, which can diffuse out and disproportionate to MPP^+ and MPTP [178], and MPP^+, which leaks out of glia. MPP^+ in extracellular space enters nearby DA neurons by the active DA transporter. This is not the complete story because different DA neurons differ in their susceptibility to MPTP neurotoxicity, viz. SN neurons are more sensitive than ventral tegmental DA neurons.

3.1.2. Role of VMAT2

A number of other factors contribute to DA neuron neurotoxicity susceptibility. An additional system regulating MPP^+ distribution was identified at the synaptic vesicle level with the demonstration of the neuronal-specific vesicular monoamine transporter (VMAT2) [137]. VMAT can concentrate endogenous monoamine neurotransmitters, as well as abnormal, toxic substances, such as MPP^+, from the cytoplasm into synaptic vesicles and thus serve to isolate and protect toxin-sensitive intracellular components [137,151]. The densities of radiolabeled MPP^+ (after systemic labeled MPTP injection) in autoradiograms and VMAT2 immunoreactivity in the midbrain vary greatly among monoamine nuclei, but they both follow a similar pattern and one consistent with neuronal sensitivity to MPTP neurotoxicity [167]. LC neurons containing higher VAMT2 levels are relatively spared, while SN neurons with much lower VMAT2 are affected, supporting the fact that vesicular sequestration of MPP^+ is neuroprotective. Transgenic mice lacking VMAT2 are more sensitive to MPTP [56]. Furthermore, inhibitors of VMAT2 block this vesicular uptake, resulting in higher cytoplasmic concentrations and enhanced MPP^+ neurotoxicity [170]. After MPTP treatment, VMAT2 immunostaining in monkey caudate and putamen is markedly reduced, reflecting DA cell loss, while the hypothalamus and amygdala are relatively spared [124]. In addition, they showed a similar pattern of VMAT2 immunostaining in PD brains. Using analogs of the VMAT2 inhibitor, tetrabenazine, in vivo PET imaging studies revealed decreased binding in PD [53] and MPTP-treated primate [39] brain regions, which paralleled other neurochemical and behavioral data.

3.1.3. Iron and melanin involvement

If not bound to carrier proteins, transition metals, such as iron, copper and zinc, can form reactive oxygen species (ROS; see below) and thereby damage cellular components, including enzymes, membrane lipoproteins and even DNA. DA neurons contain relatively high iron [190] and species-dependent melanin [9] concentrations, which interact with each other and exogenous compounds such as MPTP and MPP^+, to produce ROS, making DA neurons more vulnerable and enhancing MPTP neurotoxicity [10,178]. In addition, reactive iron facilitates DA auto-oxidation, also forming ROS and melanin, which binds iron, MPTP and MPP^+ [38]. These investigators showed that melanin binds MPP^+ with high affinity, influencing its distribution and serving as a high concentration toxin release depot in species forming neuromelanin. Examining competitors of MPP^+ binding, chloroquine was potent and treatment with it prior to MPTP attenuated neurotoxicity [37]. Iron chelators also antagonize MPTP neurotoxicity [10] and ROS formation. Since iron is increased in PD SN [64,81,88], it might contribute to the enhanced oxidative burden on affected neuron in this region. Attempts to find differences in the iron carrier protein, ferritin, in PD tissues have been equivocal [190].

3.1.4. MPTP analogs

With a better understanding of the potent effects of one environmental toxin, investigators looked at compounds structurally related to MPTP for possible similar effects and association with PD. Various MPTP analogs existing in the environment were delineated [166], although few have been rigorously tested for neurotoxicity. This group found that 4-phenylpyridine (4PP; a constituent of some teas and an industrial chemical) produced moderate inhibition of DA synthesis in vivo and in vitro. 4PP and related pyridine derivatives also inhibited mitochondrial NADH dehydrogenase to a similar extent as MPP^+ [145]. In addition, 4PP might be metabolized to MPP^+ by a broad nicotinamide-N-methyltransferase (a similar pathway is involved for salsolinol-related derivatives; see Naoi, this volume), expressed in some brain neurons [148]. The formation, localization and actions of the endogenous monoamine and aldehyde-derived condensation or metabolic products related to MPTP are discussed in other papers in this volume (see Naoi; Collins).

Ten of numerous MPTP analogs have been cited to be neurotoxic [74], with the $2'$-substituted analogs being equal or more potent than MPTP [192]. Because a number of epidemiologic reports found an association of PD with rural living, wellwater consumption and pesticide exposure [11,139] studies on some compounds in use were carried out. MPP^+ itself has been field-tested as a herbicide (trade name Cyberquat) but neurotoxicity was not reported [104]. Paraquat is also a widely used herbicide and, with two N-methyl pyridinium moieties, is structurally similar to MPP^+. It impairs mitochondrial respiration [42], however, because its charge does not readily enter the brain after peripheral administration, but where it can cause lung damage. Direct brain injections of paraquat produce neurotoxicity, but not specific to DA neurons [35].

3.2. Actions on the electron transport system

Because of the important role of MAO in metabolizing MPTP to its presumptive toxic product, MPP^+, as well as the considerable oxidative "load" on DA neurons [82], the participation of oxidative stress in MPTP and related neurotoxins effects came under intensive scrutiny. MPP^+ and MPTP cause massive DA release [155], which is then metabolized by oxidative degradative processes, including MAO, forming ROS, such as H_2O_2 and superoxide radicals [33], which themselves can be toxic. Multiple antioxidant ROS-neutralizing systems exist, including superoxide dismutase (SOD), glutathione peroxidase and catalase, but they can be overloaded or compromised by genetic variations in their protein components. Enhanced ROS production after MPTP and MPP^+ has been detected in vivo using radical spin-trapping agents [32]. In striatal slices and brain mitochondrial preparations, MPP^+ inhibits NADH-linked oxidation, without affecting succinate oxidation [131,184], and parts of the mitochondrial electron transport cascade that generates ATP, but also generates ROS. MPTP was inactive in the mitochondrial preparation, but inhibited oxidation in brain slices, where it could be metabolized to MPP^+. These studies established that MPP^+ is a specific inhibitor of Complex I (NADH dehydrogenase or NADH ubiquinone oxidoreductase) [144] at the ubiquinone-binding site. Mitochondria have an energy-dependent active uptake system that can concentrate MPP^+ [146,147]. The MPP^+ concentrations required to inhibit mitochondrial respiration are relatively high (0.1–0.5 mM) [184], but combined active neuronal plasma and mitochondrial membrane transporters might achieve more than 20 mM in mitochondria [143,147].

SN Complex I activity is decreased by about 30% in PD, but the rest of the respiratory chain is not affected [159]. In addition, PD platelets also show a small but consistent Complex I deficiency [134,158], as yet insufficient to be of diagnostic value. Unfortunately, this field is hampered by technical variability, largely dependent on mitochondrial preparation purity [173], but the SN findings are consistent; the platelet data less so. Multiple factors might underlie impairments in Complex I activity; defects would produce more ROS, which would further impair activity. The genetics of human Complex I encoding is indeed complex: composed of 41 protein subunits, of which seven are encoded by genes on mitochondrial DNA and the remainder by nuclear DNA. Susceptible to ROS damage, mtDNA is dependent on nuclear DNA for replication, translation and repair [164]. In addition, the copy number of mitochondrial DNA per mitochondrion varies with tissue: platelets have one copy, while brain has more than five, making considerable diversity possible [173]. Finally, exogenous or endogenous toxins could be inhibiting Complex I.

Manipulating antioxidant systems modulates MPTP neurotoxicity. Antioxidants, such as ascorbate, α-tocopherol, N-acetylcysteine, etc., protect against MPTP neurotoxicity in mice [136]. Glutathione normally inactivates H_2O_2, but inhibiting its synthesis or depleting it with drugs potentiated MPTP or MPP$^+$ toxicity in the mouse [2,187]. Conversely, increasing brain glutathione attenuates MPTP neurotoxicity [185]. Furthermore, MPTP treatment depletes brainstem glutathione, which is prevented by antioxidant pretreatment [191]. Mice deficient in glutathione peroxidase exhibit increased vulnerability to MPTP (and other neurotoxins) [94]. Enhancing SOD reduces MPTP neurotoxicity in transgenic mice [142], while reducing SOD makes mice more vulnerable to several neurotoxins, including MPTP [5]. Because copper is a prosthetic group for several antioxidant enzymes, mice were pretreated with copper sulfate prior to MPP$^+$, which reduced lipid peroxide formation and produced a marked increase in SOD activity [4], suggesting that suboptimal SOD activity might contribute to neurotoxicity.

With this greater appreciation for MPTP acting on the electron transport chain, the potent Complex I inhibitor (without any apparent structural similarities to MPTP) and widely used pesticide, rotenone, was tested for DA neuron toxicity. Stereotaxic rotenone injections into rat striatum produced greater mitochondrial respiratory inhibition than and comparable DA depletion to 100-fold higher doses of MPP$^+$ [77]. Recently, after chronic systemic treatment to rats, rotenone produced a parkinsonian syndrome, including brain DA neurotoxicity with LB-like inclusions [14], which has renewed interest in the role of environmental toxins in brain neurotoxicity.

It has been difficult to establish the primary step in the oxidative pathway causing neuronal death, since MPTP and MPP$^+$ themselves alter many of its components and manipulation of many of these components in turn alters many neurotoxin endpoints. ROS formation along several affected metabolic pathways does not help neuronal health. However, the key effect of electron transport inhibition at mitochondrial Complex I sets in motion numerous processes that seemingly send a neuron into a downward death spiral (see below). Decreased ATP formation from MPP$^+$ or MPTP exposure [126] would then jeopardize the neuron's ability to maintain critical energy-dependent concentration gradients. For example, compromised Na$^+$/K$^+$ ATPase function could cause increased intracellular Ca^{2+}, protease activation and neuronal depolarization. Such depolarization can remove Mg$^+$ blockade of N-methyl-D-aspartate (NMDA) receptors, allowing receptor activation by its agonists glutamate and NMDA, with resultant further increases in intracellular Ca^{2+}, nitric oxide synthase (NOS) and NO [80]. Lamotrigine, an anticonvulsant that inhibits glutamate release, reduced MPTP neurotoxicity [92]. Although not universal [49], some laboratories have reported that NMDA receptor

blockade with MK-801 attenuates MPTP and MPP$^+$ neurotoxicity [180].

NOS activity is high in the nigrostriatal system, but attempts to show differences in PD brains have been inconclusive. MPTP induces striatal and SN NOS activity, probably in activated glia, where highly diffusible NO can act on already impaired DA neurons [106]. NO reacts with superoxide, forming peroxynitrite, which is highly reactive and can damage lipids, proteins and DNA. Tyrosine hydroxylase, the rate-limiting step in DA synthesis, can be inactivated by peroxynitrite tyrosine nitration after MPTP treatment [6]. NOS inhibitor treatments (7-nitroindazole [160]; S-methylthiocitrulline [118]) and transgenic mice deficient in neuronal NOS [41,141] resist MPTP neurotoxicity.

These processes, no doubt, operate to some degree in many cells, but the DA neuron's susceptibility results from the potential fatal flaw of its plasma transporter's ability to concentrate MPP$^+$-related toxins, followed by high oxidative metabolic activity and iron content, which all enhance ROS formation, pushing antioxidant defense mechanisms to the limit. Adding possible mutations in any proteins handling metabolic, structural or transport functions, age-related DA neuron losses and possible exposure to other environmental toxins like MPTP could result in a situation that the DA neuron might not survive. Studies on MPTP neurotoxicity over the past 20 years have made us well aware of the complexities of neuronal homeostatic systems and have given insights into equally complex systems likely underlying PD-related neurodegenerative diseases.

Synthesizing the results of both in vivo and in vitro experiments from a large number of laboratories up to this point, a working hypothesis that emerges involves the rapid appearance of administered MPTP in the brain, where MAO B, probably in glial cells, forms MPDP$^+$ or the active toxin, MPP$^+$, which diffuse or are secreted into extracellular space. The active DA uptake transporter moves MPP$^+$ into these DA neurons selectively, where toxin that is not sequestered in synaptic vesicles by VMAT2 can exert its toxic effects on intracellular components, as described in Section 3.3.

3.3. Molecular death signaling

As a better understanding of the metabolism and subcellular targets of MPTP and MPP$^+$ emerged, investigators in turn focused on the sequelae of neuronal death signaling triggered by MPTP exposure, and how these toxins might provide more generalizable principles of neuronal death. Cell death can occur by necrosis or apoptosis, also call programmed cell death. As a result of a gross, acute insult, necrotic cells exhibit cytoplasmic and nuclear swelling and membrane breakdown, with loss of contents, inducing an immune reaction. On the other hand, apoptosis is part of normal cellular development and homeostasis, although it is present in some disease states, including Alzheimer's disease [99,130] and perhaps PD [69]. Apoptosis involves a complex cascade of enzymatic checkpoints (for details, see figure: www.cellsignal.com/References/Pathways/mito-chon_ctrl.html) and mitochondria play a critical role. Apoptotic cells show membrane blebbing, nuclear chromatin condensation and DNA laddering from DNA breakdown, but no immune reaction. However, apoptosis and necrosis are two different but interconnected pathways. If ATP concentrations are reduced sufficiently, an apoptotic stimulus cannot completely activate the biochemical apparatus, but there will be cell necrosis [161]. This switching might then adversely affect adjacent neurons by triggering an inflammatory response.

Apoptosis is regulated by the Bcl-2 gene family, caspase enzyme activity and p53 protein. In response to death signals, proapoptotic Bax, Bad, Bid and Bim proteins migrate from the cytosol into the mitochondria to release cytochrome c (which functionally connects Complex III and IV in the electron transport cascade) into the cytosol, where it binds with Apaf1, and this complex activates caspases [66], the enzymatic executioners. Bcl-2 and Bcl-xl proteins are in the outer mitochondrial wall and are anti-apoptotic, inhibiting cytochrome c release. DNA damage activates p53, which in turn induces Bax transcription. Apoptosis results from upregulation of the death promoting pathways, the downregulation of death antagonistic pathways and their net effect. Transgenic mice expressing human Bcl-2 in neurons are resistant to MPTP treatment [133], whereas decreasing Bcl-2 and Bcl-xl enhance MPP^+-induced neuronal death [83]. After MPTP exposure in mice, SN Bax mRNA and protein are increased [71, 181], while Bax-deficient animals are resistant [179,181] to MPTP neurotoxicity.

Oxidative stress and interference with mitochondrial respiration, such as from MPTP treatment, can initiate apoptosis [98]. Inhibition of mitochondrial respiration causes release of cytochrome c, presumably through the voltage-dependent anion channel or pore, and activation of the apoptotic death cascade, similar to what other cell death-inducing treatments do acting further upstream [24]. Complex I inhibitors block electron transport and ATP synthesis, inducing dose-dependent apoptosis in PC12 cells [68]. Therefore, MPTP and MPP^+ effects on ATP formation are critical, and the degree of ATP alteration might underlie the controversy in the literature as to which form of neuronal death they produce. With brain ischemia and hypoxia, neurons die by both mechanisms, depending on the severity of the insult, and the balance might be similar with MPTP neurotoxicity [132].

Numerous studies report that MPP^+ can induce apoptosis in vitro in a variety of cell lines, including monoamine-related cells [16], using morphological or biochemical criteria. In vivo, SN DNA fragmentation in mice treated with MPTP has been reported [55,168,175], but this is not absolute proof. Drugs that block the mitochondrial pore attenuate [157] and those that keep it open enhance [24]

MPP^+-induced apoptosis. As with other apoptotic stimuli, released cytochrome c activates caspases and their activities are elevated in mouse SN after MPTP treatment [183]. They further showed that caspase inhibition reduced MPTP neurotoxicity. Blockade of the protein in vitro with antisense treatment prevents apoptosis. After MPTP treatment that would produce apoptosis, pretreatment with a caspase inhibitor in vitro did not neuroprotect, but resulted in neuronal necrosis [70]. Nonetheless, there are a number of reports [16] that do not show apoptosis after MPTP or MPP^+, possibly due to dosage, cell line or endpoint differences. Prostate apoptosis response-4, an important protein in neuronal apoptosis, is increased in primates and mice in SN neurons and the striatum after MPTP [44].

4. Emerging new players

4.1. Genetic susceptibility

Several genetic modulators of subcellular parkinsonian processes have been described recently. They do not appear to cause PD, per se, but interact with other systems to contribute to the PD neuropathology.

4.1.1. α-Synuclein

With great initial excitement, mutations in the α-synuclein (syn) gene (at chromosome 4q21) in several large family studies were reported to be associated with an early-onset PD, with autosomal dominant inheritance [138]. However, a number of subsequent prospective studies for syn mutations in familial PD [27,47] found that they are rare. Nonetheless, the prominent localization of syn to PD LBs and other intracellular inclusions is not only of neuropathological interest, but mechanistically is important. Native and especially mutant syn isoforms readily form filamentous aggregates in vitro, and specific syn antibodies label LBs and some neurites in PD and diffuse LB disease, most Alzheimer's disease senile plaques and at axon terminals, in proximity to synaptic vesicles [62]. LBs are also immunoreactive to ubiquitin and neurofilament antibodies, suggesting that syn interacts with these proteins. In transfected cells, syn is degraded by the proteosome–ubiquitin mechanism and syn mutations prolong its half-life [12]. PD-related syn mutations show reduced vesicular binding, which might contribute to the pathophysiology [90]. However, the mutated alleles could be transcribed at a much lower level, resulting in reduced expression, instead of a pathological gain of function [116]. Elevated ubiquitin in PD LBs suggests a defect in proteosomal breakdown, supported by the finding of missense mutations in the ubiquitin carboxy-terminal hydrolase 1 gene in PD [105], normally functioning as a member of the deubiquitinating enzyme family. In addition, there is now evidence for increased PD susceptibility in individuals with altered syn and apolipoprotein E genotypes [97], showing that a single polymorphism might not have a strong association, but can interact with others to create the pathological phenotype.

After an MPTP protocol that kills DA neurons by apoptosis, syn expression was upregulated, in a time course paralleling DA neuron degeneration [182]. There was a dramatic increase in syn immunoreactivity exclusively in SN neurons, but without any apoptotic death features. This increased staining seems to be due to a redistribution of syn immunoreactivity from the terminals to the cell body after MPTP treatment of primates [96] or mice [109] and correlated with greater neurotoxic severity, perhaps reflecting the formation of aggregates in degenerating neuronal cell bodies or a compensatory response. A recent study addresses interactions of syn with paraquat, the herbicide and inhibitor of cellular respiration mentioned above, which illustrates how genetic and environmental factors might interact in PD. The in vitro syn fibril formation was enhanced, while mice showed thioflavin-positive SN aggregates after paraquat treatment [113]. In syn-transfected cells or in vivo in mouse striatum and midbrain, MPTP treatment caused a specific syn nitration [140], which has a deleterious effect on other proteins, although this exact consequence is not known. On the other hand, mice transfected with mutant *syn* exhibited no difference in MPTP neurotoxicity than wild-type controls [150].

4.1.2. Parkin

An autosomal recessive juvenile form of PD was first described in Japan [189], which mapped to chromosome 6 and was named parkin. It is a large gene (>1 Mb) and includes a coding sequence moderately homologous to that for ubiquitin. The protein has about 465 amino acids, with a metal-binding terminal domain that can also interact with DNA and other proteins. Multiple small gene exon deletions and truncating mutations prevent the full protein synthesis, while missense mutations cause amino acid substitutions. The protein is expressed in many tissues and abundant in brain, but its function is unknown, except for possible ubiquitin-like proteosomal protein breakdown. In affected individuals, there is extensive SN and LC depopulation, but no LBs or parkin immunostaining in brain regions [125], suggesting a loss-of-function defect.

4.1.3. Cytochrome P450 isoenzymes

The cytochrome P450 enzymes are a protein family that metabolize a wide variety of endogenous and exogenous substances, showing tissue distribution diversity and relative amounts, as well as genetic polymorphisms. Of special interest are reports of associations of the CYP 2D6 subfamily with PD, which might predispose individuals to less efficient detoxification or alternative metabolism to form some toxin [28]. Liver CYP 2D6 can metabolize MPTP to nontoxic products [172], compared to its metabolism in brain via MAO B to the toxin, MPP^+. Various specific substrates were tested for their P450 N-demethylation of MPTP in vitro in yeast and human microsomes and pro-

vided evidence that CYP 2D6, 1A2 and, to a lesser extent, 3A4 isoforms could play a role in in vivo MPTP metabolism [34]. Diverse investigations over the world show some associations of CYP 2D6 with PD, but which vary greatly with ethnic groups. Variants in this gene combined with those for apolipoprotein E show an association in PD [18].

4.2. Inflammatory processes

In response to toxins, including MPTP, and neuronal death, glia become activated and can upregulate production of a broad array of inflammatory cytokines, as well as ROS, which can contribute to neurotoxicity of DA neurons in culture [120]. Since glial cells are critically involved in MPTP oxidative metabolism, multiple glial pathways are also activated, including superoxide formation, glutamate release, growth factors, etc. As brain-supportive cells, glia provide trophic factors, including glial cell line-derived neurotrophic factor (GDNF) and brain-derived neurotrophic factor (BDNF) [186], and MPP^+ impairs GDNF release in primary astrocyte cultures [122]. Fibroblasts modified to express BDNF and implanted prevent MPP^+-induced DA neuron toxicity in rats [54]. BDNF stimulates antioxidant enzymes [30], which might underlie this neuroprotective effect. On the other hand, intranigral injections of proinflammatory lipopolysaccharide activated glia and caused striatal and nigral DA depletion, with loss of SN DA neurons [26]. Mouse strain differences in MPTP neurotoxicity might be related to glia, since cocultured DA neurons are more resistant to MPP^+ toxicity when associated with glia cultures from resistant strains than those from susceptible strains [165].

Glial proliferation and reactive microglia have been described in the striatum and SN of PD [188] and of humans exposed to MPTP years previously [103]. Glial cells expressing interleukins, interferon and tumor necrosis factor α were seen in PD SN, but not control tissue [85]. These reactions are not specific to PD, as they are seen in other neurodegenerative diseases, and so are probably not causative.

Mice treated with MPTP show transient SN glial activation and elevated interleukin-6 [95]. Pretreatment with the antiinflammatory, cyclosporin, or the immunosuppressant, FK-506, inhibited DA depletion caused by MPTP [93]. Aspirin and meloxicam—a specific cyclooxygenase-2 inhibitor—pretreatment comparably protected mice from MPTP DA neurotoxicity [176]. Similarly, aspirin or salicylate protected mice from MPTP-induced DA depletion, but this was attributed to ROS scavenging and not their antiinflammatory actions [8]. A more specific radical spin-trapping compound, α-phenyl-N-tert-butyl nitrone (PNB) that has shown protection against ischemic damage, protected against MPTP toxic effects [48], but without altering striatal hydroxyl radical formation, suggesting additional properties of PNB. The free radical spin traps, MDL 101002 and tempol, gave dose-dependent neuroprotection from toxins, including MPTP, attenuating striatal DA depletions and nitrotyrosine formation [117].

5. Therapeutic strategies

The use of MPTP/MPP$^+$-induced neuronal damage in animal studies has served as a backdrop to test compounds or protocols that attenuate or reverse toxic actions, and thus might be clinically useful in PD or other neurodegenerative diseases. Based on the many anatomical and biochemical similarities of MPTP neurotoxicity to PD neuropathology, it does not seem unreasonable to assume that similar fundamental mechanisms of neuronal death are operating. The MPTP animal model of PD is useful to evaluate new therapies, imaging, transplantation and other surgical approaches. However, most studies have used high neurotoxin doses, evaluated after relatively short exposures and examining short-term effects. In contrast, the current hypothesis posits that PD is not a single disease [22] and might be due, in part, to very long duration exposure to low levels of toxin(s) that interact with multiple factors [153], including age-related losses of DA neurons and individual genetic polymorphisms, which all synergize to produce the related phenotypes. Therefore, neuroprotective approaches that are useful for acute exposure give insights into mechanisms operating over chronic exposure, but might suggest therapies that are harder to implement over long-term, low-level exposure. Identification of risk factors, including various genetic polymorphisms and the development of early or presymptomatic diagnostic tests, such as brain imaging [52,171] or other biological markers [134], are essential for initiating neuroprotective therapy before there has been excess neuronal loss.

A number of treatments, which have been mentioned previously, antagonize neurotoxic features of MPTP or MPP$^+$, some acting very early in toxin formation, necessitating an early response to acute, rapidly acting high-level toxin exposure. These include blockade of MPP$^+$ formation with MAO B inhibitors or uptake with DA transporter inhibitors, problematic outside of experimental studies. Modulating SN iron levels with chelators has specificity and brain permeability problems. An innovative approach is the use of prodrugs, which enter the brain and are converted to the active drug, after stimuli such as oxidative stress [57]. Competition for melanin binding with chloroquine-related drugs has not been exploited clinically. These have remained intriguing laboratory exercises, but some drug interventions exist.

5.1. Antioxidants/ROS scavenging

Oxidative stress is a key aspect of MPTP neurotoxicity that provides multiple points of potential therapeutic intervention. Augmenting antioxidant mechanisms with combined deprenyl/antioxidant had some success in clinical trials, and based on animal studies seemed a very rational approach. The Deprenyl and Tocopherol Antioxidative Therapy of Parkinsonism (DATATOP) study was a large, double-blind, multicenter, placebo-controlled, long-term clinical trial undertaken in the late 1980s to test this oxidative stress hypothesis with deprenyl and α-tocopherol, alone or in combination. Initial and long-term follow-up results [162,163] indicated a slowing of PD symptom progression. Although the use of L-DOPA has been the most significant PD therapy, problems with fluctuating efficacy and dyskinesias, as well as questions of possible deleterious effects of greater DA synthesis [123], stimulated the use of DA agonists. Some newer, more specific agonists have been postulated to also have ROS-scavenging properties [149], but have had only limited clinical trials. A recent clinical trial comparing L-DOPA and ropinirol, a new agonist, done with PET visualization was encouraging [152]. Another new DA agonist, SKF-38393, attenuated MPTP DA depletion and blocked glutathione reduction, partly by altering MAO B activity [128]. The prototypical DA agonist, apomorphine, exhibits antioxidant/scavenging properties at low doses in vitro, but higher concentrations are cytotoxic [59], limiting its potential clinical usefulness. Notwithstanding, some doses of apomorphine protect mice from MPTP neurotoxicity [67], via antioxidant, iron chelation and MAO inhibition actions.

In addition to the diverse neuroendocrine functions of melatonin, antioxidant properties have been described [154], acting to scavenge hydroxyl and peroxyl radicals, and in some systems more potent than glutathione. Melatonin, administered prior to and after MPTP, protected against striatal DA neurotoxicity and lipid peroxidation [1]. It also reversed DA neurotoxicity produced by intranigral MPP$^+$ injections [91]. There is growing in vitro, animal and epidemiologic evidence that estrogen (es) is neuroprotective, with neurotrophic and antiinflammatory effects, apart from its traditional steroidal actions. Es attenuates striatal DA neurotoxicity from toxins, including MPTP. 17-β-es (and progesterone, but not 17-α-es) blunted the MPTP-induced loss in DA transporter mRNA [21]. Striatal DA depletion following methamphetamine is antagonized by es, which is reversed by the anti-es compound, tamoxifen [58].

Ebselen, an antioxidant with glutathione peroxidase-like activity that has been used successfully in stroke, prevents MPTP neurotoxicity in primates, when administered before, during or after MPTP [127]. A tetracycline that also was neuroprotective in ischemic injury, minocycline, prevented SN DA neurodegeneration from MPTP in mice [43]. Using cultures of mesencephalic or cerebellar granule neurons, minocycline inhibited MPP$^+$-mediated NOS induction and NO toxicity, but only in the presence of glia. It also inhibited NO-induced phosphorylation of p38 MAP kinase, which might be an important neuroprotective mechanism.

5.2. Epidemiologic clues

Epidemiologic analyses have shown decreased PD incidence in smokers and recently a protective role for modest coffee consumption. This inverse association with smoking has been a consistent finding [46,79] and attempts to find alternative explanations have been unsuccessful, but the exact cause remains unclear. However, compounds in tobacco smoke inhibit MPP^+ and DA uptake into synaptosomes [23]. Recently, a quinone present in tobacco and its smoke were isolated and prevented MPTP-induced striatal DA depletion [25]. Nicotine pretreatment also antagonizes MPTP neurotoxicity [108].

Two recent, large studies showed a reduced PD risk associated with moderate coffee or caffeine consumption [7,156]. An adenosine-2A antagonist protected monkeys from MPTP toxicity [65]. The effects of treatment with caffeine and specific adenosine antagonists were tested in MPTP-treated mice [29]. Caffeine and adenosine-2A antagonists, but not adenosine-1 antagonists, attenuated striatal DA neurotoxicity. Furthermore, adenosine-2A knockout mice were more resistant to MPTP than wild-type littermates.

5.3. Genetic probes

cDNA expression microarrays were utilized to look for candidate gene expression changes in brains of mice treated with MPTP [110]. As might be expected when examining 1200 gene fragments, many changes were found, including altered expression of 50 genes related to oxidative stress, inflammation, cytokines, NMDA receptors, neurotrophic factors, neuronal NOS, several caspases, cell cycle regulators, signal transduction, etc. They further examined the effects of apomorphine on MPTP actions, which prevented the expression changes in a number of genes, including those involved in cell death.

6. Summary and conclusions

In the two decades since MPTP was identified as producing a clinical parkinsonian syndrome, not only have its mechanisms of action been unraveled, but it has revealed important features of PD neuropathology. The key role of oxidative stress, its subcellular components contributing to DA neuronal susceptibility and potential biological markers have emerged from this body of work. The next level of endeavor will be the identification of reliable diagnostic markers and risk factors, so that neuroprotective therapeutic interventions can be instituted early enough to at least slow down pathologic brain DA neuron losses.

References

[1] D. Acuna-Castroviejo, A. Coto-Montes, M. Gaia Monti, G.G. Ortiz, R.J. Reiter, Melatonin is protective against MPTP-induced striatal and hippocampal lesions, Life Sci. 60 (1997) PL23–PL29.

[2] J.D. Adams Jr., N. Odunze, Biochemical mechanisms of 1-methyl-4-phenyl-1,2,3,6-tetrahydropyridine toxicity. Could oxidative stress be involved in the brain? Biochem. Pharmacol. 41 (1991) 1099–1105.

[3] Y. Agid, F. Javoy-Agid, Peptides and Parkinson's disease, Trends Neurosci. 8 (1985) 30–35.

[4] M. Alcaraz-Zubeldia, S. Montes, C. Rios, Participation of manganese-superoxide dismutase in the neuroprotection exerted by copper sulfate against 1-methyl-4-phenylpyridinium neurotoxicity, Brain Res. Bull. 55 (2001) 277–279.

[5] O.A. Andreassen, R.J. Ferrante, A. Dedeoglu, D.W. Albers, P. Klivenyi, E.J. Carlson, C.J. Epstein, M.F. Beal, Mice with a partial deficiency of manganese superoxide dismutase show increased vulnerability to the mitochondrial toxins malonate, 3-nitropropionic acid and MPTP, Exp. Neurol. 167 (2001) 189–195.

[6] J. Ara, S. Przedborski, A.B. Naini, V. Jackson-Lewis, R.R. Trifiletti, J. Horwitz, H. Ischiropoulos, Inactivation of tyrosine hydroxylase by nitration following exposure to peroxynitrite and 1-methyl-4-phenyl-1,2,3,6-tetrahydropyridine (MPTP), Proc. Natl. Acad. Sci. USA 95 (1998) 7659–7663.

[7] A. Ascherio, S.M. Zhang, M.A. Hernan, I. Kawachi, G.A. Colditz, F.E. Speizer, W.C. Willett, A prospective study of caffeine consumption and risk of Parkinson's disease in men and women, Ann. Neurol. 50 (2001) 56–63.

[8] N. Aubin, O. Curet, A. Deffois, C. Carter, Aspirin and salicylate protect against MPTP-induced dopamine depletion in mice, J. Neurochem. 71 (1998) 1635–1642.

[9] M.J. Bannon, M. Goedert, B. Williams, The possible relation of glutathione, melanin and 1-methyl-4-phenyl-1,2,5,6-tetrahydropyridine (MPTP) to Parkinson's disease, Biochem. Pharmacol. 33 (1984) 2697–2698.

[10] A. Barbeau, J. Poirier, L. Dallaire, E. Rucinska, N.T. Buu, J. Donaldson, Studies on MPTP, MPP^+ and paraquat in frogs and in vitro, in: S.P. Markey, et al. (Eds.), MPTP: A Neurotoxin Producing a Parkinsonian Syndrome, Academic Press, Orlando, 1986, pp. 85–103.

[11] A. Barbeau, M. Roy, G. Bernier, G. Campanella, S. Paris, Ecogenetics of Parkinson's disease: Prevalence and environmental aspects in rural areas, Can. J. Neurol. Sci. 14 (1987) 36–41.

[12] M.C. Bennett, J.F. Bishop, Y. Leng, P.B. Chock, T.N. Chase, M.M. Mouradian, Degradation of α-synuclein by proteosome, J. Biol. Chem. 274 (1999) 33855–33858.

[13] H. Bernheimer, W. Birkmayer, O. Hornykiewicz, K. Jellinger, F. Seitelberger, Brain dopamine and the syndromes of Parkinson and Huntington. Clinical, morphological and neurochemical correlations, J. Neurol. Sci. 20 (1973) 415–455.

[14] R. Betarbet, T.B. Sherer, G. Macenzie, M. Garcia-Osuna, A.V. Panov, J.T. Greenamyre, Chronic systemic pesticide exposure reproduces features of Parkinson's disease, Nat. Neurosci. 3 (2000) 1301–1306.

[15] E. Bezard, C.E. Gross, M.C. Fournier, S. Dovero, B. Bloch, B. Jaber, Absence of MPTP-induced neuronal death in mice lacking the dopamine transporter, Exp. Neurol. 155 (1999) 268–273.

[16] D. Blum, S. Torch, N. Lambeng, M.-F. Nissou, A.-L. Benabid, R. Sadoul, J.-M. Verna, Molecular pathways involved in the neurotoxicity of 6-OHDA, dopamine and MPTP: Contribution to the apoptotic theory in Parkinson's disease, Prog. Neurobiol. 65 (2001) 135–172.

[17] B. Bogerts, J. Hantsch, M. Herzer, A morphometric study of the dopamine-containing cell groups in the mesencephalon of normals, Parkinson patients, and schizophrenics, Biol. Psychiatry 18 (1983) 951–969.

[18] M.A. Bon, E.N. Jansen-Steur, R.A. de Vos, I. Vermes, Neurogenetic correlates of Parkinson's disease: Apolipoprotein-E and cytochrome P450 2D6 genetic polymorphism, Neurosci. Lett. 266 (1999) 149–151.

[19] R.S. Burns, C.C. Chiueh, S.P. Markey, M.H. Ebert, D.M. Jacobowitz, I.J. Kopin, A primate model of parkinsonism: Selective destruction of dopaminergic neurons in the pars compacta of the

substantia nigra by N-methyl-4-phenyl-1,2,3,6-tetrahydropyridine, Proc. Natl. Acad. Sci. USA 80 (1983) 4546–4550.

[20] R.S. Burns, C.C. Chiueh, J. Parisi, S. Markey, I.J. Kopin, Biochemical and pathological effects of MPTP in the rhesus monkey, in: S. Fahn, et al. (Eds.), Recent Developments in Parkinson's Disease, Raven Press, New York, 1986, pp. 127–136.

[21] S. Callier, M. Morissette, M. Grandbois, D. Pelaprat, T. Di Paola, Neuroprotective properties of 17beta-estradiol, progesterone, and raloxifene in MPTP C57BL/6 mice, Synapse 41 (2001) 131–138.

[22] D.B. Calne, Parkinson's disease is not one disease, Parkinsonism Relat. Disord. 7 (2001) 3–7.

[23] L.A. Carr, J.K. Basham, B.K. York, P.P. Rowell, Inhibition of 1-methyl-4-phenylpyridinium ion and dopamine in striatal synaptosomes by tobacco smoke components, Eur. J. Pharmacol. 215 (1992) 285–287.

[24] D.S. Cassarino, J.K. Parks, W.D. Parker Jr., J.P. Bennett Jr., The parkinsonian neurotoxin MPP^+ opens the mitochondrial permeability transition pore and releases cytochrome c in isolated mitochondria via an oxidative mechanism, Biochim. Biophys. Acta 1453 (1999) 49–62.

[25] K.P. Castagnoli, S.J. Steyn, J.P. Petzer, C.J. Van der Schyf, N. Castagnoli Jr., Neuroprotection in the MPTP parkinsonian C57BL/6 mouse model by a compound isolated from tobacco, Chem. Res. Toxicol. 14 (2001) 523–527.

[26] A. Castano, A.J. Herrera, J. Cano, A. Machado, Lipopolysaccharide intranigral injection induces inflammatory reaction and damage in nigrostriatal dopaminergic system, J. Neurochem. 70 (1998) 1584–1592.

[27] P. Chan, Jiang, L.S. Forno, D.A. DiMonte, C.M. Tanner, J.W. Langston, Absence of mutations in the coding region of the α-synuclein gene in pathologically proven Parkinson's disease, Neurology 50 (1998) 1136–1137.

[28] H. Checkoway, F.M. Farin, P. Costa-Mallen, S.C. Kirchner, L.G. Costa, Genetic polymorphisms in Parkinson's disease, Neurotoxicology 19 (1998) 635–643.

[29] J.-F. Chen, K. Xu, J.P. Petzer, R. Staal, Y.-H. Xu, M. Beilstein, P.K. Sonsalla, K. Castagnoli, N. Castagnoli Jr., M.A. Schwarzschild, Neuroprotection by caffeine and A2A adenosine receptor inactivation in a model of Parkinson's disease, J. Neurosci. 21 (2001) 1–6 (RC143).

[30] B. Cheng, M.P. Mattson, NT-3 and BDNF protect CNS neurons against metabolic/excitotoxic insults, Brain Res. 640 (1994) 56–67.

[31] K. Chiba, A. Trevor, N. Castagnoli Jr., Metabolism of the neurotoxic tertiary amine, MPTP, by brain monoamine oxidase, Biochem. Biophys. Res. Commun. 120 (1984) 574–578.

[32] C.C. Chiueh, R.M. Wu, K.P. Mohanakumar, L.M. Sternberger, G. Krishna, T. Obata, D.L. Murphy, In vivo generation of hydroxyl radicals and MPTP-induced dopaminergic toxicity in the basal ganglia, Ann. NY Acad. Sci. 738 (1994) 25–36.

[33] G. Cohen, Monoamine oxidase and oxidative stress at dopaminergic synapses, J. Neural Transm. Suppl. 32 (1990) 229–238.

[34] T. Coleman, S.W. Ellis, I.J. Martin, M.S. Lennard, G.T. Tucker, 1-Methyl-4-phenyl-1,2,3,6-tetrahydropyridine (MPTP) is N-demethylated by cytochromes P450 2D6, 1A2 and 3A4-implications for susceptibility to Parkinson's disease, J. Pharmacol. Exp. Ther. 277 (1996) 685–690.

[35] M.T. Corasaniti, M.C. Strongoli, D. Rotiroti, G. Bagetta, G. Nistico, Paraquat: A useful tool for the in vivo study of mechanisms of neuronal cell death, Pharmacol. Toxicol. 83 (1998) 1–7.

[36] J.L. Cummings, Understanding Parkinson disease, JAMA, J. Am. Med. Assoc. 281 (1999) 376–378.

[37] R.J. D'Amato, G.M. Alexander, R.J. Schwartzman, C.A. Kitt, D.L. Price, S.H. Snyder, Evidence for neuromelanin involvement in MPTP-induced neurotoxicity, Nature 327 (1987) 324–326.

[38] R.J. D'Amato, Z.P. Lipman, S.H. Snyder, Selectivity of the parkinsonian neurotoxin MPTP: Toxic metabolite MPP^+ binds to neuromelanin, Science 231 (1986) 987–989.

[39] J.N. DaSilva, M.R. Kilbourn, E.F. Domino, In vivo imaging of monoaminergic nerve terminals in normal and MPTP-lesioned primate brain using positron emission tomography (PET) and ^{11}C-tetrabenazine, Synapse 14 (1993) 128–131.

[40] G.C. Davis, A.C. Williams, S.P. Markey, M.H. Ebert, E.D. Caine, C.M. Reichert, I.J. Kopin, Chronic parkinsonism secondary to intravenous injection of meperidine analogues, Psychiatr. Res. 1 (1979) 249–254.

[41] T. Dehmer, J. Lindenau, S. Haid, J. Dichgans, J.B. Schulz, Deficiency of inducible nitric oxide synthase protects against MPTP toxicity in vivo, J. Neurochem. 74 (2000) 2213–2216.

[42] Q. Ding, J.N. Keller, Proteosome inhibition in oxidative stress neurotoxicity: Implications for heat shock proteins, J. Neurochem. 77 (2001) 1010–1017.

[43] Y. Du, Z. Ma, S. Lin, R.C. Dodel, F. Gao, K.R. Bales, L.C. Triarhou, E. Chernet, K.W. Perry, D.L. Nelson, S. Luecke, L.A. Phebus, F.P. Bymaster, S.M. Paul, Minocycline prevents nigrostriatal dopaminergic neurodegeneration in the MPTP model of Parkinson's disease, Proc. Natl. Acad. Sci. USA 98 (2001) 14669–14674.

[44] W. Duan, Z. Zhang, D.M. Gash, M.P. Mattson, Participation of Par-4 in degeneration of dopaminergic neurons in models of Parkinson's disease, Ann. Neurol. 46 (1999) 587–597.

[45] R.C. Duvoisin, Is Parkinson's disease acquired or inherited? Can. J. Neurol. Sci. 11 (1984) 151–155.

[46] R. Eldridge, W.A. Rocca, S.E. Ince, Parkinson's disease: Evidence against a toxic etiology and for an alternative theory, in: S.P. Markey, et al. (Eds.), MPTP: A Neurotoxin Producing a Parkinsonian Syndrome, Academic Press, Orlando, 1986, pp. 355–367.

[47] M. Farrer, F. Wavrant-De Vrieze, R. Crook, L. Boles, J. Perez-Tur, J. Hardy, W.G. Johnson, J. Steele, D. Maraganore, K. Gwinn, T. Lynch, Low frequency of α-synuclein in Parkinson's disease, Ann. Neurol. 43 (1998) 394–397.

[48] B. Ferger, P. Teismann, C.D. Earl, K. Kuschinsky, W.H. Oertel, The protective effects of PBN against MPTP toxicity are independent of hydroxyl radical trapping, Pharmacol., Biochem. Behav. 65 (2000) 425–431.

[49] F. Finiels-Marlier, A.M. Marini, P. Williams, S.M. Paul, The N-methyl-D-aspartate antagonist MK-801 fails to protect dopaminergic neurons from 1-methyl-4-phenylpyridinium toxicity in vitro, J. Neurochem. 60 (1993) 1968–1971.

[50] L.S. Forno, L.E. DeLanney, I. Irwin, J.W. Langston, Similarities and differences between MPTP-induced parkinsonism and Parkinson's disease. Neuropathologic considerations, Adv. Neurol. 60 (1993) 600–608.

[51] L.S. Forno, J.W. Langston, L.E. DeLanney, I. Irwin, G.A. Ricaurte, Locus ceruleus lesions and eosinophilic inclusions in MPTP-treated monkeys, Ann. Neurol. 20 (1986) 449–455.

[52] K.A. Frey, R.A. Koeppe, M.R. Kilbourne, Imaging the vesicular monoamine transporter, in: D. Calne, S.M. Calne (Eds.), Parkinson's Disease: Advances in Neurology, vol. 86, Lippincott Williams and Wilkins, Philadelphia, 2001, pp. 237–247.

[53] K.A. Frey, R.A. Koeppe, M.R. Kilbourne, T.M. Vander Borght, R.L. Albin, S. Gilman, D.E. Kuhl, Presynaptic monoaminergic vesicles in Parkinson's disease and normal aging, Ann. Neurol. 40 (1996) 873–884.

[54] D.M. Frim, T.A. Uhler, W.R. Galpern, M.F. Beal, X.O. Breakefield, O. Isacson, Implanted fibroblasts genetically engineered to produce brain-derived neurotrophic factor prevent 1-methyl-4-phenylpyridinium toxicity to dopaminergic neurons in the rat, Proc. Natl. Acad. Sci. USA 91 (1994) 5104–5108.

[55] T. Fukuda, J. Tanaka, Nucleosomal DNA fragmentation in 1-methyl-4-phenyl-1,2,3,6,-tetrahydropyridine-treated mice, Soc. Neurosci. Abstr. 26 (2000) 1027.

[56] R.R. Gainetdinov, F. Fumagalli, Y.M. Wang, S.R. Jones, A.I. Levey, G.W. Miller, M.G. Caron, Increased MPTP neurotoxicity in vesicular monoamine transporter 2 heterozygote knockout mice, J. Neurochem. 70 (1998) 73–78.

[57] J.B. Galey, J. Dumats, I. Beck, B. Fernandez, M. Hocquaux, N, N'-bis-dibenzyl ethylenediaminediacetic acid (DBED): A site-specific hydroxyl radical scavenger acting as an "oxidative stress activatable" iron chelator in vitro, Free Radical Res. 22 (1995) 67–86.

[58] X. Gao, D.E. Dluzen, Tamoxifen abolishes estrogen's neuroprotective effect upon methamphetamine neurotoxicity of the nigrostriatal dopaminergic system, Neuroscience 103 (2001) 385–394.

[59] M. Gassen, M.B.H. Youdim, Free radical scavengers: Chemical concepts and clinical relevance, J. Neural Transm. Suppl. 56 (1999) 193–210.

[60] T. Gasser, Molecular genetics of Parkinson's disease, in: D. Calne, S.M. Calne (Eds.), Parkinson's Disease: Advances in Neurology, vol. 86, Lippincott Williams and Wilkins, Philadelphia, 2001, pp. 23–32.

[61] D.C. German, K. Manaye, W.K. Smith, D.J. Woodward, C.D. Saper, Midbrain dopaminergic cell loss in Parkinson's disease: Computer visualization, Ann. Neurol. 26 (1989) 507–514.

[62] B.I. Giasson, C.A. Wilson, J.Q. Trojanowski, V.M. Lee, Tau and α-synuclein in neurodegenerative diseases, in: M.-F. Chesselet (Ed.), Molecular Mechanisms of Neurodegenerative Diseases, Humana Press, Totowa, NJ, 2001, pp. 151–176.

[63] A. Giovanni, B.A. Sieber, R.E. Heikkila, P.K. Sonsalla, Correlation between the neostriatal content of the 1-methyl-4-phenylpyridinium species and dopaminergic neurotoxicity following 1-methyl-4-phenyl-1,2,3,6,-tetrahydropyridine administration to several strains of mice, J. Pharmacol. Exp. Ther. 257 (1991) 691–697.

[64] P.F. Good, C.W. Olanow, D.P. Perl, Neuromelanin-containing neurons of the substantia nigra accumulate iron and aluminum in Parkinson's disease: A LAMMA study, Brain Res. 593 (1992) 343–346.

[65] R. Grondin, P.J. Bedard, A.H. Tahar, L. Gregoire, A. Mori, H. Kase, Antiparkinsonian effect of a new selective adenosine A-2A receptor antagonist in MPTP-treated monkeys, Neurology 52 (1999) 1673–1677.

[66] A. Gross, J.M. McDonnell, S.J. Korsmeyer, Bcl-2 family members and the mitochondria in apoptosis, Genes Dev. 13 (1999) 1899–1911.

[67] E. Grunblatt, S. Mande, G. Maor, M. Youdim, Effects of R- and S-apomorphine on MPTP-induced nigrostriatal dopamine neuronal loss, J. Neurochem. 77 (2001) 146–156.

[68] A. Hartley, J.M. Stone, C. Heron, J.M. Cooper, A.H. Schapira, Complex I inhibitors induce dose-dependent apoptosis in PC12 cells: Relevance to Parkinson's disease, J. Neurochem. 63 (1994) 1987–1990.

[69] A. Hartmann, E.C. Hirsch, Parkinson's disease: The apoptosis hypothesis revisited, in: D. Calne, S.M. Calne (Eds.), Parkinson's Disease: Advances in Neurology, vol. 86, Lippincott Williams and Wilkins, Philadelphia, 2001, pp. 143–153.

[70] A. Hartmann, J.D. Troadec, S. Hunot, K. Kikly, B.A. Faucheux, A. Mouatt-Prigent, M. Ruberg, Y. Agid, E.C. Hirsch, Caspase-8 is an effector in apoptotic death of dopaminergic neurons in Parkinson's disease, but pathway inhibition results in neuronal necrosis, J. Neurosci. 21 (2001) 2247–2255.

[71] I. Hasouna, H. Wickert, M. Zimmermann, F. Gillardon, Increase in bax expression in substantia nigra following MPTP treatment of mice, Neurosci. Lett. 204 (1996) 85–88.

[72] R. Hassler, Zur pathologischen anatomie des senilen und des parkinsonistischen tremor, J. Psychol. Neurol. 49 (1939) 13–55.

[73] R.E. Heikkila, Differential neurotoxicity of 1-methyl-4-phenyl-tetrahydropyridine (MPTP) in Swiss–Webster mice from different sources, Eur. J. Pharmacol. 117 (1985) 1451–1453.

[74] R.E. Heikkila, 1-Methyl-4-phenyl-1,2,3,6-tetrahydropyridine (MPTP), a dopaminergic neurotoxin, RBI Neurotransm. 4 (1) (1988) 1–4.

[75] R.E. Heikkila, A. Hess, R.C. Duvoisin, Dopaminergic neurotoxicity of 1-methyl-4-phenyl-1,2,5,6-tetrahydropyridine in mice, Science 224 (1984) 1451–1453.

[76] R.E. Heikkila, L. Manzino, F.S. Cabbat, R.C. Duvoisin, Inhibition of monoamine oxidase produces protection against the dopaminergic neurotoxicity of 1-methyl-4-phenyl-1,2,5,6-tetrahydropyridine, Nature 311 (1984) 467–469.

[77] R.E. Heikkila, W.J. Nicklas, I. Vyas, R.C. Duvoisin, Dopaminergic toxicity of rotenone and the 1-methyl-4-phenylpyridinium ion after their stereotaxic administration to rats: Implication for the mechanism of 1-methyl-4-phenyl-1,2,3,6-tetrahydropyridine toxicity, Neurosci. Lett. 62 (1985) 389–394.

[78] R.E. Heikkila, W.J. Nicklas, R.C. Duvoisin, Studies on the mechanism of MPTP-MPP$^+$-induced neurotoxicity in rodents, in: S.P. Markey, et al. (Eds.), MPTP: A Neurotoxin Producing a Parkinsonian Syndrome, Academic Press, Orlando, 1986, pp. 69–83.

[79] W. Hellenbrand, A. Seidler, B.P. Robra, P. Vieregge, W.H. Oertel, J. Joerg, P. Nischan, E. Schneider, G. Ulm, Smoking and Parkinson's disease: A case-controlled study in Germany, Int. J. Epidemiol. 26 (1997) 328–339.

[80] R.C. Henneberry, A. Novelli, J.A. Cox, P.G. Lysko, Neurotoxicity at the N-methyl-D-aspartate receptor in energy-compromised neurons. A hypothesis for cell death in aging and disease, Ann. NY Acad. Sci. 568 (1989) 225–233.

[81] E.C. Hirsch, J.P. Brandel, P. Galle, F. Javoy-Agid, Y. Agid, Iron and aluminum increase in the substantia nigra of patients with Parkinson's disease: An X-ray microanalysis, J. Neurochem. 56 (1991) 446–451.

[82] E.C. Hirsch, B. Faucheux, P. Damier, A. Mouatt-Prigent, Y. Agid, Neuronal vulnerability in Parkinson's disease, J. Neural. Transm. Suppl. 50 (1997) 79–88.

[83] A. Hochman, H. Sternin, S. Gorodin, S. Korsmeyer, I. Ziv, E. Offen, D. Offen, Enhanced oxidative stress and altered antioxidants in brains of Bcl-2-deficient mice, J. Neurochem. 71 (1998) 741–748.

[84] O. Hornykiewicz, Brain dopamine in Parkinson's disease and other neurological disturbances, in: A.S. Horn, et al. (Eds.), The Neurobiology of Dopamine, Academic Press, New York, 1979, pp. 633–654.

[85] S. Hunot, N. Dugas, B. Faucheux, A. Hartmann, M. Tardieu, P. Agid, Y. Agid, B. Dugas, E.C. Hirsch, Fc epsilon RII/CD23 is expressed in Parkinson's disease and induces, in vitro, production of nitric oxide and tumor necrosis factor-alpha in glial cells, J. Neurosci. 19 (1999) 3440–3447.

[86] M.F. Jarvis, G.C. Wagner, Age-dependent effects of 1-methyl-4-phenyl-tetrahydropyridine (MPTP), Neuropharmacology 24 (1985) 581–583.

[87] J.A. Javitch, R.J. D'Amato, S.M. Strittmatter, S.H. Snyder, Parkinsonism-inducing neurotoxin, N-methyl-4-phenyl-1,2,3,6-tetrahydropyridine: Uptake of the metabolite N-methyl-4-phenylpyridine by dopamine neurons explains selective toxicity, Proc. Natl. Acad. Sci. USA 82 (1985) 2173–2177.

[88] K. Jellinger, E. Kienzl, G. Rumpelmair, P. Riederer, H. Stachelberger, D. Ben-Shachar, M.B. Youdim, Iron–melanin complex in substantia nigra of parkinsonian brains: An X-ray microanalysis, J. Neurochem. 59 (1992) 1168–1171.

[89] P. Jenner, N.M. Rupniak, S. Rose, E. Kelly, G. Kilpatrick, A. Lees, C.D. Marsden, 1-Methyl-4-phenyl-1,2,3,6-tetrahydropyridine-induced parkinsonism in the common marmoset, Neurosci. Lett. 50 (1984) 85–90.

[90] P.H. Jensen, M.S. Nielsen, R. Jakes, C.G. Dotti, M. Goedert, Binding of α-synuclein to brain vesicles is abolished by familial Parkinson's disease mutation, J. Biol. Chem. 273 (1998) 26292–26294.

[91] B.K. Jin, D.Y. Shin, M.Y. Jeong, M.R. Gwag, H.W. Baik, K.S. Yoon, Y.H. Cho, W.S. Joo, Y.S. Kim, H.H. Baik, Melatonin protects nigral dopaminergic neurons from 1-methyl-4-phenylpyridinium ion (MPP$^+$) neurotoxicity in rats, Neurosci. Lett. 245 (1998) 61–64.

[92] S.A. Jones-Humble, P.F. Morgan, B.R. Cooper, The novel anticonvulsant lamotrigine prevents dopamine depletion in C57 black mice in the MPTP animal model of Parkinson's disease, Life Sci. 54 (1994) 245–252.

[93] Y. Kitamura, Y. Itano, T. Kubo, Y. Nomura, Suppressive effect of FK-506, a novel immunosuppressant, against MPTP-induced dopamine depletion in the striatum of young C57BL/6 mice, J. Neuroimmunol. 50 (1994) 221–224.

[94] P. Klivenyi, O.A. Andreassen, R.J. Ferrante, A. Dedeoglu, G. Lancelot, E. Lancelot, M. Bogdanov, J.K. Andersen, D. Jiang, M.F. Beal, Mice deficient in cellular glutathione peroxidase show increased vulnerability to malonate, 3-nitropropionic acid and 1-methyl-4-phenyl-1,2,5,6-tetrahydropyridine, J. Neurosci. 20 (2000) 1–7.

[95] M. Kohutnicka, E. Lewandowska, I. Kurkowska-Jastrzebska, A. Czlonkowski, A. Czlonkowska, Microglial and astrocytic involvement in a murine model of Parkinson's disease induced by 1-methyl-4-phenyl-1,2,3,6-tetrahydropyridine (MPTP), Immunopharmacology 39 (1998) 167–180.

[96] N.W. Kowall, P. Hantraye, E. Brouillet, M.F. Beal, A.C. McKee, R.J. Ferrante, MPTP induces alpha-synuclein aggregation in the substantia nigra of baboons, NeuroReport 11 (2000) 211–213.

[97] R. Kruger, A.M. Vieira-Saecker, W. Kuhn, D. Berg, T. Muller, N. Kuhnl, G.A. Fuchs, A. Storch, M. Hungs, D. Woitalla, H. Przuntek, J.T. Epplen, L. Schols, O. Reiss, Increased susceptibility to sporadic Parkinson's disease by a certain combined alpha-synuclein/apolipoprotein-E genotype, Ann. Neurol. 45 (1999) 611–617.

[98] I. Kruman, Q. Guo, M.P. Mattson, Calcium and reactive oxygen species mediate staurosporine-induced mitochondrial dysfunction and apoptosis in PC12 cells, J. Neurosci. Res. 51 (1998) 293–308.

[99] F.M. LaFerla, J.C. Troncoso, D.K. Strickland, C.H. Kawas, G. Jay, Neuronal cell death in Alzheimer's disease correlates with apoE uptake and intracellular Aβ stabilization, J. Clin. Invest. 100 (1997) 310–320.

[100] J.W. Langston, MPTP: Current concepts and controversies, Clin. Neuropharmacol. 9 (1986) 485–507.

[101] J.W. Langston, P.A. Ballard, Parkinson's disease in a chemist working with MPTP, N. Engl. J. Med. 309 (1983) 310.

[102] J.W. Langston, P. Ballard, J.W. Tetrud, I. Irwin, Chronic parkinsonism in humans due to a product of meperidine-analog synthesis, Science 219 (1983) 979–980.

[103] J.W. Langston, L.S. Forno, J. Tetrud, A.G. Reeves, J.A. Kaplan, D. Karluk, Evidence of active nerve cell degeneration in the substantia nigra of humans years after 1-methyl-4-phenyl-1,2,3,6-tetrahydropyridine exposure, Ann. Neurol. 46 (1999) 598–605.

[104] N.N. Lermontiva, L.S. Soliakov, S.O. Bachurin, S.E. Tkachenko, T.P. Serkova, Evaluation of the capability of 1-methyl-4-phenyl-1,2,3,6-tetrahydropyridine and pyridine derivatives to evoke parkinsonism, Biul. Eksp. Biol. Med. 107 (1989) 239–244 (In Russian).

[105] E. Leroy, R. Boyer, G. Auburger, B. Leube, G. Ulm, E. Mezey, G. Harta, M.J. Brownstein, S. Jonnalagada, T. Chernova, A. Dehejia, C. Lavedan, T. Gasser, P.J. Steinbach, K.D. Wilkinson, M.H. Polymeropoulos, The ubiquitin pathway in Parkinson's disease, Nature 395 (1998) 451–452.

[106] G.T. Liberatore, V. Jackson-Lewis, S. Vukosavic, A.S. Mandir, M. Vila, W.G. McAuliffe, V.L. Dawson, T.M. Dawson, S. Przedborski, Inducible nitric oxide synthase stimulates dopaminergic neurodegeneration in the MPTP model of Parkinson disease, Nat. Med. 5 (1999) 1403–1409.

[107] A.H. Lockwood, Pesticides and Parkinsonism: Is there an etiological link? Curr. Opin. Neurol. 13 (2000) 687–690.

[108] R. Maggio, M. Riva, F. Vaglini, F. Fornai, R. Molteni, M. Armogida, G. Racagni, G.U. Corsini, Nicotine prevents experimental parkinsonism in rodents and induces striatal increase of neurotrophic factors, J. Neurochem. 71 (1998) 2439–2446.

[109] S.C. Mahoney, R.J. Ferrante, A. Dedeoglu, A.C. McKee, N.W. Kowall, Alpha synuclein is selectively redistributed in the substantia nigra of MPTP-treated mice, J. Neuropathol. Exp. Neurol. 60 (2001) 549.

[110] S. Mandel, E. Grunblatt, M. Youdim, cDNA microarray to study gene expression of dopaminergic neurodegeneration and neuroprotection in MPTP and 6-hydroxydopamine models: Implication for idiopathic Parkinson's disease, J. Neural Transm. Suppl. 60 (2000) 117–124.

[111] D.M.A. Mann, P.O. Yates, Pathological basis for neurotransmitter changes in Parkinson's disease, Neuropathol. Appl. Neurobiol. 9 (1983) 3–19.

[112] D.M.A. Mann, P.O. Yates, J. Hawkes, The pathology of the human locus coeruleus, Clin. Neuropathol. 2 (1983) 1–7.

[113] A.B. Manning-Bog, A.L. McCormack, J. Li, V.N. Uversky, A.L. Fink, D.A. Di Monte, The herbicide paraquat causes up-regulation and aggregation of alpha-synuclein in mice, J. Biol. Chem. 277 (2002) 1641–1644.

[114] S.P. Markey, N. Castagnoli, A.J. Trevor, I.J. Kopin (Eds.), MPTP: A Neurotoxin Producing a Parkinsonian Syndrome, Academic Press, Orlando, 1986.

[115] S.P. Markey Jr., J.N. Johannessen, C.C. Chiueh, R.S. Burns, M. Herkenham, Intraneuronal generation of a pyridinium metabolite may cause drug-induced parkinsonism, Nature 311 (1984) 464–467.

[116] K. Markopoulou, Z.K. Wszolek, R.F. Pfeiffer, B.A. Chase, Reduced expression of the G209A alpha-synuclein allele in familial Parkinsonism, Ann. Neurol. 46 (1999) 374–381.

[117] R.T. Matthews, P. Klivenyi, G. Mueller, L. Yang, M. Wermer, C.E. Thomas, M.F. Beal, Novel free radical spin traps protect against malonate and MPTP neurotoxicity, Exp. Neurol. 157 (1999) 120–126.

[118] R.T. Matthews, L. Yang, M.F. Beal, S-Methylthiocitrulline, a neuronal nitric oxide synthase inhibitor, protects against malonate and MPTP neurotoxicity, Exp. Neurol. 143 (1997) 282–286.

[119] M.M. Matzuk, C.B. Saper, Preservation of hypothalamic dopaminergic neurons in Parkinson's disease, Ann. Neurol. 18 (1985) 552–555.

[120] P.L. McGeer, K. Yasojima, E.G. McGeer, Inflammation in Parkinson's disease, in: D. Calne, S.M. Calne (Eds.), Parkinson's Disease: Advances in Neurology, vol. 86, Lippincott Williams and Wilkins, Philadelphia, 2001, pp. 83–89.

[121] K.St.P. McNaught, Isoquinoline derivatives as endogenous neurotoxins in the aetiology of Parkinson's disease, Biochem. Pharmacol. 56 (1998) 921–933.

[122] K.St.P. McNaught, P. Jenner, Dysfunction of rat forebrain astrocytes in culture alters cytokine release and neurotrophic factor release, Neurosci. Lett. 285 (2000) 61–65.

[123] M.A. Mena, B. Pardo, M.J. Casarejos, S. Fahn, J.G. de Yebenes, Neurotoxicity of levodopa on catecholamine rich neurons, Mov. Disord. 7 (1992) 23–31.

[124] G.W. Miller, J.D. Erickson, J.T. Perez, S.N. Penland, D.C. Mash, D.B. Rye, A.I. Levey, Immunochemical analysis of vesicular monoamine transporter (VMAT2) protein in Parkinson's disease, Exp. Neurol. 154 (1999) 138–148.

[125] Y. Mizuno, N. Hattori, T. Kitada, H. Matsumine, H. Mori, H. Shimura, S. Kubo, H. Kobayashi, S. Aakawa, S. Minoshima, N. Shimizu, Familial Parkinson's disease: α-Synuclein and parkin, in: D. Calne, S.M. Calne (Eds.), Parkinson's Disease: Advances in Neurology, vol. 86, Lippincott Williams and Wilkins, Philadelphia, 2001, pp. 13–21.

[126] Y. Mizuno, K. Suzuki, N. Sone, T. Saitoh, Inhibition of ATP synthesis by 1-methyl-4-phenylpyridinium ion (MPP+) in isolated mitochondria from mouse brains, Neurosci. Lett. 81 (1987) 204–208.

[127] S. Moussaoui, M.C. Obinu, N. Daniel, M. Reibaud, V. Blanchard, A. Imperato, The antioxidant ebselen prevents neurotoxicity and clinical symptoms in a primate model of Parkinson's disease, Exp. Neurol. 166 (2000) 235–245.

[128] D. Muralikrishnan, M. Ebadi, SKF-38393, a dopamine receptor agonist, attenuates 1-methyl-4-phenyl-1,2,3,6-tetrahydropyridine-induced neurotoxicity, Brain Res. 892 (2001) 241–247.

[129] C. Mytilineou, Mechanism of MPTP neurotoxicity, in: J. Segura-Aguilar (Ed.), Mechanisms of Degeneration and Protection of the Dopaminergic System, Graham Publishing, Johnson City, TN, 2001, pp. 131–148.

[130] T. Nakagawa, H. Zhu, N. Morishima, E. Li, J. Xu, B.A. Yankner, J. Yuan, Caspase-12 mediates endoplasmic-reticulum-specific apoptosis and cytotoxicity by amyloid-beta, Nature 403 (2000) 98–103.

[131] W.J. Nicklas, I. Vyas, R.E. Heikkila, Inhibition of NADH-linked

oxidation in brain mitochondria by 1-methyl-4-phenylpyridine, a metabolite of the neurotoxin, 1-methyl-4-phenyl-1,2,3,6-tetrahydropyridine, Life Sci. 36 (1985) 2503–2508.

[132] A. Nicotra, S.H. Parvez, Cell death induced by MPTP, a substrate for monoamine oxidase B, Toxicology 153 (2000) 157–166.

[133] D. Offen, P.M. Beart, N.S. Cheung, C.J. Pascoe, A. Hochman, S. Gorodin, E. Melamed, R. Bernard, O. Bernard, Transgenic mice expressing human Bcl-2 in their neurons are resistant to 6-hydroxydopamine and 1-methyl-4-phenyl-1,2,3,6-tetrahydropyridine neurotoxicity, Proc. Natl. Acad. Sci. USA 95 (1998) 5789–5794.

[134] W.D. Parker, S.J. Boyson, J.K. Parks, Abnormalities of the electron transport chain in idiopathic Parkinson's disease, Ann. Neurol. 26 (1989) 719–723.

[135] J. Parkinson, An Essay on the Shaking Palsy, Sherwood, Neely and Jones, London, 1817.

[136] T.L. Perry, V.W. Yong, R.M. Clavier, K. Jones, J.M. Wright, J.G. Foulks, R.A. Wall, Partial protection from the dopaminergic neurotoxin N-methyl-4-phenyl-1,2,3,6-tetrahydropyridine by four different antioxidants in the mouse, Neurosci. Lett. 60 (1985) 109–114.

[137] D. Peter, Y. Liu, C. Sternini, R. deGiorgio, N. Brecha, R.H. Edwards, Differential expression of two vesicular monoamine transporters, J. Neurosci. 15 (1995) 6179–6188.

[138] M.H. Polymeropoulos, C. Lavedan, E. Leroy, S.E. Ide, A. Dehejia, A. Dutra, B. Pike, H. Root, J. Rubenstein, R. Boyer, E.S. Stenroos, S. Chandrasekharappa, A. Athanassiadou, T. Papapetropoulos, W.G. Johnson, A.M. Lazzarini, R.C. Duvoisin, G. Di Iorio, L.I. Golbe, R.L. Nussbaum, Mutations in the α-synuclein gene identified in families with Parkinson's disease, Science 276 (1997) 2045–2047.

[139] A. Priyadarshi, S.A. Khuder, E.A. Schaub, S.S. Priyadarshi, Environmental risk factors and Parkinson's disease: A meta-analysis, Environ. Res. 86 (2001) 122–127.

[140] S. Przedborski, Q. Chen, M. Vila, B.I. Giasson, R. Djaldatti, S. Vukosavic, J.M. Souza, V. Jackson-Lewis, V.M. Lee, H. Ischiropoulos, Oxidative post-translational modifications of alpha-synuclein in the 1-methyl-4-phenyl-1,2,3,6-tetrahydropyridine (MPTP) mouse model of Parkinson's disease, J. Neurochem. 76 (2001) 637–640.

[141] S. Przedborski, V. Jackson-Lewis, R. Yokoyama, T. Shibata, V.L. Dawson, T.M. Dawson, Role of neuronal nitric oxide in 1-methyl-4-phenyl-1,2,3,6-tetrahydropyridine (MPTP)-induced dopaminergic neurotoxicity, Proc. Natl. Acad. Sc. USA 93 (1996) 4565–4571.

[142] S. Przedborski, V. Kostic, V. Jackson-Lewis, A.B. Naini, S. Simonetti, S. Fahn, E. Carlson, C.J. Epstein, J.L. Cadet, Transgenic mice with increased Cu/Zn-superoxide dismutase activity are resistant to N-methyl-4-phenyl-1,2,3,6,-tetrahydropyridine-induced neurotoxicity, J. Neurosci. 12 (1992) 1658–1667.

[143] R.R. Ramsay, J. Dadgar, A. Trevor, T.P. Singer, Energy-driven uptake of N-methyl-4-phenylpyridine by brain mitochondria mediates the neurotoxicity of MPTP, Life Sci. 39 (1986) 581–588.

[144] R.R. Ramsay, A.T. Kowal, M.K. Johnson, J.I. Salach, T.P. Singer, The inhibition site of MPP+, the neurotoxic bioactivation product of 1-methyl-4-phenyl-1,2,3,6,-tetrahydropyridine is near the Q-binding site of NADH dehydrogenase, Arch. Biochem. Biophys. 259 (1987) 645–649.

[145] R.R. Ramsay, J.I. Salach, J. Dadgar, T.P. Singer, Inhibition of mitochondrial NADH dehydrogenase by pyridine derivatives and its possible relation to experimental and idiopathic parkinsonism, Biochem. Biophys. Res. Commun. 135 (1986) 269–275.

[146] R.R. Ramsay, J.I. Salach, T.P. Singer, Uptake of the neurotoxin 1-methyl-4-phenylpyridinium (MPP+) by mitochondria and its relation to the inhibition of the mitochondrial oxidation of NAD+ linked substrates by MPP+, Biochem. Biophys. Res. Commun. 134 (1986) 743–748.

[147] R.R. Ramsay, T.P. Singer, Energy-dependent uptake of N-methyl-4-phenylpyridinium, the neurotoxic metabolite of 1-methyl-4-phenyl-1,2,3,6-tetrahydropyridine, by mitochondria, J. Biol. Chem. 261 (1986) 7585–7587.

[148] D.B. Ramsden, R.B. Parsons, S.-L. Ho, T. Xie, R.H. Waring, A.C.

Williams, Further studies in xenobiotic metabolism and Parkinson's disease, in: D. Calne, S.M. Calne (Eds.), Parkinson's Disease: Advances in Neurology, vol. 86, Lippincott Williams and Wilkins, Philadelphia, 2001, pp. 105–113.

[149] O. Rascol, J.J. Ferreira, C. Thalamus, M. Galitsky, J.-L. Montastruc, Dopamine agonists: Their role in the management of Parkinson's disease, in: D. Calne, S.M. Calne (Eds.), Parkinson's Disease: Advances in Neurology Volume, Lippincott Williams and Wilkins, Philadelphia, 2001, pp. 301–309.

[150] S. Rathke-Hartlieb, P.J. Kahle, M. Neumann, L. Ozmen, S. Haid, M. Okochi, C. Haass, J.B. Schulz, Sensitivity to MPTP is not increased in Parkinson's disease-associated mutant alpha-synuclein transgenic mice, J. Neurochem. 77 (2001) 1181–1184.

[151] J.F. Reinhard Jr., E.J. Diliberto Jr., O.H. Viveros, A.J. Daniels, Subcellular compartmentalization of 1-methyl-4-phenylpyridinium with catecholamines in adrenal medullary chromaffin vesicles may explain the lack of toxicity to adrenal chromaffin cells, Proc. Natl. Acad. Sci. USA 84 (1987) 8160–8164.

[152] P. Riederer, J. Sian, M. Gerlach, Is there neuroprotection in Parkinson syndrome? J. Neurol. 247 (Suppl. 4) (2000) 8–11.

[153] O. Riess, R. Kruger, Parkinson's disease — A multifactorial neurodegenerative disorder, J. Neural Transm. Suppl. 56 (1999) 113–125.

[154] C. Rodriguez, R.M. Sainz, J.C. Mayo, Neurotoxicity by L-DOPA and dopamine: Neuroprotective role of melatonin, in: J. Segura-Aguilar (Ed.), Mechanisms of Degeneration and Protection of the Dopaminergic System, Graham Publishing, Johnson City, TN, 2001, pp. 239–275.

[155] H. Rollema, G. Damsma, A.S. Horn, J.B. DeVries, B.H.C. Westerink, Brain dialysis in conscious rats reveals an instantaneous massive release of striatal dopamine in response to MPP+, Eur. J. Pharmacol. 126 (1986) 345–346.

[156] G.W. Ross, R.D. Abbott, H. Petrovitch, D.M. Morens, A. Tung, K.H. Tung, C.M. Tanner, K.H. Masaki, P.L. Blanchette, J.D. Curb, J.S. Popper, L. White, Association of coffee and caffeine intake with the risk of Parkinson disease, JAMA, J. Am. Med. Assoc. 283 (2000) 2674–2679.

[157] T.A. Seaton, J.M. Cooper, A.H. Schapira, Cyclosporin inhibition of apoptosis induced by mitochondrial complex I toxins, Brain Res. 809 (1998) 12–17.

[158] A.H.V. Schapira, Evidence for mitochondrial dysfunction in Parkinson's disease — A critical appraisal, Mov. Disord. 9 (1994) 125–138.

[159] A.H.V. Schapira, J.M. Cooper, D. Dexter, Mitochondrial complex I deficiency in Parkinson's disease, Ann. Neurol. 26 (1989) 122–123.

[160] J.B. Schulz, R.T. Matthews, M.M.K. Muqit, S.E. Browne, M.F. Beal, Inhibition of neuronal nitric oxide synthase by 7-nitroindazole protects against MPTP-induced neurotoxicity in mice, J. Neurochem. 64 (1995) 936–939.

[161] J.B. Schulz, P. Nicotera, Introduction: Targeted modulation of neuronal apoptosis: A double-edged sword? Brain Pathol. 10 (2000) 273–275.

[162] I. Shoulson, Deprenyl and Tocopherol Antioxidative Therapy of Parkinsonism (DATATOP). Parkinson study group, Acta Neurol. Scand. Suppl. 126 (1989) 171–175.

[163] I. Shoulson, DATATOP: A decade of neuroprotective inquiry. Parkinson study group. Deprenyl and Tocopherol Antioxidative Therapy of Parkinsonism, Ann. Neurol. Suppl. 44 (1998) S160–S166.

[164] M.T. Silva, A.H.V. Schapira, Parkinson's disease, in: M.P. Mattson (Ed.), Pathogenesis of Neurodegenerative Disorders, Humana Press, Totowa, NJ, 2001, pp. 53–79.

[165] M. Smeyne, O. Goloubeva, R.J. Smeyne, Strain-dependent susceptibility to MPTP and MPP+-induced parkinsonism is determined by glia, Glia 34 (2001) 73–80.

[166] S.H. Snyder, R.J. D'Amato, MPTP: A neurotoxin relevant to the pathophysiology of Parkinson's disease, Neurology 36 (1986) 250–258.

[167] S.G. Speciale, C.-L. Liang, P.K. Sonsalla, R.H. Edwards, D.C. German, The neurotoxin 1-methyl-4-phenylpyridinum is sequestered

within neurons that contain the vesicular monoamine transporter, Neuroscience 84 (1998) 1177–1185.

[168] W.P. Spooren, C. Gentsch, C. Wiessner, TUNEL-positive cells in the substantia nigra of C57BL/6 mice after a single dose of 1-methyl-4-phenyl-1,2,3,6,-tetrahydropyridine, Neuroscience 85 (1998) 649–651.

[169] R.G. Staal, K.A. Hogan, C.L. Liang, D.C. German, P.K. Sonsalla, In vitro studies of striatal vesicles containing the vesicular monoamine transporter (VMAT2): Rat versus mouse differences in sequestration of 1-methyl-4-phenylpyridinium, J. Pharmacol. Ther. 293 (2000) 329–335.

[170] R.G. Staal, P.K. Sonsalla, Inhibition of brain vesicular monoamine transporter (VMAT2) enhances 1-methyl-4-phenylpyridinum neurotoxicity in vivo in rat striata, J. Pharmacol. Exp. Ther. 293 (2000) 336–342.

[171] A.J. Stoessl, Neurochemical and neuroreceptor imaging with PET in Parkinson's disease, in: D. Calne, S.M. Calne (Eds.), Parkinson's Disease: Advances in Neurology, vol. 86, Lippincott Williams and Wilkins, Philadelphia, 2001, pp. 215–223.

[172] H.W. Strobel, Cytochrome P450 in neurodegeneration and Parkinson's disease, in: J. Segura-Aguilar (Ed.), Mechanisms of Degeneration and Protection of Dopaminergic System, Graham Publishing, Johnson City, TN, 2001, pp. 149–163.

[173] R.H. Swerdlow, Mitochondria and Parkinson's disease, in: M.-F. Chesselet (Ed.), Molecular Mechanisms of Neurodegenerative Diseases, Humana Press, Totowa, NJ, 2001, pp. 233–270.

[174] C.M. Tanner, R. Ottman, S.M. Goldman, J. Ellenberg, P. Chan, R. Mayeux, J.W. Langston, Parkinson disease in twins: An etiologic study, JAMA, J. Am. Med. Assoc. 281 (1999) 341–346.

[175] N.A. Tatton, S.J. Kish, In situ detection of apoptotic nuclei in the substantia nigra compacta of 1-methyl-4-phenyl-1,2,3,6,-tetrahydro-pyridine-treated mice using terminal deoxynucleotidyl transferase labelling and acridine orange staining, Neuroscience 77 (1997) 1037–1048.

[176] P. Teismann, B. Ferger, Inhibition of the cyclooxygenase isoenzymes COX-1 and COX-2 provide neuroprotection in the MPTP-mouse model of Parkinson's disease, Synapse 39 (2001) 167–174.

[177] K.F. Tipton, T.P. Singer, Advances in our understanding of the mechanisms of the neurotoxicity of MPTP and related compounds, J. Neurochem. 61 (1993) 1191–1206.

[178] A.J. Trevor, K. Chiba, E.Y. Yu, P.S. Caldera, K.P. Castagnoli, N. Castagnoli Jr., L.A. Peterson, J.I. Salach, T.P. Singer, Metabolism of MPTP in vitro: The intermediate role of 2,3-MPDP$^+$ and studies on its chemical and biochemical reactivity, in: S.P. Markey, et al. (Eds.), MPTP: A Neurotoxin Producing a Parkinsonian Syndrome, Academic Press, Orlando, 1986, pp. 161–172.

[179] P.A. Trimmer, T.S. Smith, A.B. Jung, J.P. Bennett Jr., Dopamine neurons from transgenic mice with a knockout of the p53 gene resist MPTP neurotoxicity, Neurodegeneration 5 (1996) 233–239.

[180] L. Turski, K. Bressler, K.J. Rettig, P.A. Loschmann, H. Wachtel, Protection of substantia nigra from MPP$^+$ neurotoxicity by *N*-methyl-D-aspartate antagonists, Nature 349 (1991) 414–418.

[181] M. Vila, V.V. Jackson-Lewis, S. Vukosavic, R. Djaldetti, G. Liberatore, D. Offen, S.J. Korsmeyer, S. Przedborski, Bax ablation prevents dopaminergic neurodegeneration in the 1-methyl-4-phenyl-1,2,3,6-tetrahydropyridine mouse model of Parkinson's disease, Proc. Natl. Acad. Sci. USA 98 (2001) 2837–2842.

[182] M. Vila, S. Vukosavic, V. Jackson-Lewis, M. Neystat, M. Jakowec, S. Przedborski, α-Synuclein up-regulation in substantia nigra dopaminergic neurons following administration of the parkinsonian toxin MPTP, J. Neurochem. 74 (2000) 721–729.

[183] V. Viswanath, L. Larsen, J.K. Andersen, Attenuation of 1-methyl-4-phenyl-1,2,3,6-tetrahydropyridine toxicity in mice expressing the baculoviral caspase inhibitor p35, Soc. Neurosci. Abstr. 26 (2000) 11.

[184] I. Vyas, R.E. Heikkila, W.J. Nicklas, Studies on the neurotoxicity of 1-methyl-4-phenyl-1,2,3,6-tetrahydropyridine: Inhibition of NAD-linked substrate oxidation by its metabolite, 1-methyl-4-phenylpyridinium, J. Neurochem. 46 (1986) 1501–1507.

[185] H.L. Weiner, A. Hashim, A. Lajtha, H. Shershen, (−)-2-Oxo-4-thiazolidine carboxylic acid attenuates 1-methyl-4-phenyl-1,2,3,6-tetrahydropyridine induced neurotoxicity, Res. Commun. Subst. Abuse 9 (1988) 53–60.

[186] G.P. Wilkin, C. Knott, A curtain riser, in: G. Stern (Ed.), Advances in Neurology, vol. 80, Lippincott Williams and Wilkins, Philadelphia, 1999, pp. 3–7.

[187] U. Wullner, P.A. Loschmann, J.B. Schulz, A. Schmid, R. Dringen, F. Eblen, L. Turski, T. Klockgether, Glutathione depletion potentiates MPTP and MPP$^+$ toxicity in nigral dopaminergic neurones, NeuroReport 7 (1996) 921–923.

[188] T. Yamada, P.L. McGeer, E.G. McGeer, Relationship of complement-activated oligodendrocytes to reactive microglia and neuronal pathology in neurodegenerative disease, Dementia 2 (1991) 71–77.

[189] Y. Yamamura, I. Sobue, K. Ando, M. Iida, T. Yanagi, Paralysis agitans of early onset with marked diurnal fluctuation of symptoms, Neurology 23 (1973) 239–244.

[190] F. Yantiri, J.K. Andersen, The role of iron in Parkinson disease and 1-methyl-4-phenyl-1,2,3,6-tetrahydropyridine toxicity, IUBMB Life 48 (1999) 139–141.

[191] V.W. Yong, T.L. Perry, A.A. Krisman, Depletion of glutathione in brainstems of mice by *N*-methyl-4-phenyl-1,2,3,6-tetrahydropyridine is prevented by antioxidant pretreatment, Neurosci. Lett. 63 (1986) 56–60.

[192] S.K. Youngster, R.C. Duvoisin, A. Hess, P.K. Sonsalla, M.V. Kindt, R.E. Heikkila, 1-Methyl-4-(2′-methylphenyl)-1,2,3,6-tetrahydropyridine (2′CH$_3$-MPTP) is a more potent dopaminergic neurotoxin than MPTP in mice, Eur. J. Pharmacol. 122 (1986) 283–287.

[193] A. Zuddas, F. Fascetti, G.U. Corsini, M.P. Piccardi, In brown Norway rats, MPP$^+$ is accumulated in the nigrostriatal dopaminergic terminals but it is not neurotoxic: A model of natural resistance to MPTP toxicity, Exp. Neurol. 127 (1994) 54–61.

Neurotoxicology and Teratology 24 (2002) 621–628

NEUROTOXICOLOGY

AND

TERATOLOGY

www.elsevier.com/locate/neutera

Review article

Influence of neuromelanin on oxidative pathways within the human substantia nigra

K.L. Double[a,*], D. Ben-Shachar[b], M.B.H. Youdim[c], L. Zecca[d], P. Riederer[e], M. Gerlach[f]

[a]*Prince of Wales Medical Research Institute, Barker Street, Randwick, Sydney, NSW 2031, Australia*
[b]*Laboratory of Psychobiology, Department of Psychiatry, Rambam Medical Centre, B. Rappaport Faculty of Medicine, Technion, Haifa, Israel*
[c]*Department of Pharmacology, B. Rappaport Faculty of Medicine, Eve Topf Neurodegenerative and National Parkinson Foundation Centres, Technion, Haifa, Israel*
[d]*Institute of Advanced Biomedical Technologies-CNR, Segrate, Italy*
[e]*Clinical Neurochemistry, Department of Psychiatry and Psychotherapy, University of Würzburg, 97080 Würzburg, Germany*
[f]*Clinical Neurochemistry, Department of Child and Youth Psychiatry and Psychotherapy, University of Würzburg, Würzburg, Germany*

Received 30 July 2001; accepted 24 January 2002

Abstract

Neuromelanin (NM) is a dark-coloured pigment produced in the dopaminergic neurons of the human substantia nigra (SN). The function of NM within the pigmented neurons is unknown but other melanins are believed to play a protective role via attenuation of free radical damage. Experimental evidence suggests that NM may also exhibit this characteristic, possibly by directly inactivating free radical species or via its ability to chelate transition metals, such as iron. Increased tissue iron, however, may saturate iron-chelating sites on NM and a looser association between iron and NM may result in an increased, rather than decreased, production of free radical species. The death of NM-pigmented neurons in Parkinson's disease (PD) is associated with both a measurable increase in tissue iron concentrations and indices of free radical mediated damage, suggesting that NM is involved in the aetiology of this disorder. As yet, it is unknown whether NM in the parkinsonian brain differs to that found in healthy tissue and thus may fulfil a different role within this tissue. © 2002 Elsevier Science Inc. All rights reserved.

Keywords: Neuromelanin; Iron; Substantia nigra; Human; Parkinson's disease; Oxidative stress

1. Introduction

The substantia nigra (SN) is a small nucleus located in the ventral midbrain. Its primary physiological function appears to be the control of motor function and it forms a part of what is commonly known as the "motor circuit" [22]. The efferent neurons contained in this nucleus are dopaminergic in nature and, in humans, are characterised by the presence of a dark-coloured pigment called neuromelanin (NM). The dark appearance of the SN viewed macroscopically results from the presence of this pigment and for which it was named (lat. SN, black body, Fig. 1). Despite

this nomenclature, the dopaminergic neurons of the SN of almost all other species either fail to produce this pigment entirely, or produce significantly less NM than humans [47]. The presence of large quantities of NM thus identifies this region as unique in humans.

2. Origin of NM

NM is of uncertain origin and, to date, no active synthesis pathway has been identified. The association of NM with the primary catecholaminergic pathways in the brain implicated these substances in its biosynthesis but the exact pathway resulting in NM formation is unclear. The two major forms of NM are generally regarded as the result of the spontaneous autoxidation of dopamine and noradrenaline in the SN and locus coeruleus, respectively. In the SN, NM appears to exist as a large and complex co-

---

Abbreviations: MPTP, 1-methyl-4-phenyl-1,2,3,6-tetrahydropyridine; NM, neuromelanin; PD, Parkinson's disease; SN, substantia nigra.
  * Corresponding author. Tel.: +61-2-9382-2935; fax: +61-2-9382-2681.
  E-mail address: k.double@unsw.edu.au (K.L. Double).

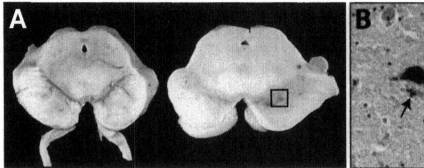
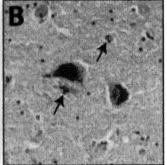

Fig. 1. (A) Macroscopic appearance of the normal human midbrain (left) demonstrating the dark pigmentation of the SN in this tissue compared with the relative pallor of the parkinsonian midbrain (right). (B) Photomicrograph of the pigmented cells of the parkinsonian midbrain demonstrating surviving pigment-filled neurons and extraneuronal pigment (arrows) released from dying neurons.

polymer incorporating both dopamine and a variety of dopaminergic metabolic products. We have recently reviewed what is known regarding the structure and possible synthesis of NM [19,20].

3. Physiological function of NM

The assumption that NM is formed spontaneously gave rise to the belief that NM plays no functional role within the cells in which it appears. Indeed, historically, it has been assumed that NM represents an inert waste product of dopamine metabolism. More recently, however, a more active role for NM within the cell has been suggested. One hypothesis suggests that the formation of NM from the metabolic products of dopamine protects the cell against toxic quinone and semiquinone species produced during the metabolism of dopamine and also from the spontaneous oxidation of dopamine by the polymerisation of these productions into the NM macromolecule [58,59]. In melanised tissues outside the central nervous system, for example, in the skin and eye, melanin protects against the effects of free radicals via its ability to scavenge and thus inactivate these potentially damaging molecules [55]. A case in point are relative rates of membrane damage measurable in the pigmented versus unpigmented liver of the frog. Lipid peroxidation induced by ferrous irons is lower in pigmented, compared with unpigmented, liver tissue of the frog [57]. Pigmented frog liver contains lower reduced glutathione levels and higher levels of polyunsaturated fatty acids than the unpigmented liver and would thus be considered to be more susceptible to free radical activity than the unpigmented tissue; the fact that this is not the case argues for a protective role for melanin in this tissue. By analogy, the regional production of an apparently similar pigment in the brain suggests that this pigment may also reflect an active protective role in the cell. In recent years, NM has gained attention because of its apparent, yet ambiguous, role in Parkinson's disease (PD) and other related neurodegenerative disorders of the basal ganglia.

4. Role of NM in PD

The loss of the dopamine-producing cells of the SN is one of most severely affected areas of degeneration in PD and corresponds with the ensuing loss of NM as the pigment is released from dying cells. The subsequent pallor of the SN is one of the most striking pathological features of PD and is usually the first indication of pathological confirmation (Fig. 1).

Indirect evidence suggests that the pattern of pigment loss is not a random event in PD and several conflicting reports have argued that NM plays an important role in the pathogenesis of PD. In 1988, Hirsch et al. [35] reported that the NM-containing cells of the SN appear particularly vulnerable in PD. This work reported a direct correlation between cell loss in the nigral dopaminergic cell groups in PD and the percentage of NM-positive cells usually found in them [35], a relationship the authors interpreted as suggesting that the development of NM may have a *negative influence* on the cells which contain it. Certainly, the oxidation of catecholamines thought to result in NM production involves the production of a variety of active intermediate products, including quinones, semiquinones and superoxide radicals, thus the presence of NM in a cell may be indicative of a cell subjected to an unusually high oxidative load. These cells are also suggested to suffer from a relative lack of glutathione peroxidase [14] and may consequently be poorly protected against free radical damage. Clearly, neuronal loss in PD is not restricted to the pigmented dopaminergic neurons of the SN. Nonmelaninised nuclei, such as the nucleus basalis of Meynert, subregions of the thalamus and cortical regions, are also affected [10,34], but this does not exclude the possible importance of this polymer in cells in which it is produced.

Contrasting with the school of thought proposing that NM has a negative influence upon the cell, Gibb [26] reported that the more vulnerable ventral tier cells of the SN contain less NM than the more heavily pigmented and relatively preserved cells of the dorsal tier. This ventral cell group is the first to degenerate in PD and the region in which cell loss is greatest [30]. This finding was interpreted as indicating that NM may *confer an advantage* on the cells in which it is found. This view is supported by the observation by Kastner et al. [38] that very heavily melanised neurons are relatively resistant to the pathological process in PD.

The conclusion drawn by Gibb [26] assumes, of course, that NM in the parkinsonian brain is identical to that found in the healthy brain. One recent report using X-band electron paramagnetic spectroscopy suggested that NM isolated from the parkinsonian brain differs in structure to that isolated from the healthy brain [45]. It is also of interest that Kastner et al. [38] report that in PD, pigmented cells contain less NM that that found in healthy brains, a suggestion also made by Aime et al. [2]. This point is of interest because should NM indeed represent a yet unrecognised protective system peculiar to this population of dopaminergic neurons, a relative lack of this polymer may result in increased vulnerability to any subsequent cellular stress. As yet, however, too little information is available to assess whether NM in the parkinsonian brain differs from that in the healthy brain.

5. Changes in iron regulation in PD

Another factor thought to contribute to the vulnerability of nigral dopaminergic neurons in PD are changes in iron homeostasis within this tissue. We have recently reviewed reports describing a range of changes in iron regulation systems in the brain and their relevance to PD [21]. A consistent finding in the parkinsonian brain, possibly as a consequence of apparent changes in iron regulation, has been an increase in tissue iron content in the SN. The increase in tissue iron is significant in magnitude (increases up to 255% of control have been reported [62]) and is observable both in the postmortem parkinsonian brain [18,29,54,62] and, using imaging techniques in the living patient [7,28,56]. PD is not the only neurodegenerative disease to be characterised by this phenomenon. Increased regional brain iron has also been variously reported in trinucleotide repeat disorders, such as Huntington's disease [6,17] in dementia syndromes, including Alzheimer's disease [12,15,46,60] and in the normal aged brain [12,48]. It is interesting, however, that in each of these disorders, the topographical distribution of the increased iron parallels the pattern of degeneration. In PD, iron is selectively increased only in the SN pars compacta, the nigral region containing pigmented dopaminergic neurons [62]. The selectivity of the increased iron in PD (and other neurodegenerative diseases) remains unexplained but the possible consequences of this phenomenon for a cellular system carrying a high oxidative load may be significant.

Iron is a potential danger to biological systems because the catalysation of the so-called Haber–Weiss reaction by iron is considered to be the major mechanism by which the hydroxyl radical (HO) is generated in biological systems. This reaction is a mechanism by which two less reactive free radical species, the superoxide ion and hydrogen peroxide (H_2O_2), react to form the more active HO radical in the presence of redox-active iron. The importance of redox-active iron in this reaction is of interest as increases in tissue iron are usually accompanied by an appropriate increase in the major iron-binding protein ferritin. Increased ferritin expression in response to an increase in tissue iron is reported not to occur in PD [13,17]. This suggests that free iron concentrations, and thus the potential for iron-stimulated oxidative damage, are increased in this tissue. The contribution of various free radical species to cell damage is unknown but the highly active nature of HO enables it to interact with a range of cellular constituents. Given that the half-life of HO is in the vicinity of a nanosecond [39], it is likely that significant HO-mediated cellular damage can only be stimulated if high local concentrations of this radical are produced. Alternatively, a local increase in iron-stimulated radical production may simply exhaust cellular defense mechanisms, and thus increase the deleterious effects of a subsequent noxious stimulus.

6. Interaction between NM and iron

One of the marked features of the parkinsonian nigra is the abundance of activated microglial cells compared with the healthy brain which stain positive for the primary iron-binding protein ferritin [37,49]. It might be suggested that the measured increase in iron in the parkinsonian SN merely reflects the increase in ferritin-containing cells in this region, a predictable response of cellular defense systems to an area of tissue damage. This cell population undoubtedly contributes to total tissue iron measures but iron has also been directly measured to be bound to pigment granules within the SN [3,27,36]. NM has the ability to bind to a variety of metals. The human SN contains levels of iron, copper, zinc, lead, manganese and titanium at concentrations up to four times higher than those found in unpigmented brain tissue and substantial amounts of paramagnetic metal iron, especially iron, are associated with NM [67]. The relative amount of iron bound to NM also appears to be increased in the parkinsonian brain [36], suggesting that the increased iron concentration reported in the parkinsonian nigra, at least in part, is bound to NM. The capacity of the NM macromolecule to bind iron appears to be considerable, a recent estimate suggesting that 10–20% of the total iron contained within the SN of healthy subjects is bound within NM [66]. Iron appears to be bound to NM in the ferric form as polynuclear oxy–hydroxy aggregates in a manner similar to that found in the endogenous iron-binding proteins ferritin and haemosiderin [25], although some differences in binding structures have been suggested [1]. These data suggest that NM may represent an endogenous iron-binding substance anchoring this metal within dopaminergic neurons.

7. Physiological consequences of the NM/iron complex

The binding of iron within the core of the major endogenous iron-binding protein ferritin has the advantage of creating a readily available store of iron which is, however, chemically inactive. This point is important because, as described above, redox-active iron represents a danger when exposed to a tissue. The binding of iron to NM has also been suggested to represent a function analogous to ferritin; that is, iron (and other redox-active metals such as copper) are effectively inactivated by their interaction with the polymer [8,64]. Interneuronal NM is suggested to be only 50% loaded with iron, the remaining chelating potential representing a mechanism by which potentially damaging free metal ions can be removed from the cellular milieu [61]. Such a role would support the hypothesis that NM plays a protective role within dopaminergic neurons. Experimental data support this hypothesis. Using direct electron spin resonance and a model synthetic dopamine melanin, Zareba et al. [65] demonstrated that at low concentrations of iron, most iron ions are bound to the melanin polymer and no free radical signal can be detected. Further, the NM-bound iron was resistant to reduction by mild physiological reductants, such as ascorbate [65].

Where iron is plentiful, however, a different picture emerges. Zareba et al. [65] reported that at intermediate iron concentrations, melanin was still able to sequester a proportion of the iron ions added to the system, but free radical activity became apparent. At high concentrations of iron, the metal-ion binding capacity of the melanin molecule was exceeded by free radical activity, which increased with increasing iron content [65] (Fig. 2).

Thus, in the presence of high iron concentrations, instead of attenuating free radical production, melanin acts as an effective pro-oxidant, resulting in a measurably increased production of hydroxyl radicals measured using electron spin resonance [53]. Krol and Liebler [41] have described a similar phenomenon for melanin in the skin. Both eumelanin and pheomelanin normally exert antioxidant effects against ultraviolet-induced lipid peroxidation. Following complexation with iron, however, the antioxidant effect of eumelanin is abolished, whereas iron-complexed pheomelanin acted as a pro-oxidant stimulating the production of radical oxidants [41]. These authors concluded that the balance of antioxidant–pro-oxidant actions of skin-derived melanins is highly dependent upon the degree of iron binding to these pigments. It was suggested that the binding of iron to these melanins could reduce antioxidant capacity by the oxidation of reducing moieties, such as hydroquinones and semiquinones, or by the formation of a melanin–iron complex that directly promotes ultraviolet-induced lipid peroxidation. The significance of NM within the dopaminergic cells of the SN may therefore reflect not so much the mere presence of the pigment itself but the status of the environment in which the pigment exists.

These data are supported by our own work in which the effect of NM varies, depending upon the chemical environment. Iron-induced lipid peroxidation in the rat brain homogenate measured in vitro was significantly attenuated by the addition of isolated NM pigment [19]. The mechanism of this apparent protective effect is unclear but might result from either the removal of small amounts of endogenous free iron from the homogenate by binding to NM or the direct inactivation of free radicals via the high reactivity of these molecules with aromatic-based molecules, such as the NM polymer [53,65]. When the NM was saturated with iron prior to adding it to the tissue, however, lipid peroxidation was found to be greater than twice that induced by the isolated pigment alone. Moreover, the addition of the iron chelator desferoxamine reduced peroxidation to levels seen in tissue treated with iron-free NM [19] demonstrating the importance of the iron in the tissue damage. Similar results had been noted previously by Youdim et al. [9,64]. In an attempt to explain this conundrum, these authors described NM as a double-edged sword; a type of cellular Dr. Jekyll and Mr. Hyde which can change its behaviour depending upon the cellular environment [64].

Cell culture systems have also been used to investigate the effect of NM on cell survival. Synthetic dopamine melanin is phagocytosed into PC12 cells in culture [51], and further ingestion of synthetic melanin induces significant apoptotic cell death in these cultures [52]. These results suggest that the intraneuronal presence of melanin per se may be toxic but does not concur with the presence of heavily pigmented cells in the normal brain, or with reports that pigmented cells are not lost during normal aging [42] or the apparent absence of apoptotic cell death in PD [5]. Apoptotic cell death was, however, seen only when large concentrations (0.45 mg/ml and above) of synthetic dopamine melanin were used, but not at lower, more physiological concentrations [52]. In

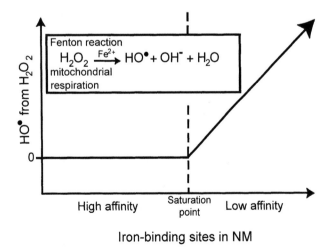

Fig. 2. Effect of dual binding sites for iron on NM for HO· formation. At low iron concentrations, iron is chelated by binding to the high-affinity binding sites on NM, and HO· production is low. At increasing iron concentrations, high-affinity binding sites are saturated and HO· production is stimulated by nonchelated iron bound to low-affinity binding sites on NM.

contrast, our recently unpublished observations suggest that incubation of synthetic dopamine melanin with either human-derived neuroblastoma or glioblastoma cells attenuates lactate dehydrogenase activity in a dose-dependent manner. These results argue for a protection effect rather than toxicity. These results concur with the finding of Mochizuki et al. [50] who report that low concentrations of synthetic dopamine melanin (6 μg/ml) had no cytotoxic effects on primary rat nigrostriatal cocultures. As the structure of human NM differs to that of the synthetically produced dopamine melanin [20], the relevance of much of the literature based upon the use of the synthetic melanin is unclear. Few studies into the role of NM are available in which NM itself, rather than the synthetic pigment, has been used. An exception to this is the 1996 paper by Aimi and McGeer [4] in which NM extracts from both healthy and parkinsonian human brain were injected into the rat SN. No nigral cell loss was reported in either case, nor was significant activation of microglia noted. It is unclear, however, whether the NM contained within the extract was able to interact with the host tissue, as the preparation of the injected suspension consisted of a simple homogenising process which may have resulted in brain-derived materials, such as lipids, coating the individual NM granules, rather than the separation of NM from the surrounding tissue.

The contradictions posed by the available data reflect the apparent ambiguous nature of NM. On one hand, NM appears to act as a protective substance within the dopaminergic cells. The mechanism of this protection is unclear but might be mediated via two mechanisms. Firstly, the NM molecule itself may attenuate oxidative damage by directly interacting with, and inactivating, free radical species. Secondly, the ability of NM to bind potentially damaging species such as transition metals may represent another mechanism by which it is neuroprotective. This capacity, however, appears to be true at most low levels of such species. At higher concentrations of metals, such as iron, the consequences of this binding for the cell appear to be negative, rather than positive. This hypothesis is supported by the available evidence. Offen et al. [52] also reported that the coincubation of ferrous iron with synthetic melanin significantly increased the toxicity above that seen for melanin alone. Further, this additive effect was abolished by the addition of the iron chelator desferrioxamine [52]. Similarly, Mochizuki et al. [50] reported that a complex of iron and synthetic melanin induced significantly greater cell death than that induced by an equivalent amount of iron alone, an effect prevented by desferrioxamine. Taken together, these results indicate that in physiological amounts, NM (or synthetic melanin) alone is not cytotoxic and may even be protective. It appears to be the overloading of NM with iron which is potentially cytotoxic.

The mechanism by which this proposed iron-mediated toxicity occurs is unclear. In our work, the abolition of membrane damage in the presence of iron-loaded NM by an iron chelator [19] suggests that the damage is iron-mediated. Iron bound to many endogenous iron-binding compounds, such as the transport protein transferrin, the neurotrophil-derived protein lactoferrin, the iron-storage proteins ferritin and haemosiderin, the oxygen-binding proteins hemoglobin and myoglobin, is not redox-active. Thus, in general, iron-dependent radical production in the brain is thought to be mediated by the small proportion of ferritin-bound iron, which can be reductively mobilised, and by a small pool of mobile iron within the cell [31,32]. It has been suggested that NM can reduce ferric iron, even when most of its ion-binding sites have been saturated with an excess of ferric iron [53]. This suggests that the ion-exchange and electron-exchange properties of this molecule may be independent. Iron-mediated damage may also be stimulated by the release of iron from its binding sites on NM. The release of iron from its binding site in NM into the cellular milieu would thus increase the concentration of free redox-active iron within the cell, resulting in the production of free radical species. It is possible that iron bound to NM may remain redox-active. Although ferrous irons bound to synthetic melanin are reported to be relatively inefficient for decomposing H_2O_2 in vitro [53], redox-active iron has recently been reported on NM granules in both the normal and the parkinsonian brain [11]. This study used immunohistochemical techniques and thus, the relative amounts of redox-active iron on NM granules on parkinsonian, compared with normal, brain could not be determined. Interestingly, these authors also report the presence of redox-active iron associated with Lewy bodies in the parkinsonian midbrain and hypothesise that these inclusion bodies sequester iron for the purpose of neuroprotection, rather than as a signature of impending cell death.

The available evidence suggests that iron binds to NM in the ferric form [1,25,36,54,59]. NM's capacity for iron storage is unknown. Studies on the iron-binding characteristics of synthetic dopamine melanin have demonstrated that this molecule contains both high- and low-affinity binding sites [8]. Our recent unpublished observations also suggest the presence of dual iron-binding sites in human NM. It is possible that iron bound to NM at the low-affinity site might remain redox-active (Fig. 3). This parallels the first phase of the incorporation of iron into ferritin, where iron is first loosely bound in the ferrous form, prior to being oxidised and stored in the ferric form within the iron core [32]. Interestingly, Lopiano et al. [45] suggested that NM isolated from the parkinsonian brain may have a decreased ability to bind iron. This same group has also reported structural changes in NM isolated from the parkinsonian brain which the authors suggest might reflect a decreased iron-sequestering ability [2]. A reduction in NM's iron-binding ability might be expected to increase the availability of redox-free iron within dopaminergic neurons.

In either event, free radical production is hypothesised to centre around the iron-saturated NM granules [40]. To date, the lack of NM in commonly used laboratory animals has made it difficult to investigate these hypothesised events experimentally, although increases in nigral iron via un-

Low Iron Environment

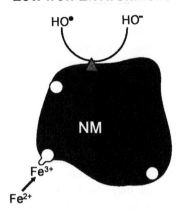

High Iron Environment

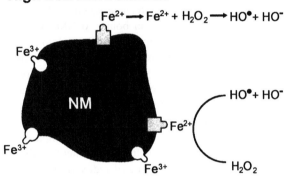

Iron exchange sites

○ high affinity iron binding sites

☐ low affinity iron binding sites

▲ electron exchange site

Fig. 3. Hypothesised interactions between NM and the cellular milieu. NM may reduce free radical levels by the direct inactivation of free radicals at electron-exchange sites upon the NM macromolecule. At low iron concentrations, ferrous iron is oxidised and chelated by high-affinity binding sites on NM, effectively reducing iron-stimulated damage. In the presence of high concentrations of iron, however, the high-affinity iron-binding sites are saturated and iron bound to low-affinity iron-binding sites may remain redox-active or may alternately be released to stimulate the local production of free radicals.

known mechanisms have been demonstrated in a variety of animal models of PD [24,33,63].

The source of the increased tissue iron is also unknown, although several mechanisms by which tissue iron might increase have been suggested (reviewed in Ref. [21]). Irrespective of the mode by which tissue iron is increased in degenerating tissue in PD, it appears that the presence of NM pigment in the dopaminergic neurons of the SN reflects a peculiar vulnerability to iron-stimulated damage because of the interaction between iron and NM. Interestingly, several lines of evidence suggest that in vivo increased brain iron per se does not stimulate significant degeneration and that this

event may be a secondary phenomenon, rather than a primary trigger of degeneration. Lan and Jiang [44] reported that mice exhibiting high brain iron concentrations following a high iron diet produced a significantly greater rate of HO⋅ production, accompanied by increased levels of oxidised glutathione. High tissue iron alone, however, was not associated with decreases in striatal dopamine concentrations or with membranal damage; a single injection of 1-methyl-4-phenyl-1,2,3,6-tetrahydropyridine (MPTP) in these animals subsequently induced significant neurochemical changes indicative of neurodegeneration. Similarly, evidence of behavioural deficients and significantly reduced measures of dopamine function was also reported in iron-loaded MPTP-treated mice by Fredriksson et al. [23]. This is also supported by findings in humans where an increase in SN iron can be neither be detected in incidental Lewy body disease, a syndrome characterised by mild nigral loss and Lewy bodies, suggested to be preclinical PD [16], nor in "mild" PD assessed semiquantitatively [54]. These data suggest that increased tissue iron alone, and perhaps, even an increase in oxidative load, is insufficient to trigger neurodegeneration, but may reflect a system less able to defend itself against additional noxious stimulus. Nevertheless, the development of strategies to attenuate tissue iron or the resultant production of free radicals in humans, an approach which has been reported to be successful in animal models [43], may have important therapeutic implications not only for PD, but for a range of neurodegenerative diseases.

Acknowledgments

K.L.D. was the recipient of an R.D. Wright Research Fellowship from the National Health and Medical Research Council of Australia. M.G. and P.R. were supported by a grant from the Deutsche Forschungsgemeinschaft (SFB 581 "Molekulare Modelle für Erkrankungen des Nervensystems"). M.G. held a grant from the Deutsche Forschungsgemeinschaft (BE1774/41). This research was completed within "The National Parkinson Foundation Center of Excellence Research Laboratories" at the Clinic and Policlinic for Psychiatry and Psychotherapy of the University of Würzburg (awarded to P.R.) and at the Prince of Wales Medical Research Institute, Sydney, Australia.

References

[1] S. Aime, B. Bergamasco, D. Biglino, G. Digilio, M. Fasano, E. Giamello, L. Lopiano, EPR investigations of the iron domain in neuromelanin, Biochem. Biophys. Acta 1361 (1997) 49–58.

[2] S. Aime, B. Bergamasco, M. Casu, G. Digilio, M. Fasano, S. Giraudo, L. Lopiano, Isolation and ^{13}C-NMR characterization of an insoluble proteinaceous fraction from substantia nigra of patients with Parkinson's disease, Mov. Disord. 15 (2000) 977–981.

[3] S. Aime, M. Fasano, B. Bergamasco, L. Lopiano, G. Quattrocolo, Nuclear magnetic resonance spectroscopy characterization and iron

content determination of human mesencephalic neuromelanin, in: L. Battistini (Ed.), Advances in Neurology, Lippencott-Raven Publishers, Philadelphia, 1996, pp. 263–270.

[4] Y. Aimi, P.L. McGeer, Lack of toxicity of human neuromelanin to rat brain dopaminergic neurons, Parkinsonism Relat. Disord. 2 (1996) 69–74.

[5] R.B. Banati, S.E. Daniel, S.B. Blunt, Glial pathology but absence of apoptotic nigral neurons in long-standing Parkinson's disease, Mov. Disord. 13 (1998) 221–227.

[6] G. Bartzokis, T. Tishler, MRI evaluation of basal ganglia ferritin iron and neurotoxicity in Alzheimer's and Huntington's disease, Cell. Mol. Biol. (Paris) 46 (2000) 821–833.

[7] G. Becker, J. Seufert, U. Bogdahn, H. Reichman, K. Reiners, Degeneration of substantia nigra in chronic Parkinson's disease visualized by transcranial color-coded real-time sonography, Neurology 45 (1995) 182–184.

[8] D. Ben-Shachar, P. Riederer, M.B.H. Youdim, Iron–melanin interaction and lipid peroxidation: Implications for Parkinson's disease, J. Neurochem. 57 (1991) 1609–1614.

[9] D. Ben-Shachar, M.B. Youdim, Selectivity of melanized nigra-striatal dopamine neurons to degeneration in Parkinson's disease may depend on iron–melanin interaction, J. Neural. Transm., Suppl. 29 (1990) 251–258.

[10] H. Braak, E. Braak, D. Yilmazer, C. Schultz, R. De Vos, E. Jansen, Nigral and extanigral pathology in Parkinson's disease, J. Neural. Transm., Suppl. 46 (1995) 15–32.

[11] R. Castellani, S. Siedlak, G. Perry, M. Smith, Sequestration of iron by Lewy bodies in Parkinson's disease, Acta Neuropathol. 100 (2000) 111–114.

[12] J.R. Connor, B.S. Synder, J.L. Beard, R.E. Fine, E.J. Mufson, Regional distribution of iron and iron-regulatory proteins in the brain in aging and Alzheimer's disease, J. Neurosci. Res. 31 (1992) 327–335.

[13] J.R. Connor, B.S. Snyder, P. Arosio, D.A. Loeffler, P. LeWitt, A quantitative analysis of isoferritins in select regions of aged, parkisonian and Alzheimer's diseased brains, J. Neurochem. 65 (1995) 717–724.

[14] P. Damier, E. Hirsch, F. Javoy-Agid, P. Zhang, F. Agid, Glutathione peroxidase, glial cells and Parkinson's disease, Neuroscience 52 (1993) 1–6.

[15] D. Dexter, J. Sian, P. Jenner, C. Marsden, Implications of alterations in trace element levels in brain in Parkinson's disease and other neurological disorders affecting the basal ganglia, in: H. Narabayashi (Ed.), Advances in Neurology, Raven Press, New York, 1993, pp. 273–281.

[16] D. Dexter, J. Sian, S. Rose, J. Hindmarsh, V. Mann, J. Cooper, F. Wells, S. Daniel, A. Lees, A. Schapira, Indices of oxidative stress and mitochondrial function in individuals with incidental Lewy body disease, Ann. Neurol. 35 (1994) 38–44.

[17] D.T. Dexter, A. Carayon, F. Javoy-Agid, Y. Agid, F.R. Wells, S.E. Daniel, A.J. Lees, P. Jenner, C.D. Marsden, Alterations in the levels of iron, ferritin and other trace metals in Parkinson's disease and other neurodegenerative diseases affecting the basal ganglia, Brain 114 (1991) 1953–1975.

[18] D.T. Dexter, F.R. Wells, A.J. Lees, F. Agid, Y. Agid, P. Jenner, C.D. Marsden, Increased nigral iron content and alterations in other metal ions occurring in brain in Parkinson's disease, J. Neurochem. 52 (1989) 1830–1836.

[19] K. Double, P. Riederer, M. Gerlach, The significance of neuromelanin in Parkinson's disease, Drug News Dev. 12 (1999) 333–340.

[20] K. Double, L. Zecca, P. Costo, M. Mauer, C. Greisinger, S. Ito, D. Ben-Shachar, G. Bringmann, R.G. Fariello, P. Riederer, M. Gerlach, Structural characteristics of human substantia nigra neuromelanin and synthetic dopamine melanins, J. Neurochem. 75 (2000) 2583–2589.

[21] K.L. Double, M. Gerlach, M.B.H. Youdim, P. Riederer, Impaired iron homeostasis in Parkinson's disease, in: P. Riederer (Ed.), Advances in Research on Neurodegeneration, Springer, Vienna, 2000, pp. 37–58.

[22] P. Foley, P. Riederer, The motor circuit of the human basal ganglia reconsidered, J. Neural. Transm. 58 (2000) 97–110.

[23] A. Fredriksson, N. Schroder, P. Eriksson, I. Izquierdo, T. Archer, Neonatal iron exposure induces neurobehavioural dysfunctions in adult mice, Toxicol. Appl. Pharmacol. 159 (1999) 25–30.

[24] A. Fredriksson, N. Schroumlautder, P. Eriksson, I. Izquierdo, T. Archer, Neonatal iron potentiates adult MPTP-induced neurodegenerative and functional deficits, Parkinsonism Relat. Disord. 7 (2001) 97–105.

[25] M. Gerlach, A.X. Trautwein, L. Zecca, M.B.H. Youdim, P. Riederer, Mössbauer spectroscopic studies of purified human neuromelanin isolated from the substantia nigra, J. Neurochem. 65 (1995) 923–926.

[26] W. Gibb, Melanin, tyrosine hydroxylase, calbindin and substance P in the human midbrain and substantia nigra in relation to nigrastriatal projections and differential neuron susceptibility in Parkinson's disease, Brain Res. 581 (1992) 283–291.

[27] P. Good, C. Olanow, D. Perl, Neuromelanin-containing neurons of the substantia nigra accumulate iron and aluminum in Parkinson's disease: A LAMMA study, Brain Res. 593 (1992) 343–346.

[28] J.M. Gorell, R.J. Ordidge, G.G. Brown, J.C. Deniau, N.M. Buderer, J.A. Helpern, Increased iron-related MRI contrast in the substantia nigra in Parkinson's disease, Neurology 45 (1995) 1138–1143.

[29] P.D. Griffiths, B.R. Dobson, G.R. Jones, D.T. Clarke, Iron in the basal ganglia in Parkinson's disease. An in vitro study using extended X-ray absorption fine structure and cryo-electron microscopy, Brain 122 (1999) 667–673.

[30] G. Halliday, D. McRitchie, H. Cartwright, R. Pamphlett, M. Hely, J. Morris, Midbrain neuropathology in idiopathic Parkinson's disease and diffuse Lewy body disease, J. Clin. Neurosci. 3 (1996) 52–60.

[31] B. Halliwell, Reactive oxygen species and the central nervous system, J. Neurochem. 59 (1992) 1609–1623.

[32] P.M. Harrison, P. Arosio, The ferritins: Molecular properties, iron storage function and cellular regulation, Biochem. Biophys. Acta 1275 (1996) 161–203.

[33] Y. He, T. Lee, S. Leong, Time course of dopaminergic cell death and changes in iron, ferritin and transferrin levels in the rat substantia nigra after 6-hydroxydopamine (6-OHDA) lesioning, Free Radical Res. 13 (1999) 103–112.

[34] J. Henderson, K. Carpenter, H. Cartwright, G. Halliday, Loss of thalamic intralaminar nuclei in progressive supranuclei palsy and Parkinson's disease, Brain 123 (2000) 1410–1421.

[35] E. Hirsch, A. Graybiel, Y. Agid, Melanized dopamine neurons are differentially susceptible to degeneration in Parkinson's disease, Nature (1988) 345–348.

[36] K. Jellinger, E. Kienzel, G. Rumpelmair, P. Riederer, H. Stachellberger, D. Ben-Shachar, M.B.H. Youdim, Iron–melanin complex in substantia nigra of parkinsonian brains: An X-ray microanalysis, J. Neurochem. 59 (1992) 1168–1171.

[37] K. Jellinger, W. Paulus, I. Grundke-Iqbal, P. Riederer, M.B.H. Youdim, Brain iron and ferritin in Parkinson's disease and Alzheimer's disease, J. Neural. Transm.: Parkinson's Dis. Dementia Sect. 2 (1990) 327–340.

[38] A. Kastner, E. Hirsch, O. Lejeune, F. Javoy-Agid, O. Rascol, Y. Agid, Is the vulnerability of neurons in the substantia nigra of patients with Parkinson's disease related to their neuromelanin content? J. Neurochem. 59 (1992) 1080–1089.

[39] J. Kehrer, The Haber–Weiss reaction and mechanisms of toxicity, Toxicology 149 (2000) 43–50.

[40] E. Kienzl, K. Jellinger, H. Stachelberger, W. Linert, Iron as a catalyst for oxidative stress in the pathogenesis of Parkinson's disease? Life Sci. 65 (1999) 1973–1976.

[41] E.S. Krol, D.C. Liebler, Photoprotective actions of natural and synthetic melanins, Chem. Res. Toxicol. 11 (1998) 1434–1440.

[42] N. Kubis, B. Faucheux, G. Ransmayr, P. Damier, C. Duyckaerts, D. Henin, B. Forette, Y. Le Charpentier, J. Hauw, Y. Agid, E. Hirsch, Preservation of midbrain catecholaminergic neurons in very old human subjects, Brain 123 (2000) 366–373.

[43] J. Lan, D. Jiang, Desferrioxamine and vitamin E protect against iron and MPTP-induced neurodegeneration in mice, J. Neural. Transm. 104 (1997) 469–481.

[44] J. Lan, D. Jiang, Excessive iron accumulation in the brain: A possible potential source of neurodegeneration in Parkinson's disease, J. Neural. Transm. 104 (1997) 649–660.

[45] L. Lopiano, M. Chiesa, D. Digilio, G. Giraudo, B. Bergamasco, M. Fasano, Q-band EPR investigations of neuromelanin in control and Parkinson's disease patients, Biochem. Biophys. Acta 1500 (2000) 306–312.

[46] W. Markesbery, Oxidative stress hypothesis in Alzheimer's disease, Free Radical Biol. Med. 23 (1997) 134–147.

[47] C.D. Marsden, Pigmentation in the nucleus substantia nigra of mammals, J. Anat. 95 (1961) 256–261.

[48] W. Martin, F. Ye, P. Allen, Increasing striatal iron content associated with normal aging, Mov. Disord. 13 (1998) 281–286.

[49] P. McGeer, S. Itagaki, H. Akiyama, E. McGeer, Rate of cell death in Parkinson's disease indicates an active neuropathological process, Ann. Neurol. 24 (1988) 574–576.

[50] H. Mochizuki, K. Nishi, Y. Mizuno, Iron–melanin complex is toxic to dopaminergic neurons in nigrostriatal co-culture, Neurodegeneration 2 (1993) 1–7.

[51] D. Offen, S. Gorodin, E. Melamed, J. Hanania, Z. Malik, Dopamine-melanin is actively phagocytized by PC12 cells and cerebellar granular cells: Possible implications for the etiology of Parkinson's disease, Neurosci. Lett. 260 (1999) 101–104.

[52] D. Offen, I. Ziv, A. Barzilai, S. Gorodin, E. Glater, A. Hochman, E. Melamed, Dopamine-melanin induces apoptosis in PC12 cells: Possible implications for the etiology of Parkinson's disease, Neurochem. Int. 31 (1997) 207–216.

[53] B. Pilas, T. Sarna, B. Kalyanaraman, H. Swartz, The effect of melanin on iron associated decomposition of hydrogen peroxide, Free Radical Biol. Med. 4 (1988) 285–293.

[54] P. Riederer, E. Sofic, W.D. Rausch, B. Schmidt, G.P. Reynolds, K. Jellinger, M.B. Youdim, Transition metals, ferritin, glutathione, and ascorbic acid in parkinsonian brains, J. Neurochem. 52 (1989) 515–520.

[55] M. Rozanowska, T. Sarna, E. Land, T. Truscott, Free radical scavenging properties of melanin interaction of eu- and pheo-melanin models with reducing and oxidising radicals, Free Radical Biol. Med. 26 (1999) 518–525.

[56] P. Ryvlin, E. Broussolle, H. Piollet, F. Viallet, Y. Khalfallah, G. Chazot, Magnetic resonance imaging evidence of decreased putamenal iron content in idiopathic Parkinson's disease, Arch. Neurol. 52 (1995) 583–588.

[57] M. Scalia, E. Geremia, C. Corsaro, C. Santoro, D. Baratta, G. Sichel, Lipid peroxidation in pigmented and unpigmented liver tissues: Protective role of melanin, Pigm. Cell Res. 3 (1990) 115–119.

[58] X. Shen, F. Zhang, G. Dryhurst, Oxidation of dopamine in the presence of cysteine: Characterization of new toxic products, Chem. Res. Toxicol. 10 (1997) 147–155.

[59] T. Shima, T. Sarna, H. Swartz, A. Stroppolo, R. Gerbasi, L. Zecca, Binding of iron to neuromelanin of human substantia nigra and synthetic melanin: An electron paramagnetic resonance spectroscopy study, Free Radical Biol. Med. 23 (1997) 110–119.

[60] M. Smith, P. Harris, L. Sayre, G. Perry, Iron accumulation in Alzheimer disease is a source of redox-generated free radicals, Proc. Natl. Acad. Sci. U. S. A. (1997) 9866–9868.

[61] J. Smythies, On the function of neuromelanin, Proc. R. Soc. London, Ser. B 263 (1996) 487–489.

[62] E. Sofic, W. Paulus, K. Jellinger, P. Riederer, M. Youdim, Selective increase of iron in substantia nigra zona compacta of parkinsonian brains, J. Neurochem. 56 (1991) 978–982.

[63] J. Temlett, J. Landsberg, F. Watt, G. Grime, Increased iron in the substantia nigra compacta of the MPTP-lesioned hemiparkisonian African Green monkey: Evidence from proton microprobe elemental microanalysis, J. Neurochem. 62 (1994) 134–146.

[64] M.B.H. Youdim, D. Ben-Shachar, P. Riederer, The enigma of neuromelanin in Parkinson's disease substantia nigra, J. Neural. Transm., Suppl. 43 (1994) 113–122.

[65] M. Zareba, A. Bober, W. Korytowski, L. Zecca, T. Sarna, The effect of a synthetic neuromelanin on yield of free hydroxyl radicals generated in model systems, Biochem. Biophys. Acta 1271 (1995) 343–348.

[66] L. Zecca, M. Gallorini, V. Schümann, A. Trautwein, M. Gerlach, P. Riederer, P. Vezzoni, D. Tampellini, Iron, neuromelanin and ferritin content in substantia nigra of normal subjects at different ages. Consequences for iron storage and neurodegenerative processes, J. Neurochem. 76 (2001) 1766–1773.

[67] L. Zecca, H.M. Swartz, Total and paramagnetic metals in human substantia nigra and its neuromelanin, J. Neural Transm.: Parkinson's Dis. Dementia Sect. 5 (1993) 203–213.

ELSEVIER

Neurotoxicology and Teratology 24 (2002) 629–638

NEUROTOXICOLOGY

AND

TERATOLOGY

www.elsevier.com/locate/neutera

Review article

DOPA causes glutamate release and delayed neuron death by brain ischemia in rats

Yoshimi Misu[a,b,*], Nobuya Furukawa[a,c], Nobutaka Arai[d], Takeaki Miyamae[a], Yoshio Goshima[a], Kiyohide Fujita[c]

[a]Department of Pharmacology, Yokohama City University School of Medicine, Yokohama 236-0004, Japan
[b]Shinobu Hospital, Fukushima 960-1101, Japan
[c]Department of Oral and Maxillofacial Surgery, Yokohama City University School of Medicine, Yokohama 236-0004, Japan
[d]Department of Clinical Neuropathology, Tokyo Metropolitan Institute of Neuroscience, Tokyo 183-8526, Japan

Received 2 April 2001; accepted 24 January 2002

Abstract

DOPA seems to be a neuromodulator in striata and hippocampal CA1 and a neurotransmitter of the primary baroreceptor afferents terminating in the nucleus tractus solitarii (NTS) and baroreflex pathways in the caudal ventrolateral medulla and rostral ventrolateral medulla in the brainstem of rats. DOPA recognition sites differ from dopamine (DA) D_1 and D_2 and ionotropic glutamate receptors. Via DOPA sites, DOPA stereoselectively releases by itself neuronal glutamate from in vitro and in vivo striata. In the cultured neurons, DOPA and DA cause neuron death via autoxidation. In addition, DOPA causes autoxidation-irrelevant neuron death via glutamate release. Furthermore, DOPA released by four-vessel occlusion seems to be an upstream causal factor for glutamate release and resultant delayed neuron death by brain ischemia in striata and hippocampal CA1. Glutamate has been regarded as a neurotransmitter of baroreflex pathways. Herein, we propose a new pathway that DOPA is a neurotransmitter of the primary aortic depressor nerve and glutamate is that of secondary neurons in neuronal microcircuits of depressor sites in the NTS. DOPA seems to release unmeasurable, but functioning, endogenous glutamate from the secondary neurons via DOPA sites. A common following pathway may be ionotropic glutamate receptors–nNOS activation–NO production–baroreflex neurotransmission and delayed neuron death. However, we are concerned that DOPA therapy may accelerate neuronal degeneration process especially at progressive stages of Parkinson's disease. © 2002 Elsevier Science Inc. All rights reserved.

Keywords: DOPA; Competitive DOPA antagonists; Brain ischemia; Glutamate release; Delayed neuron death; Neurotransmitter candidate of baroreflex pathways

1. Introduction

Since the 1950s, L-3,4-dihydroxyphenylalanine (DOPA) has been believed to be an inert amino acid that affects Parkinson's disease via conversion to dopamine (DA) by aromatic L-amino acid decarboxylase (AADC) [2]. Meanwhile, since 1986 [11,24], we have proposed that DOPA is a neurotransmitter and/or neuromodulator as well as precursor for DA [23,26,28]. Several criteria must be satisfied before acceptance as a neurotransmitter. There exist neurons showing immunocytochemically tyrosine hydroxylase (TH)-(+),

AADC-(−), DOPA-(+) and DA-(−) reactivity in various regions of the CNS including the nucleus tractus solitarii (NTS) [26,51,58]. These neurons may contain DOPA as an end product. DOPA is released by neuronal activities from in vitro [12] and in vivo [34] striata and in vivo NTS [58]. DOPA seems to be released from some cytoplasmic compartment other than DA-containing vesicles in striata [12,23] and from neurons that may have DOPA as an end product in the NTS [58].

DOPA methyl ester (ME), a prodrug for DOPA, is the first competitive antagonist for exogenously applied [13] and endogenously released DOPA [58]. Competitive antagonism suggests existence of a recognition site and/or sites for DOPA, which differ(s) from catecholamine receptors. DOPA and DOPA ME do not displace selective binding of

* Corresponding author. Shinobu Hospital, Fukushima 960-1101, Japan. Tel.: +81-24-546-3311; fax: +81-24-546-9467.
E-mail address: ymisu@zd5.so-net.ne.jp (Y. Misu).

DA D_1, D_2 and $\beta[^3H]$ ligands with negligible inhibition of that of an $\alpha_2[^3H]$ ligand in rat brain membranes [8,13]. DOPA ME, however, is effective, only when supplied continuously in brain slices [13–15] or within several minutes after being microinjected into depressor and presser sites of the lower brainstem [19,32,37,55,56,58]. DOPA ME is easily hydrolyzed. Thus, we screened a series of DOPA esters with bulky ring structures against esterases. Among candidates, DOPA cyclohexyl ester (CHE) is the most potent and relatively stable competitive antagonist [8,25]. Furthermore, DOPA recognition sites differ from ionotropic glutamate receptors because DOPAergic agonists and antagonists do not interact on these receptors: among binding sites labeled with ionotropic glutamate $[^3H]$ ligands, DOPA acts only on AMPA receptors with low affinity, whereas DOPA ME and DOPA CHE act only on NMDA ion channel domain with millimolar IC_{50} [31]. The DOPA ME- and DOPA CHE-sensitive DOPA recognition sites differ from ionotropic glutamate receptors.

No immunocytochemical evidence for neurons having DOPA as an end product is proved in striata and hippocampal CA1 pyramidal cell layers [26]. Notwithstanding, in striata, DOPA is released in a transmitter-like manner [12,34] and acts as a neuromodulator. Sensitivity of immunocytochemical analysis to find DOPAergic neurons seems to be lower, compared to that of biochemical or functional approach [26]. Exogenously applied DOPA potentiates activities of presynaptic β-adrenoceptors [27] to facilitate DA release [14,24]. Exogenously applied [35] and endogenously released [57] DOPA potentiate activities of postsynaptic D_2 receptors related to locomotor activities. Applied DOPA inhibits acetylcholine release in a rat Parkinson's model [52]. These responses [14,24,35,52,57] and, of course, conversion to DA [2] may cooperate in the effectiveness for Parkinson's disease. Meanwhile, micromolar DOPA by itself releases neuronal glutamate from slices [15]. This glutamate may be related to neuroexcitatory side effects at least such as dyskinesia encountered during chronic therapy of Parkinson's disease [17]. All of these responses [14,15,24,35,52,57] are not due to conversion to DA, but due to DOPA itself, because responses are seen under inhibition of central AADC with 3-hydroxybenzylhydrazine (NSD-1015) or DA does not mimic DOPA.

In this review, we will survey two topics. One is a neuromodulator role of DOPA as a causal factor for neuron death in striata [3,7,22] and hippocampal CA1 pyramidal cell layers [1]. The other is a neurotransmitter role of DOPA in baroreflex pathways and central regulation of blood pressure in the NTS [19,58,59], caudal ventrolateral medulla (CVLM) [32,33,55] and rostral ventrolateral medulla (RVLM) [37,56,59]. Glutamate has been regarded as a neurotransmitter of baroreflex pathways [43–45,50]. Finally, we propose a new pathway that DOPA is a neurotransmitter of the primary aortic depressor nerve (ADN) and glutamate is that of secondary neurons in neuronal microcircuits of depressor sites in the NTS [16].

2. DOPA-induced neurotoxicity in rat striata in vitro

Neurons in the dorsolateral striata and hippocampal CA1 pyramidal cell layers are vulnerable against ischemia, precipitating into cell death [41,42,46]. Excitotoxicity due to increases in extracellular glutamate is proposed to be a causal factor for neuron death [38,39] and implicated in a final common pathway for neurologic disorders [4,20]. Increased glutamate elicits excessive Ca^{2+} entry [4,48] via activation of postsynaptic non-NMDA [47] and NMDA [4,9,60] receptors. One of following pathways seems to be nitric oxide (NO) production [5], leading to delayed neuron death [18].

For neuron death in striata, clear relationships are seen between DOPA and glutamate. DOPA releases stereoselectively neuronal glutamate with ED_{50} of 140 μM from slices [15]. This release is seen under inhibition of AADC and is antagonized by DOPA ME in a competitive manner. Meanwhile, a corresponding concentration of DA (300 μM) releases no glutamate. DOPA and DA elicit differential toxicity in primary neuron cultures [3]. DA elicits neuron death in 3- and 10-day cultures. It is protected by ascorbic acid, an antioxidant, but not by non-NMDA and NMDA antagonists. Many studies show that hydroxyl radicals play an important role in toxicity induced by DA [3,6]. DOPA elicits toxicity via two different pathways. One is antioxidant-sensitive toxicity by DOPA in 3-day cultures. D-DOPA also elicits toxicity. DOPA itself or converted DA produces hydroxyl radicals (Fig. 1). In addition, DOPA auto-oxidizes to toxic 3,4,6-trihydroxyphenylalanine (TOPA, 6-hydroxy-DOPA) and TOPA quinones [36]. The other is stereoselective toxicity in 10-day cultures. It is protected by CNQX, a non-NMDA antagonist, MK-801, a NMDA ion channel domain antagonist, Mg^{2+} addition and Ca^{2+} removal, but not by ascorbic acid [3]. It is mediated by release of neuronal glutamate [22]. This type of neuron death is explained as follows. DOPA releases neuronal glutamate [15,22], which acts on non-NMDA receptors to depolarize neurons. These neurons facilitate the relief of NMDA ion channel function from Mg^{2+} block. Glutamate in turn acts on NMDA receptors [22] to elicit Ca^{2+} influx [4,48]. Ca^{2+}, forming the complex with calmodulin, activates neuronal NO synthase (nNOS), which produces NO with a detrimental property to cellular and mitochondrial membranes [5].

(1) DOPA → dopamine → autoxidation → hydroxyl radicals

(2) DOPA → autoxidation → TOPA and TOPA quinones

(3) DOPA → glutamate release → nNOS activation → NO production

Fig. 1. Three types of DOPA-induced neuron death in cultured neurons.

3. Endogenously released DOPA is a causal factor for glutamate release and resultant delayed neuron death by transient ischemia in striata and hippocampal CA1 in conscious rats

At first, we attempted to clarify whether endogenous DOPA is released by 10-min ischemia due to four-vessel occlusion during striatal microdialysis and, further, whether DOPA, when released, functions to cause glutamate release and resultant delayed neuron death by ischemia [7]. If it is the case, AADC inhibition should exaggerate these events because it markedly increases extracellular DOPA [34], whereas DOPA CHE, a potent and relatively stable competitive antagonist [8,25], ought to protect these events.

Extracellular DOPA, DA and glutamate were stabilized 2–3 h after perfusion with Ringer's solution. Ischemia increased extracellular DOPA, DA and glutamate (Fig. 2). An increase in DOPA in the first sample during ischemia was apparently slower than that in DA and glutamate. Peak increases in these substances were seen in the second sample immediately after ischemia. Some parallelism was seen between DOPA and glutamate because the peak release ratio of DOPA:DA:glutamate was 6:220:9. This ischemia released the 16-fold-higher amount of glutamate, compared to a basal level.

Compared to sham ischemia (Fig. 3a and e), the dorsolateral striatum (Fig. 3b) and the hippocampal CA1 pyramidal cell layers (Fig. 3f) showed tissue damages such as reactive astrocytosis, atrophic acidophilic cytoplasm and pyknotic nuclei 96 h after reperfusion. Ischemic damage was mild to moderate in the striatum and severe in the hippocampus. By a quantitative analysis, respectively, 20% and 50% neurons showed cell damage in the striatum and hippocampus (Table 1).

DOPA released may function to cause glutamate release and resultant delayed neuron death by ischemia. Inhibition of intrastriatal AADC with perfusion of NSD-1015 (30 μM) markedly increased basal DOPA, tripled the peak glutamate release with a tendency of decrease in DA release by ischemia [7], exaggerated delayed neuron death (Fig. 3c) and increased the density of ischemic neurons (Table 1). Furthermore, intrastriatal perfusion of 30–100 nM DOPA CHE decreased by 80% the peak glutamate release without modification of DA release by ischemia. The antagonist at 100 nM elicited a slight ceiling effect on the decreases in glutamate release [7], protected neuron death (Fig. 3d) and decreased the density of ischemic neurons. In the hippocampus, the degree and density of ischemic neurons were not modified by intrastriatal perfusion of either drug (Fig. 3g–h). There is concern that DOPA therapy may accelerate neuronal degeneration process especially at progressive stages of Parkinson's disease [26]. DA seems to be not related to glutamate release and resultant delayed neuron death by ischemia in this system, although it has been implicated in these events [10,40].

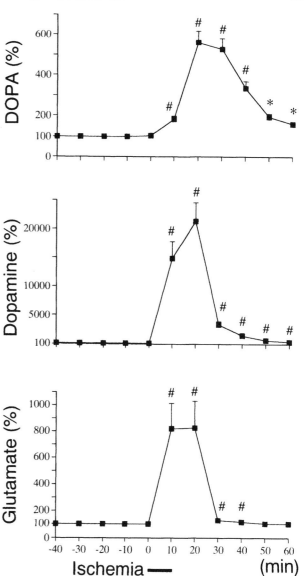

Fig. 2. Time course of DOPA, DA and glutamate released by 10-min transient ischemia due to four-vessel occlusion ($n = 38–40$) during microdialysis of the right striata in conscious rats. The bilateral vertebral arteries were electrocauterized and dummy cannula was implanted into the striatum. Two to 3 days later, dummy cannula was replaced by dialysis probe. Ringer's solution was perfused at a rate of 2 μl/min. Dialysates were collected every 10 min. After extracellular basal levels became stable, the bilateral carotid arteries were occluded at a horizontal bar. Carotid clamps were removed to make postischemic reperfusion. Data are mean ± S.E.M. $*P < .05$, $^{\#}P < .01$, vs. the value immediately before ischemia (paired t test). Reproduced from Ref. [7] with permission from ISN.

Judging from time course of increases in DOPA, DA and glutamate released by ischemia, it was not clear that DOPA is a causal factor for glutamate release because increase in DOPA in the first sample was one third of the peak, while that in glutamate reached the peak. It seems likely that there is a threshold of extracellular DOPA to trigger glutamate release by ischemia and an apparently low level of DOPA cleared over the threshold. Extremely low DOPA exerts responses [14]: noneffective 3–10 pM

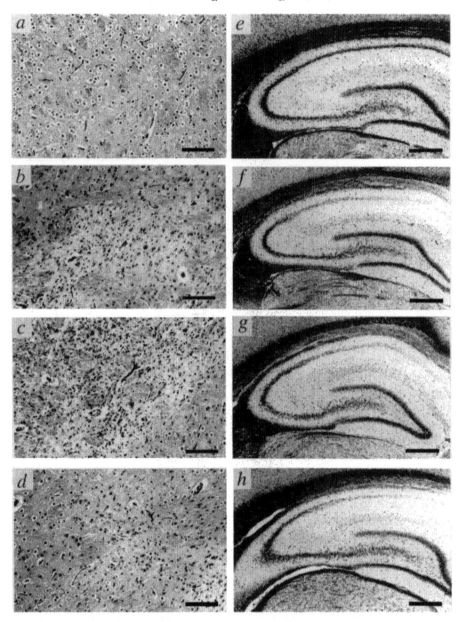

Fig. 3. Photomicrographs of the striatum (a–d) and the hippocampus (e–h): sham ischemia (a, e), typical pathological changes of delayed neuron death by transient ischemia (b, f), exaggeration of the changes by intrastriatal perfusion of 30 μM NSD-1015 10 min before ischemia until 1 h after postischemic reperfusion (c), protection by 100 nM DOPA CHE (d) in the striatum, and no modifications in the hippocampus (g, h). Ninety-six hours after ischemia, rats were decapitated and the brains were removed, fixed with 10% formalin, cut, embedded in paraffin and stained by hematoxylin–eosin and Klüver–Barrera methods. Scale bar, 100 μm in (a–d) and 500 μm in (e–h). Reproduced from Ref. [7] with permission from ISN.

DOPA potentiates facilitation of noradrenaline release by 1–3 nM isoproterenol via presynaptic β-adrenoceptors [27] in rat hypothalamic slices. We confirmed that exogenously perfused DOPA (0.3–1 mM) is a stereo-selective causal factor for glutamate release [7]. Another issue is some decline of increase in the peak glutamate release accompanied with further increase in DOPA release by ischemia. There were also clear discrepancies between continuously increased DOPA levels due to exposure to NSD-1015 and no further increases in glutamate release by ischemia. These findings may be explained by "downregulation" of the DOPA recognition

sites functioning to release glutamate due to exposure to increased DOPA levels. This idea is supported by the finding that the highest concentration of exogenously perfused DOPA (1 mM) tended to elicit desensitization of glutamate release in a time course similar to that after ischemic insult. It seems likely that NSD-1015 tripled the peak glutamate release at a critical time before occurrence of such desensitization.

Then, we tried to clarify whether DOPA CHE antagonizes glutamate release and delayed neuron death in the hippocampal CA1 most vulnerable against brain ischemia [4,38,41,46]. The antagonist is neuroprotective under a mild

Table 1
Delayed neuron death by 10-min ischemia in striatum and hippocampus CA1 and its modifications by NSD-1015 and DOPA CHE

Groups	n	Density of ischemic neurons/mm^2	
		Striatum	Hippocampus
Ischemia alone	17	34.9 ± 1.2	194.0 ± 2.0
NSD-1015 30 μM + ischemia	10	$49.4 \pm 0.8*$	195.2 ± 1.8
DOPA CHE 100 nM + ischemia	12	$22.0 \pm 1.2*$	192.0 ± 1.8

Intrastriatal perfusion of NSD-1015 or DOPA CHE was done 10 min before ischemia until 1 h after postischemic reperfusion. The brain was dissected 96 h after the acute experiments. Values are mean ± S.E.M. of ischemic neurons per millimeters squared in the dorsolateral striatum and hippocampus CA1 from *n* estimations. Intact neurons per millimeters squared in sham ischemia (*n* = 5) are 179.5 ± 8.7 in the striatum and 366.7 ± 14.0 in the hippocampus.
 * $P < .01$, vs. ischemia alone (Student's *t* test).

Table 2
Delayed neuron death by 5-min ischemia in the hippocampal CA1 and its protection by DOPA CHE

Groups	n	Density of ischemic neurons/mm^2
Ischemia alone	8	78.5 ± 7.0
DOPA CHE 100 nM + ischemia	6	$11.9 \pm 2.8*$

Intrahippocampal perfusion of 100 nM DOPA CHE was done 10 min before ischemia until 1 h after postischemic reperfusion. Other details are as in Table 1.
 * $P < .01$, vs. ischemia alone (Student's *t* test).

ischemic condition [1]. DOPA CHE is a mother compound to develop neuroprotectants. Five-to 10-min ischemia caused slight to mild glutamate release in 10-min samples during microdialysis (Fig. 4) and 20–50% [7] neuron

A. 10 min Ischemia

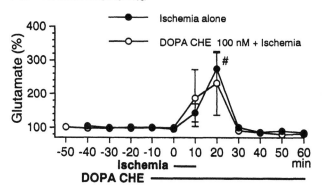

B. 5 min Ischemia

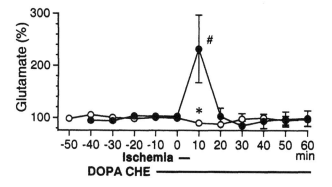

Fig. 4. Effect of intrahippocampal perfusion with 100 nM DOPA CHE on glutamate release by 10-min (A) and 5-min (B) ischemia at short horizontal bars during microdialysis of the right hippocampal CA1 in conscious rats. DOPA CHE was perfused 10 min before ischemia until 1 h after postischemic reperfusion at long horizontal bars. $^{\#}P < .05$, vs. value immediately before ischemia (paired *t* test); *$P < .01$, vs. ischemia alone (Mann–Whitney *U* test). Other details are as in Fig. 2. Reproduced from Ref. [1] with permission from Elsevier.

damages 96 h after reperfusion (Table 2). Compared to striata [7], the release is lower (1/4), but the density of ischemic neurons is higher (2.5-fold), showing the highly vulnerable property [4,38,41,46]. DOPA and DA release is under assay limit in this design, but these substances are released by 20-min ischemia. The result for DA is consistent with previous findings [38]. We cannot measure extracellular DOPA less than 0.4 nM because its assay limit is 0.05 fmol/μl and the recovery of probes used is 12% [7].

Intrahippocampal perfusion of 100 nM DOPA CHE abolished glutamate release and protected neurons from cell death by 5-min ischemia. It ought to exert neuroprotection via antagonism for DOPA. DOPA released less than 0.4 nM seems to function and to clear over a threshold to trigger glutamate release by ischemia because extremely low 3–10 pM DOPA exerts responses [14]. DOPA released less than 0.4 nM seems to be a causal factor for glutamate release and delayed neuron death by ischemia in the hippocampal CA1, suggesting also a highly vulnerable property. DOPA CHE acts on NMDA ion channel domain with millimolar IC$_{50}$ [31]. The antagonist, however, is unlikely to elicit protection via this site because the nanomolar dose was effective. Neuroprotection by systemic MK-801 is partially attributed to hypothermia [60]. DOPA CHE, however, elicits no hypothermia during and after ischemia at least by intrastriatal perfusion [7].

DOPA CHE at 100 nM does not antagonize some low DOPA released to release the twofold-higher amount of glutamate by 10-min ischemia, compared to 5-min ischemia (Fig. 4). An additional possibility is a role of DOPA converted from DOPA CHE. It still behaves as a DOPA prodrug [8]. It, at 100 nM, tends to increase basal DOPA in striata [7]. DOPA converted may weaken or exceed the competitive activity of the antagonist. These ideas are supported by the finding that 1 μM DOPA CHE increases inversely glutamate release by ischemia [1]. This dose still increases extracellular DOPA [8]. DOPA converted seems to be, by itself, a factor for glutamate release and delayed neuron death by ischemia because basal DOPA increased by AADC inhibition may function to exaggerate these events and perfused DOPA stereoselectively releases glutamate in striata in vivo [7].

We showed a delicate property of the hippocampal CA1 highly vulnerable against ischemia. An analogy was seen in the NTS [58]. DOPA ME antagonizes depressor responses to mild stimulation of the ADN, but not to the strong stimulation. DOPA release is revealed by the stronger intermittent stimulation.

4. General survey in the lower brainstem

Baroreflex is the principal neuronal mechanism by which the cardiovascular system is regulated under negative feedback control in the CNS [26,28]. Baroreceptors are located in the aortic arch and carotid sinus (Fig. 5). The primary baroreceptor afferents terminate in the NTS to carry reflex information [26,30]. The neurotransmitter has been believed to be glutamate [50] among many candidates for the primary baroreceptor afferents, microcircuit nerve terminals and interneurons within depressor sites of the NTS [28]. The NTS has depressor neurons. The CVLM receives inputs from the NTS [26,28]. Glutamate is also a neurotransmitter candidate. Vasomotor tone is reduced by neuronal activities in the CVLM, the integration of which constitutes the central pathways for baroreflex. Sympatho-inhibitory neurons in the CVLM project to the RVLM [26,28,45]. This

neurotransmitter is believed to be GABA [26,28]. The RVLM receives inputs from various regions including the NTS [45,53], CVLM [45] and posterior hypothalamic nucleus (PHN) [54]. Glutamate is a neurotransmitter candidate of neurons projecting from the NTS [26,28]. The RVLM has sympatho-excitatory neurons, which project directly to the intermediolateral cell column of the thoraco-lumbar spinal cord—the main origin of the sympathetic outflow [44,49]. The RVLM plays an important role in controlling resting and reflex integration of blood pressure. Sympatho-excitatory effect of the pathway from the NTS to RVLM may be masked by predominant effect of the pathway from the NTS to CVLM (Fig. 5) and then to RVLM, mediating depressor responses [53].

Meanwhile, neurons in the NTS and dorsal motor vagal nucleus complex have immunocytochemically DOPA as an end product [26,51,58]. Such evidence, however, is not proved in the CVLM and RVLM [26]. During microdialysis of the NTS [58,59], and even the CVLM [32,33,55] and RVLM [37,56,59] in anesthesized rats, basal DOPA release is in part Ca^{2+}-dependent and tetrodotoxin (TTX)-sensitive and is abolished by intraperitoneal α-methyl-p-tyrosine (α-MPT), a TH inhibitor. High

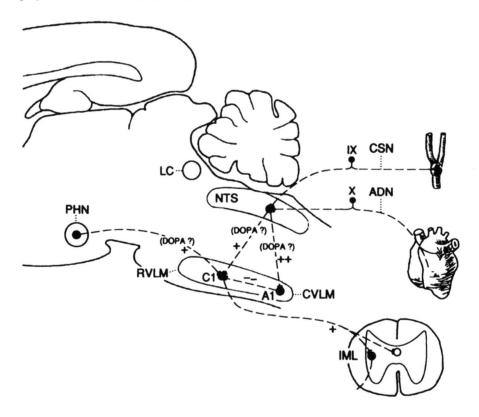

Fig. 5. Diagram of cardiovascular responses of anesthesized rats to microinjected and endogenously released DOPA in the NTS, CVLM and RVLM and of baroreflex pathways and blood pressure regulation in the lower brainstem. DOPA microinjected elicits by itself postsynaptic, stereoselective and DOPA ME-sensitive hypotension and bradycardia in the NTS and CVLM and hypertension and tachycardia in the RVLM. Endogenously released DOPA functions tonically to produce these responses in these areas. It is highly probable that DOPA is a neurotransmitter of the primary baroreceptor afferents terminating in the NTS. There may exist a DOPAergic baroreceptor–ADN–NTS–CVLM depressor relay and a PHN–RVLM pressor relay. Symbols: (+) sympathoexcitation; (−) sympathoinhibition. Abbreviations: ADN, aortic depressor nerve; CSN, carotid sinus nerve; CVLM, caudal ventrolateral medulla; IML, intermediolateral cell column; LC, locus coeruleus; NTS, nucleus tractus solitarii; PHN, posterior hypothalamic nucleus; RVLM, rostral ventrolateral medulla. Reproduced from Ref. [26] with permission from Elsevier.

K^+ releases DOPA Ca^{2+}-dependently. DOPA (10–100 ng) microinjected into depressor sites of the NTS [19,59] and CVLM [32,33,55] elicits hypotension and bradycardia. DOPA (30–600 ng) microinjected into pressor sites of the RVLM elicits hypertension and tachycardia [56,59]. Differences of these doses between the NTS or CVLM and the RVLM suggest less tonicity of the pathway from the NTS to RVLM than that from the NTS to CVLM [53]. These doses are lower approximately by two orders of magnitude, compared to the minimum effective dose of DA, noradrenaline or adrenaline microinjected (1 μg) [26]. These responses [19,32,33,55,56,59] are dose-dependent, stereoselective, seen under inhibition of central AADC with intraperitoneal NSD-1015, seen under treatment with intraventricular 6-hydroxy-DA, and antagonized by prior microinjection of DOPA ME [13]. Tonic function of DOPA recognition sites is proved by the antagonism for endogenously released DOPA: DOPA ME alone microinjected bilaterally elicits hypertension and tachycardia in the NTS [58] and CVLM [55] and hypotension and bradycardia in the RVLM [56]. These responses are abolished by intraperitoneal α-MPT.

5. Further evidence for DOPAergic system in the NTS

It is highly probable that DOPA is a neurotransmitter of the primary baroreceptor afferents terminating in the NTS to carry reflex information [19,58,59]. After denervation of the left ADN, TH-, DOPA-, but not DA- and DA-β-hydroxylase immunoreactivity decreases in the left NTS. In the left ganglion nodosum, TH-, DOPA-, but not DA immunoreactivity decreases [58]. DOPA ME microinjected into the left NTS antagonizes depressor and bradycardic responses to left electrical ADN stimulation [58] in accord with those to DOPA microinjected [19]. During microdialysis of the NTS, ADN stimulation releases TTX-sensitive DOPA [58]. Baroreceptor activation by intravenous phenylephrine elicits DOPA release and reflex bradycardia, which are abolished by denervation of the bilateral ADN and carotid sinus nerve without changes of hypertension. DOPA ME microinjected bilaterally antagonizes the reflex bradycardia.

6. Further evidence for DOPAergic system in the CVLM

It is highly probable that there exists a DOPAergic baroreceptor–ADN–NTS–CVLM depressor relay to carry baroreflex information [32]. Electrolytic lesion of the right NTS selectively decreases by 45% the tissue content of DOPA in the dissected right CVLM without decreases in catecholamines. During microdialysis of the right CVLM, intravenous phenylephrine selectively releases DOPA. This release is abolished by acute lesion of the right NTS. No catecholamines are released. ADN stimulation releases DOPA without increases in catecholamines, accompanied

with hypotension and bradycardia. Local inhibition of DOPA synthesis with α-MPT ipsilaterally microinjected gradually decreases basal DOPA, and interrupts DOPA release, reduces hypotension, but not reduces bradycardia by ADN stimulation. DOPA ME microinjected reduces hypotension, but not bradycardia by ADN stimulation. It seems likely that the main pathway of the preganglionic efferents of baro-vagal reflex does not pass through depressor sites of the CVLM [32].

7. Further evidence for DOPAergic system in the RVLM

Electrolytic lesion of the right NTS tends to decrease by 20% the tissue content of DOPA in the dissected right RVLM without decreases in catecholamines [32]. This slight tendency may be due to less tonic function of the pathway from the NTS to RVLM, compared to that from the NTS to CVLM [53]. It is highly probable that there exists a DOPAergic pressor relay from the PHN to RVLM [37]. Electrolytic lesion of the bilateral PHN selectively decreases by 50% the tissue content of DOPA in the right RVLM without decreases in catecholamines. During microdialysis of the right RVLM, electrical right PHN stimulation releases DOPA without increases in catecholamines, accompanied with hypertension and tachycardia. TTX microinjected ipsilaterally suppresses DOPA release, but partially inhibits hypertension with only slight inhibition of tachycardia. This partial inhibition may be explained mainly by a neuronal activity of the pathway through the contralateral RVLM. DOPA ME microinjected bilaterally antagonizes pressor and tachycardiac responses to unilateral PHN stimulation.

8. Further evidence for DOPAergic system in adult spontaneously hypertensive rats (SHR) and age-matched Wistar–Kyoto rats (WKY)

Furthermore, the absolute basal DOPA release is lower in the NTS [59] and CVLM [33] of adult SHR, compared to age-matched WKY. TTX reduces basal DOPA to the same levels in the two strains. TTX-sensitive tonic neuronal activity to release basal DOPA, calculated as that before TTX perfusion minus that during the perfusion, is lower in the NTS [59] and is lost in the CVLM [33] of SHR, compared to WKY. Meanwhile, basal DOPA release is higher in the RVLM of SHR, compared to WKY [59]. TTX-sensitive tonic neuronal activity to release basal DOPA in the RVLM is enhanced in SHR, compared to WKY. Additional components in the RVLM are a decrease in AADC activity in the dissected area and an increase in the sensitivity of pressor responses to lower dose ranges of DOPA microinjected in SHR, compared to WKY. These alterations may be involved in the maintenance of hypertension in SHR.

9. Is DOPA a neurotransmitter of the primary ADN and is glutamate that of secondary neurons in neuronal microcircuits of the NTS?

DOPA seems to be a neurotransmitter of the primary baroreceptor afferents terminating in the NTS [19,58,59]. Glutamate has been regarded as the most probable neurotransmitter candidate [43–45,50]. Thus, it is important to determine whether or not there are interactions between DOPA and glutamate. DOPA ME does not antagonize hypotension and bradycardia by glutamate microinjected in the NTS [19] and CVLM [55] and hypertension and tachycardia in the RVLM [56]. Kynurenate, a broad-spectrum antagonist for ionotropic glutamate receptors, however, abolishes depressor and bradycardic responses to ADN stimulation, but not those to glutamate microinjected [26]. There has been an idea that an unknown excitatory amino acid, the responses to which are antagonized by kynurenate, is a primary neurotransmitter [26]. Kynurenate prior microinjected (114 ng) into depressor sites of the NTS reduces by 80% depressor responses to 60 ng of DOPA [16], showing a major pathway to be mediated by ionotropic glutamate receptors. Herein, we propose a new pathway that DOPA is a neurotransmitter of the primary ADN and glutamate is that of secondary neurons or interneurons in neuronal microcircuits of the NTS (Fig. 6). Additional possibilities are that DOPA recognition sites are located on or near the terminals of glutamate neurons as the primary ADN and that DOPA activates directly depressor neurons in the NTS. DOPA recognition sites may differ from kynurenate-sensitive ionotropic glutamate receptors because DOPA esters do not displace kynurenate-sensitive binding of [^{3}H] ligands for NMDA glycine site, NMDA binding site and kainate receptor [31]. A following pathway seems to be production of NO, a neuromodulator in the NTS [21], because the responses to DOPA are reduced by 90% stereoselectively by a nNOS inhibitor, L-NMMA (25 μg), and by 70% by nNOS antisense oligos (28 μg). Sense and scrambled oligos elicit no effect. Kynurenate fails to modify depressor responses to a NO precursor, L-arginine (6.3 μg). These findings suggest that DOPA is a neurotransmitter of the primary ADN, DOPA may release unmeasurable but functioning glutamate from the second-order neurons via DOPA recognition sites, and NO production via kynurenate-sensitive postsynaptic ionotropic glutamate receptors is involved in depressor responses to DOPA. Endogenous basal glutamate release is measurable during microdialysis of the NTS. Evoked glutamate release over the basal level, however, is not apparently seen upon ADN and baroreceptor stimulation under the same experimental conditions to DOPA release [26,29,58]. This is consistent with the findings by the other study groups [26]. Based on the assumption that glutamate is a second-order neurotransmitter in neuronal microcircuits of the NTS, the unmeasurable evoked release might be reasonable.

In conclusion, DOPA seems to be a neuromodulator as an upstream causal factor for glutamate release and resultant

Baroreceptors → primary DOPAergic aortic depressor nerve → postsynaptic DOPA recognition sites → glutamate release from secondary glutamatergic neurons → kynurenate-sensitive postsynaptic ionotropic glutamate receptors → nNOS activation → NO production → depressor responses

Fig. 6. A proposed new pathway for DOPAergic neurons as the primary ADN and second-order glutamatergic neurons in neuronal microcircuits of depressor sites in the NTS.

delayed neuron death by brain ischemia in striata and hippocampal CA1 pyramidal cell layers. DOPA seems to be a neurotransmitter of the primary baroreceptor afferents terminating in the NTS and baroreflex pathways in the CVLM and RVLM. Herein, we propose that DOPA is a neurotransmitter of the primary ADN and glutamate is that of second-order neurons in neuronal microcircuits of depressor sites in the NTS. A following common pathway seems to be glutamate release–ionotropic glutamatergic receptors–nNOS activation–NO production–delayed neuron death and baroreflex neurotransmission. However, we are concerned that DOPA therapy may accelerate neuronal degeneration process especially at progressive stages of Parkinson's disease.

Acknowledgments

This study was mainly supported by Grants-in-Aid for Developmental Scientific Research (no. 06557143), Scientific Research (nos. 61480119, 03454146, 07407003, 09877022, 09280280, 10176229, 10470026) from the Ministry of Education, Science, Sports and Culture, Japan, and by grants from The Mitsubishi Foundation, Japan, and SRF, Japan. Y.M. and Y.G. were supported, respectively, twice by The Uehara Memorial Foundation, Japan.

References

[1] N. Arai, N. Furukawa, T. Miyamae, Y. Goshima, Y. Sasaki, E. Ohshima, F. Suzuki, K. Fujita, Y. Misu, DOPA cyclohexyl ester, a competitive antagonist, protects glutamate release and resultant delayed neuron death by transient ischemia in hippocampus CA1 of conscious rats, Neurosci. Lett. 299 (2001) 213–216.

[2] A. Carlsson, M. Lindqvist, T. Magnusson, 3,4-Dihydroxyphenylalanine and 5-hydroxytryptophan as reserpine antagonists, Nature 180 (1957) 1200.

[3] N.-N. Cheng, T. Maeda, T. Kume, S. Kaneko, H. Kochiyama, A. Akaike, H. Kochiyama, Y. Goshima, Y. Misu, Differential neurotoxicity induced by L-DOPA and dopamine in cultured striatal neurons, Brain Res. 743 (1996) 278–283.

[4] D.W. Choi, Glutamate neurotoxicity and diseases of the nervous system, Neuron 1 (1988) 623–634.

[5] V.L. Dawson, T.M. Dawson, D.A. Bartley, G.R. Uhl, S.H. Snyder, Mechanisms of nitric oxide-mediated neurotoxicity in primary brain cultures, J. Neurosci. 13 (1993) 2651–2661.

[6] S. Fahn, Is levodopa toxic? Neurology 47 (Suppl. 3) (1996) S184–S195.

[7] N. Furukawa, N. Arai, Y. Goshima, T. Miyamae, E. Ohshima, F. Suzuki, K. Fujita, Y. Misu, Endogenously released DOPA is a causal factor for glutamate release and resultant delayed neuronal cell death by transient ischemia in rat striata, J. Neurochem. 76 (2001) 815–824.

[8] N. Furukawa, Y. Goshima, T. Miyamae, Y. Sugiyama, M. Shimizu, E. Ohshima, F. Suzuki, N. Arai, K. Fujita, Y. Misu, L-DOPA cyclohexyl ester is a novel potent and relatively stable competitive antagonist against L-DOPA among several L-DOPA ester compounds in rats, Jpn. J. Pharmacol. 82 (2000) 40–47.

[9] R. Gill, A.C. Foster, G.N. Woodruff, Systemic administration of MK-801 protects against ischemia-induced hippocampal neurodegeneration in the gerbil, J. Neurosci. 7 (1987) 3343–3349.

[10] M.Y.-T. Globus, R. Busto, W.D. Dietrich, E. Martinez, I. Valdes, M.D. Ginsberg, Effect of ischemia on the in vivo release of striatal dopamine, glutamate, and γ-aminobutyric acid studied by intracerebral microdialysis, J. Neurochem. 51 (1988) 1455–1464.

[11] Y. Goshima, T. Kubo, Y. Misu, Biphasic actions of L-DOPA on the release of endogenous noradrenaline and dopamine from rat hypothalamic slices, Br. J. Pharmacol. 89 (1986) 229–234.

[12] Y. Goshima, T. Kubo, Y. Misu, Transmitter-like release of endogenous 3,4-dihydroxyphenylalanine from rat striatal slices, J. Neurochem. 50 (1988) 1725–1730.

[13] Y. Goshima, S. Nakamura, Y. Misu, L-Dihydroxyphenylalanine methyl ester is a competitive antagonist of the L-dihydroxyphenylalanine-induced facilitation of the evoked release of endogenous norepinephrine from rat hypothalamic slices, J. Pharmacol. Exp. Ther. 258 (1991) 466–471.

[14] Y. Goshima, S. Nakamura, K. Ohno, Y. Misu, Picomolar concentrations of L-DOPA stereoselectively potentiate activities of presynaptic β-adrenoceptors to facilitate the release of endogenous noradrenaline from rat hypothalamic slices, Neurosci. Lett. 129 (1991) 214–216.

[15] Y. Goshima, K. Ohno, S. Nakamura, T. Miyamae, Y. Misu, A. Akaike, L-DOPA induces Ca^{2+}-dependent and tetrodotoxin-sensitive release of endogenous glutamate from rat striatal slices, Brain Res. 617 (1993) 167–170.

[16] Y. Goshima, K. Yamanashi, T. Miyamae, Y. Sasaki, M. Maeda, H. Hirano, Y. Misu, Involvement of nitric oxide production via ionotropic glutamate receptor in DOPA-induced depressor responses in the nucleus tractus solitarii of anesthetized rats, Abstracts, The 9th International Catecholamine Symposium, 2001, p. 22.

[17] J.T. Greenamyre, Glutamate–dopamine interactions in the basal ganglia: Relationship to Parkinson's disease, J. Neural Transm. 91 (1993) 255–269.

[18] T. Kirino, Delayed neuronal death in the gerbil hippocampus following ischemia, Brain Res. 239 (1982) 57–69.

[19] T. Kubo, J.-L. Yue, Y. Goshima, S. Nakamura, Y. Misu, Evidence for L-DOPA systems responsible for cardiovascular control in the nucleus tractus solitarii of the rat, Neurosci. Lett. 140 (1992) 153–156.

[20] S.A. Lipton, P.A. Rosenberg, Excitatory amino acids as a final common pathway for neurologic disorders, N. Engl. J. Med. 330 (1994) 613–622.

[21] M. Maeda, M. Inoue, S. Takao, M. Nakai, Central control mechanisms of circulation in the medulla oblongata by nitric oxide, Jpn. J. Physiol. 49 (1999) 467–478.

[22] T. Maeda, N.-N. Cheng, T. Kume, S. Kaneko, H. Kouchiyama, A. Akaike, M. Ueda, M. Satoh, Y. Goshima, Y. Misu, L-DOPA neurotoxicity is mediated by glutamate release in cultured rat striatal neurons, Brain Res. 771 (1997) 159–162.

[23] Y. Misu, Y. Goshima, Is L-DOPA an endogenous neurotransmitter? Trends Pharmacol. Sci. 14 (1993) 119–123.

[24] Y. Misu, Y. Goshima, T. Kubo, Biphasic actions of L-DOPA on the release of endogenous dopamine via presynaptic receptors in rat striatal slices, Neurosci. Lett. 72 (1986) 194–198.

[25] Y. Misu, Y. Goshima, T. Miyamae, N. Furukawa, Y. Sugiyama, Y. Okumura, M. Shimizu, E. Ohshima, F. Suzuki, L-DOPA cyclohexyl ester is a novel stable and potent competitive antagonist against L-DOPA, as compared to L-DOPA methyl ester, Jpn. J. Pharmacol. 75 (1997) 307–309.

[26] Y. Misu, Y. Goshima, H. Ueda, H. Okamura, Neurobiology of L-DOPAergic systems, Prog. Neurobiol. 49 (1996) 415–454.

[27] Y. Misu, T. Kubo, Presynaptic β-adrenoceptors, Med. Res. Rev. 6 (1986) 197–225.

[28] Y. Misu, H. Ueda, Y. Goshima, Neurotransmitter-like actions of L-DOPA, Adv. Pharmacol. 32 (1995) 427–459.

[29] Y. Misu, J.-L. Yue, Y. Okumura, T. Miyamae, M. Nishihama, F. Okumura, Endogenous L-DOPA but not glutamate or GABA is released by aortic nerve and baroreceptor stimulation in the nucleus tractus solitarii (NTS) of rats, Jpn. J. Pharmacol. 64 (Suppl. I) (1994) 342P.

[30] M. Miura, D.J. Reis, Termination and secondary projections of carotid sinus nerve in the cat brain stem, Am. J. Physiol. 217 (1969) 142–153.

[31] T. Miyamae, Y. Goshima, M. Shimizu, T. Shibata, K. Kawashima, E. Ohshima, F. Suzuki, Y. Misu, Some interactions of L-DOPA and its related compounds with glutamate receptors, Life Sci. 64 (1999) 1045–1054.

[32] T. Miyamae, Y. Goshima, J.-L. Yue, Y. Misu, L-DOPAergic components in the caudal ventrolateral medulla in baroreflex neurotransmission, Neuroscience 92 (1999) 137–149.

[33] T. Miyamae, J.-L. Yue, Y. Okumura, Y. Goshima, Y. Misu, Loss of tonic neuronal activity to release L-DOPA in the caudal ventrolateral medulla of spontaneously hypertensive rats, Neurosci. Lett. 198 (1995) 37–40.

[34] S. Nakamura, Y. Goshima, J.-L. Yue, Y. Misu, Transmitter-like basal and K^+-evoked release of 3,4-dihydroxyphenylalanine from the striatum in conscious rats studied by microdialysis, J. Neurochem. 58 (1992) 270–275.

[35] S. Nakamura, J.-L. Yue, Y. Goshima, T. Miyamae, Y. Misu, Non-effective dose of exogenously applied L-DOPA itself stereoselectively potentiates postsynaptic D_2 receptor-mediated locomotor activities of conscious rats, Neurosci. Lett. 170 (1994) 22–26.

[36] T.A. Newcomer, P.A. Rosenberg, E. Aizenman, TOPA quinone, a kainate-like agonist and excitotoxin, is generated by a catecholaminergic cell line, J. Neurosci. 15 (1995) 3172–3177.

[37] M. Nishihama, T. Miyamae, Y. Goshima, F. Okumura, Y. Misu, An L-DOPAergic relay from the posterior hypothalamic nucleus to the rostral ventrolateral medulla and its cardiovascular function in anesthetized rats, Neuroscience 92 (1999) 123–135.

[38] T.P. Obrenovitch, D.A. Richards, Extracellular neurotransmitter changes in cerebral ischaemia, Cerebrovasc. Brain Metab. Rev. 7 (1995) 1–54.

[39] J.W. Olney, L.G. Sharpe, Brain lesions in an infant rhesus monkey treated with monosodium glutamate, Science 166 (1969) 386–388.

[40] R. Prado, R. Busto, M.Y.-T. Globus, Ischemia-induced changes in extracellular levels of striatal cyclic AMP: Role of dopamine neurotransmission, J. Neurochem. 59 (1992) 1581–1584.

[41] W.A. Pulsinelli, Pathophysiology of acute ischaemic stroke, Lancet 339 (1992) 533–536.

[42] W.A. Pulsinelli, J.B. Brierley, F. Plum, Temporal profile of neuronal damage in a model of transient forebrain ischemia, Ann. Neurol. 11 (1982) 491–498.

[43] D.J. Reis, A.R. Granata, M.H. Perrone, W.T. Talman, Evidence that glutamic acid is the neurotransmitter of baroreceptor afferent terminating in the nucleus tractus solitarius (NTS), J. Auton. Nerv. Syst. 3 (1981) 321–344.

[44] D.J. Reis, D.A. Ruggiero, S.F. Morrison, The C1 area of the rostral ventrolateral medulla oblongata, Am. J. Hypertens. 2 (1989) 363S–374S.

[45] C.A. Ross, D.A. Ruggiero, D.J. Reis, Projections from the nucleus tractus solitarii to the rostral ventrolateral medulla, J. Comp. Neurol. 242 (1985) 511–534.

[46] R. Schmidt-Kastner, T.F. Freund, Selective vulnerability of the hippocampus in brain ischemia, Neuroscience 40 (1991) 599–636.

[47] M.J. Sheardown, E.Ø. Nielsen, A.J. Hansen, P. Jacobsen, T. Honoré, 2,3-Dihydroxy-6-nitro-7-sulfamoyl-benzo(F)quinoxaline: A neuroprotectant for cerebral ischemia, Science 247 (1990) 571–574.

[48] B.K. Siesjö, F. Bengtsson, Calcium fluxes, calcium antagonists, and calcium-related pathology in brain ischemia, hypoglycemia, and spreading depression: A unifying hypothesis, J. Cereb. Blood Flow Metab. 9 (1989) 127–140.

[49] M.-K. Sun, Pharmacology of reticulospinal vasomotor neurons in cardiovascular regulation, Pharmacol. Rev. 48 (1996) 465–494.

[50] W.T. Talman, M.H. Perrone, D.J. Reis, Evidence for L-glutamate as the neurotransmitter of baroreceptor afferent nerve fibers, Science 209 (1980) 813–815.

[51] F. Tison, N. Mons, S. Rouet-Kamara, M. Geffard, P. Henry, Endogenous L-DOPA in the rat dorsal vagal complex: An immunocytochemical study by light and electron microscopy, Brain Res. 497 (1989) 260–270.

[52] H. Ueda, K. Sato, F. Okumura, Y. Misu, L-DOPA inhibits spontaneous acetylcholine release from the striatum of experimental Parkinson's model rats, Brain Res. 698 (1995) 213–216.

[53] R.W. Urbansky, H.N. Sapru, Evidence for a sympathoexcitatory pathway from the nucleus tractus solitarii to the ventrolateral medullary pressor area, J. Auton. Nerv. Syst. 23 (1988) 161–174.

[54] R.P. Vertes, A.M. Crane, Descending projections of the posterior nucleus of the hypothalamus: Phaseolus vulgaris leucoagglutinin analysis in the rat, J. Comp. Neurol. 354 (1996) 607–631.

[55] J.-L. Yue, Y. Goshima, Y. Misu, Transmitter-like L-3,4-dihydroxyphenylalanine tonically functions to mediate vasodepressor control in the caudal ventrolateral medulla of rats, Neurosci. Lett. 159 (1993) 103–106.

[56] J.-L. Yue, Y. Goshima, T. Miyamae, Y. Misu, Evidence for L-DOPA relevant to modulation of sympathetic activity in the rostral ventrolateral medulla of rats, Brain Res. 629 (1993) 310–314.

[57] J.-L. Yue, S. Nakamura, H. Ueda, Y. Misu, Endogenously released L-DOPA itself tonically functions to potentiate D_2 receptor-mediated locomotor activities of conscious rats, Neurosci. Lett. 170 (1994) 107–110.

[58] J.-L. Yue, H. Okamura, Y. Goshima, S. Nakamura, M. Geffard, Y. Misu, Baroreceptor–aortic nerve-mediated release of endogenous L-3,4-dihydroxyphenylalanine and its tonic function in the nucleus tractus solitarii of rats, Neuroscience 62 (1994) 145–161.

[59] J.-L. Yue, Y. Okumura, T. Miyamae, H. Ueda, Y. Misu, Altered tonic L-3,4-dihydroxyphenylalanine systems in the nucleus tractus solitarii and the rostral ventrolateral medulla of spontaneously hypertensive rats, Neuroscience 67 (1995) 95–106.

[60] L. Zhang, A. Mitani, H. Yanase, K. Kataoka, Continuous monitoring and regulating of brain temperature in the conscious and freely moving ischemic gerbil: Effect of MK-801 on delayed neuronal death in hippocampal CA1, J. Neurosci. Res. 47 (1997) 440–448.

NEUROTOXICOLOGY

AND

TERATOLOGY

Neurotoxicology and Teratology 24 (2002) 639–653

www.elsevier.com/locate/neutera

Review article

Manganese-induced apoptosis in PC12 cells

Yoko Hirata*

Laboratory for Genes of Motor Systems, Bio-Mimetic Control Research Center, RIKEN, 2271-130, Anagahora, Shimoshidami, Moriyama, Nagoya 463-0003, Japan
Department of Biomolecular Science, Faculty of Engineering, Gifu University, Gifu 501-1193, Japan

Received 15 July 2001; accepted 24 January 2002

Abstract

Manganese has been known to induce neurological disorders similar to parkinsonisms for a long time. Dopamine deficiency has been demonstrated in Parkinson's disease and in chronic manganese poisoning, suggesting that the mechanisms underlying the neurotoxic effects of the metal ion are related to dysfunction of the extrapyramidal system. However, the details of the mechanisms have yet to be elucidated. In an effort to learn more about the toxicity of manganese, we have employed an in vitro model that uses the PC12 catecholaminergic cell line. In this model, manganese induces apoptosis in PC12 cells. In this paper, experiments conducted with this model, the cellular biochemical changes, and the mechanism of the cell death are reviewed. © 2002 Elsevier Science Inc. All rights reserved.

Keywords: Apoptosis; Manganese; Parkinson's disease; JNK; MAPK; p38 MAPK; p70 S6 kinase; Bcl-2; Caspase

1. Introduction

The hazards of manganese have been known for decades. Early reports suggested functional disabilities resulting from chronic manganese poisoning closely resembled the extrapyramidal signs and symptoms of idiopathic parkinsonism [6–8,29]. The critical pathological process in Parkinson's disease is the selective degeneration of nigrostriatal dopaminergic neurons, which results in depletion of striatal dopamine [14]. Several lines of evidence suggest that manganese is a potential dopaminergic neurotoxin in vivo and in vitro. First, dopamine is decreased in the striatum of

monkeys intoxicated with manganese [41], and the injection of manganese directly into the rodent striatum reduces the concentration of dopamine and impairs oxidative metabolism [5,55]. Clues about the locus of manganese's action were revealed when manganese was recently found to inhibit tyrosine hydroxylation, the rate-limiting step in dopamine synthesis, in rat striatal slices (Hirata et al., submitted). Finally, these results also suggest that the neurotoxic effect of manganese is similar to that of MPTP, a well-known dopaminergic neurotoxin. On the other hand, significant differences between parkinsonism and manganese poisoning in the response to L-DOPA, their clinical features, and neuropathological changes have been described [6], and despite numerous studies in the past, the mechanisms by which this metal injures the basal ganglia remain unclear.

New hypotheses have been advanced recently about the proximate causes of neurodegenerative diseases. It has been suggested that apoptosis is involved not only in physiological cell death during normal development but also in neurodegenerative diseases [34,44,47]. Furthermore, several studies indicate that apoptosis may play a role in the dopaminergic neurotoxicity associated with manganese and MPTP [11,26,40,58,63]. Open questions remain, however, whether the pathways that lead to cell death in Parkinson's disease are the same ones involved in the striatal cell death associated with manganese toxicity.

Abbreviations: ATA, aurin tricarboxylic acid; ATF, activating transcription factor; CREB, cyclic-AMP response element binding protein; DMEM, Dulbecco's modified Eagle's medium; ECL, enhanced chemiluminescence; ERK, extracellular signal-regulated kinase; JNK, c-Jun N-terminal kinase; LDH, lactate dehydrogenase (EC 1.1.1.27); MAPK, mitogen-activated protein kinase; MTT, 3-(4,5-dimethylthiazol-2-yl)-2,5-diphenyl tetrasolium bromide; NGF, nerve growth factor; PBS, phosphate-buffered saline; SAPK, stress-activated protein kinase; SDS-PAGE, sodium dodecyl sulfate–polyacrylamide gel electrophoresis; TUNEL, terminal deoxynucleotidyl transferase-mediated dUTP-biotin nick end labeling.

* Department of Biomolecular Science, Faculty of Engineering, Gifu University, Gifu 501-1193, Japan. Tel.: +81-58-293-2601; fax: +81-58-230-1893.

E-mail address: yoko@biomol.gifu-u.ac.jp (Y. Hirata).

In this review, we describe a PC12 cell model system designed to characterize neurotoxicity following acute exposure to manganese. PC12 is a clonal cell line of rat pheochromocytoma cells that respond to nerve growth factor (NGF) by extending neurites, thus acquiring the appearance of neurons [22]. Both NGF-treated and untreated cells synthesize, store, secrete, and take up dopamine by processes that are similar to those of dopaminergic neurons [23]. Using catecholaminergic PC12 cells, we recently reported that manganese induced characteristic internucleosomal DNA fragmentation, a biochemical hallmark of apoptosis [26]. It was relatively specific for manganese and blocked in Bcl-2 overexpressed PC12 cells, and mediated by activation of the caspase family of proteases. The results indicate that apoptosis may play a role in the dopaminergic neurotoxicity associated with manganese, the first metal to be reported to induce this form of cell death. Mitochondrial oxidative phosphorylation without ATP depletion is impaired early in the cascade of biochemical events that leads to cell death, and the process may require new synthesis of proteins such as c-Fos and c-Jun. In addition, manganese induces various signaling cascades including c-Jun N-terminal kinase (JNK), p38 mitogen-activated protein kinase (MAPK), extracellular signal-regulated kinase (ERK), cyclic-AMP response element binding protein (CREB)/activating transcription factor (ATF)-1, and p70 S6 kinase, and tyrosine phosphorylation of several proteins, although contributions of these pathways to apoptosis remain to be elucidated.

2. Materials and methods

2.1. Cell culture

PC12 cells were grown in Dulbecco's modified Eagle's medium (DMEM) with 7% horse serum and 4% fetal bovine serum. Bcl-2-transfected PC12 cells were kindly supplied by Drs. Yutaka Eguchi and Yoshihide Tsujimoto, Osaka University.

2.2. Analysis of DNA fragmentation

PC12 cells ($\sim 2 \times 10^7$ cells) were incubated at 37 °C for the indicated periods and then adherent cells were harvested. Soluble DNA was isolated and extracted using the method of Hockenbery et al. [28]. Approximately half of the recovered soluble DNA per condition was separated by electrophoresis in 1.2% agarose gels and visualized with an ultraviolet transilluminator. To increase the sensitivity for detecting fragmented DNA, Southern blot analysis was performed. Blots were probed with digoxigenin (DIG)-labeled total genomic PC12 cell DNA digested with Sau 3A and chemiluminescent detection was performed using CSPD as a substrate according to the manufacturer's protocol (Roche Molecular Biochemicals).

2.3. Lactate dehydrogenase (LDH) assay

Cytotoxicity was determined by the measurement of LDH released into the supernatant medium using the method of Storrie and Madden [60].

2.4. MTT assay

[3-(4,5-Dimethylthiazol-2-yl)-2,5-diphenyl tetrasolium bromide] (MTT) assay was performed with a kit according to the manufacturer's protocol (Chemicon International). PC12 cells ($\sim 3 \times 10^3$ cells/well) were cultured in 96-well culture plates for 3 days in 100 μl of medium per well. After the treatment of cells under various experimental conditions (see below), 10 μl of MTT solution was added to each well, and the plates were incubated at 37 °C for another 4 h in a CO_2 incubator. The reaction was stopped by adding 100 μl of 0.04 N HCl/isopropanol, and the results were quantified by measuring the absorbance at 570 nm with 620 nm as a reference.

2.5. Terminal deoxynucleotidyl transferase-mediated dUTP-biotin nick end-labeling (TUNEL) method

PC12 cells were cultured on polylysine-coated chamber slides, washed with phosphate-buffered saline (PBS), and fixed with 4% paraformaldehyde in PBS for 25 min at 4 °C. The cells were then permeabilized with 0.2% Triton X-100 in PBS for 5 min at 4 °C, and incubated with terminal deoxynucleotidyl transferase and biotinylated dUTP. The reaction product was visualized with avidin/biotin-conjugated horseradish peroxidase according to the method of Gavrieli et al. [20].

2.6. Lactate and ATP assay

PC12 cells were homogenized with 50 mM sodium phosphate buffer (pH 6.5) and boiled for 15 min. The lactate content was measured spectrophotometrically at 340 nm using LDH and 0.4 M hydrazine/0.5 M glycine buffer (pH 9.0). The amount of ATP was measured by the firefly luciferin–luciferase assay [61] using a Turner Designs Luminometer TD-20/20. Protein concentration was measured using the DC Protein Assay (Bio-Rad Laboratories).

2.7. Western blotting

Cells were lysed in $2\times$ sodium dodecyl sulfate–polyacrylamide gel electrophoresis (SDS-PAGE) sample buffer (62.5 mM Tris–HCl, pH 6.8, 2% SDS, 10% glycerol) and sonicated for ~ 20 s. Total cell lysates (approximately 40 μg protein) were separated by SDS-PAGE and transferred onto nitrocellulose membranes (Amersham Pharmacia Biotech). Immunoblotting was performed with the appropriate antibody using the enhanced chemiluminescence (ECL) system (Amersham Pharmacia Biotech) for detection and visualization. In some cases, blots were reprobed with different antibodies after stripping in 62.5 mM Tris–HCl (pH 6.7)/ 100 mM β-mercaptoethanol/2% SDS.

2.8. Labeling and visualization of biotinylated caspase inhibitor-labeled proteins

To identify the p17-kDa subunit of an activated member of the caspase family, we applied an affinity-labeling technique developed by Srinivasan et al. [59] using a biotinylated caspase inhibitor that reacts covalently with the catalytic site cysteine. Cell extraction and labeling were done as described [59]. PC12 cells were lysed in cell extraction buffer [100 mM HEPES (pH 7.5), 1% Triton X-100, 10 mM DTT, 1 mM PMSF, 5 μg/ml leupeptin, 1 μg/ ml pepstatin A, 1 mM EDTA] and centrifuged at $14,000 \times g$ for 10 min. The protein concentration of the supernatant was measured by DC Protein Assay (Bio-Rad Laboratories). In a 50-μl volume reaction, cell extracts (approximately 0.3 mg protein) were incubated with 2 μM of biotin-X-VAD-fmk, a biotinylated caspase inhibitor (Calbiochem) for 20 min at 25 °C. Reactions were stopped by adding 50 μl of $2 \times$ SDS-PAGE sample buffer. Biotinylated caspase inhibitor-labeled cell extracts were resolved on SDS-PAGE (15%) and transferred onto Hybond-PVDF membranes (Amersham Pharmacia Biotech). The membranes were blocked with 0.05% Tween 20/phosphate-buffered saline (T-PBS) containing 5% milk for 1 h, washed with T-PBS twice, and incubated with streptavidin–HRP (1:1500 in T-PBS) for 1 h.

The blots were washed in T-PBS three times for 15 min and visualized using the ECL system.

3. Results

3.1. Manganese-induced DNA fragmentation in PC12 cells

To examine whether manganese causes apoptosis in catecholaminergic PC12 cells, the cells were treated with manganese chloride (MnCl₂) in complete growth medium for various periods, and soluble DNA was then extracted and analyzed. Manganese induced internucleosomal DNA fragmentation, a characteristic marker for apoptosis, in a time- and concentration-dependent manner (Fig. 1A). To increase the detection sensitivity of fragmented DNA, Southern blotting using the DIG chemiluminescent detection system was performed (Fig. 1A). The degree of DNA fragmentation was much more evident in the Southern blots compared to that in gels stained with ethidium bromide. This increased sensitivity is especially notable in lanes showing DNA fragmentation after 16 h of 1 mM MnCl₂ treatment. In contrast to data shown in the Southern blot at this time point, very few DNA fragments were observed in the ethidium bromide-stained gel (Fig. 1A). DNA fragmentation is detectable by Southern blot analysis as early as 3 h after serum deprivation [2]. This indicates that the time course of appearance of DNA ladders in manganese-treated PC12 cells is much slower than that in

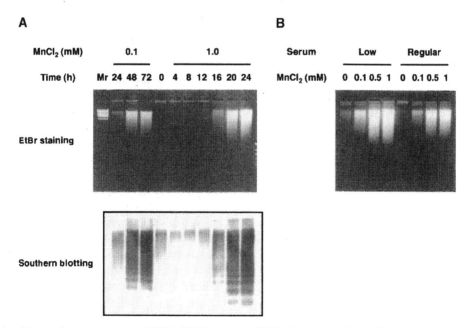

Fig. 1. Manganese induced internucleosomal cleavage of DNA. (A) Time course of DNA fragmentation induced by manganese. PC12 cells were cultured in either the absence or presence of 0.1 or 1 mM MnCl₂ for the indicated periods. Soluble DNA was analyzed by agarose gel electrophoresis and visualized under UV light after ethidium bromide (EtBr) staining or by Southern blotting using DIG-labeled total genomic PC12 cell DNA digested with Sau 3A and chemiluminescent detection as described in Materials and methods. Mr; HindIII digests of λ-DNA. (B) Effects of various concentrations of MnCl₂ on DNA fragmentation in the presence of low serum (1% horse serum, 1% fetal bovine serum) and regular serum (7% horse serum, 4% fetal bovine serum). PC12 cells were incubated with the indicated concentrations of MnCl₂ for 20 h. Soluble DNA was isolated and analyzed using agarose gel electrophoresis and visualized under UV light after EtBr staining as described in Materials and methods. Comparable results were obtained in several independent experiments.

serum-starved PC12 cells. However, the concentration of Mn^{2+} that induces DNA laddering is quite high to be of physiological significance.

To examine the influence of plasma proteins on manganese-induced apoptosis, PC12 cells were incubated under low (1% horse serum plus 1% fetal bovine serum) and regular serum conditions (7% horse serum plus 4% fetal bovine serum). These serum concentrations were used because concentrations lower than 1% induce DNA fragmentation (unpublished data). The degree of manganese-induced DNA fragmentation was more marked under low serum conditions compared to that under regular serum conditions (Fig. 1B). Because manganese is capable of binding plasma proteins such as albumin, transferrin, and α_2-macroglobulin [48], the effective concentration could be lower than 0.1 mM under experimental conditions.

The effects of manganese on cell death as assessed by LDH release is shown in Fig. 2. LDH was not detected following 20 h of manganese treatment, the time at which DNA fragmentation was already evident, in contrast to the intracellular O_2^- generator, paraquat [46], which was used as a positive control. Cell viability assessed using Trypan blue exclusion was also similar following 20 h of treatment (percent viable cells; control, 97%; $MnCl_2$, 98%), indicating that membrane integrity is preserved at this point. These results demonstrate that manganese-induced internucleosomal cleavage of DNA precedes cell death.

The effects of manganese and MPP^+ on cell viability as assessed using the MTT assay are summarized in Fig. 3. Manganese-induced cell death was observed at concentrations of 0.1 and 1 mM but not at 0.01 mM. The concentrations that caused cell death also produced DNA fragmentation, suggesting that fragmentation is closely correlated with subsequent cell death. Interestingly, while 1 mM MPP^+ induced slight DNA fragmentation (data not shown) and inhibited cell proliferation, it did not induce cell death under these experimental conditions (Fig. 3), indicating that PC12 cells are relatively resistant to MPP^+. Incidentally, the morphology of PC12 cells changed after manganese treatment with the appearance of short processes (data not shown), as already reported by Lin et al. [36].

DNA breaks in situ were confirmed by the TUNEL method, a relatively specific technique used for the in situ labeling of apoptotic cells [20]. Characteristic DNA changes were much more common in PC12 cells treated with 1 mM $MnCl_2$ for 20 h than in control cells not receiving treatment (Fig. 4). This result indicates that most of the cells are already undergoing apoptosis 20 h after initial exposure to manganese.

Various divalent cations were tested to determine whether other divalent cations also induce apoptosis or whether manganese-induced apoptosis is an ion-specific phenomenon. The Southern blots of Fig. 5 show that neither calcium, magnesium, or ferrous ions (1 mM for 22 h) nor copper, nickel, or zinc ions (0.1 mM for 46 h) induced DNA fragmentation in PC12 cells. At a concentration of 1 mM, both copper and zinc ions were severely toxic, and only 1 mM nickel ion mimicked the effect of manganese (data not shown). These results suggest that manganese-induced DNA fragmentation is quite specific, and is not a general action of divalent cations.

To determine at what time after the addition of manganese the cells are irreversibly committed to undergo apoptosis, PC12 cells incubated with manganese for 12, 14, or 16 h were washed with fresh medium, and then incubated further in regular medium for a combined total incubation of 20 h. Soluble DNA was analyzed at the 20-h time point. DNA laddering was evident after only 12 h of incubation, indicating that irreversible changes leading to DNA fragmentation must occur prior to this time point (Fig. 6).

NGF and aurin tricarboxylic acid (ATA), a general nuclease inhibitor, were reported to prevent PC12 cell death caused by serum starvation [1]. Experiments in our laboratory, however, showed that ATA (100 μM) did not inhibit or reduce manganese-induced DNA fragmentation, and NGF (50–150 ng/ml) only partially reduced DNA fragmentation when added to treatment schedules of 1 mM $MnCl_2$ for 16 and 20 h (data not shown). These findings indicate that the mechanism involved in manganese-induced apoptosis is different from that involved in serum withdrawal-induced apoptosis.

3.2. Manganese toxicity leads to dysfunction of mitochondria

Abnormalities in the mitochondrial respiratory chain have been found in Parkinson's disease and in the MPTP model

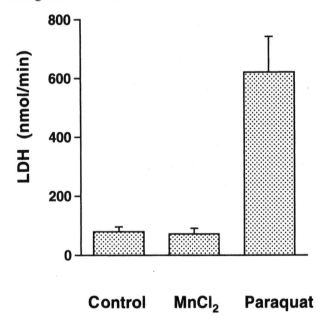

Fig. 2. PC12 cells remain viable after 20 h of $MnCl_2$ treatment. Treatment with paraquat, an intracellular free radical generator, served as a positive control. Cytotoxicity was assessed by measuring the amount of LDH released into the cell culture medium. PC12 cells (8.5×10^5/well) in 24-well plates were treated with either $MnCl_2$ (1 mM) or paraquat (0.15 mM) for 20 h, and the culture media were collected and then analyzed for LDH activity [60]. Data are means ± S.D. ($n = 4$).

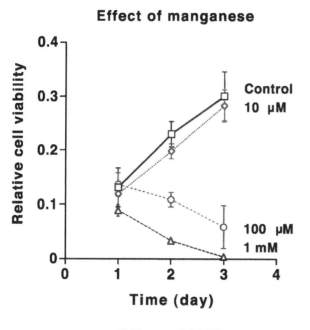

Effect of manganese

Effect of MPP+

Fig. 3. Delayed death of PC12 cells following persistent MnCl₂ treatment and relative resistance to MPP⁺, the active MPTP metabolite. PC12 cells were cultured in 96-well plates for the indicated times with various concentrations of MnCl₂ or MPP⁺ (squares, control; diamonds, 10 μM; circles, 100 μM; triangles, 1 mM). Cell viability was assessed using MTT assays that were performed according to the manufacturer's protocol (Chemicon International). Data are means ± S.D. (n = 9).

[63]. Increased lactate formation and decreased ATP synthesis are indicative of mitochondrial damage. The accumulation of lactate, for example, results from the inhibition of mitochondrial oxidation and/or subsequent stimulation of anaerobic glycolysis [65]. Thus, to determine whether manganese alters mitochondrial function in PC12 cells, intracellular lactic acid and ATP concentrations were measured

treatment, lactic acid concentration increased over twofold compared to control levels (Fig. 7A). At the 8-h time point, lactic acid levels increased fourfold and reached a plateau (Fig. 7A). Intracellular ATP levels, in contrast, were not affected by manganese treatment (Fig. 7B). These findings indicate that while manganese disrupts mitochondrial function by possibly affecting oxidative processes, managanese does not affect pathways involved in ATP production (e.g., glycolysis). The lack of ATP accumulation in our model also suggests that, in PC12 cells, the apoptotic program driven by manganese involves an energy-requiring step.

The Bcl-2 family of proteins localizes in multiple membrane compartments, including the mitochondria, and acts as anti-apoptotic agents in a variety of physiological and pathological contexts [28]. Apoptosis of PC12 cells resulting from either the removal of serum or growth factors or from an increase in intracellular Ca²⁺ concentration by using calcium ionophore A23187 was inhibited by Bcl-2 overexpression [3,37]. The Bcl-2 family of proteins also regulates the release of apoptogenic cytochrome c by

Control

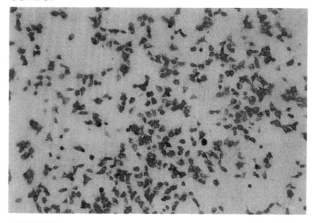

MnCl₂

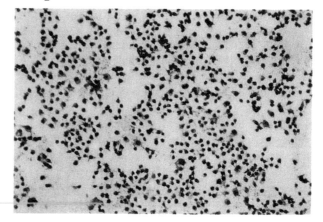

Fig. 4. TUNEL staining of apoptotic PC12 cells. PC12 cells were cultured on polylysine-coated chamber slides and treated with 1 mM MnCl₂ for 20 h

Y. Hirata / Neurotoxicology and Teratology 24 (2002) 639–653

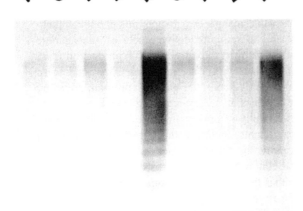

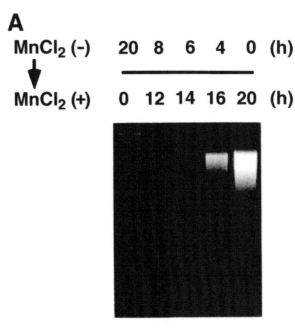

though the effects of these macromolecular synthesis inhibitors were not straightforward, it appears that mRNA and protein synthesis are indeed required for the induction of manganese-associated apoptosis.

Consistent with these findings are data from Western blots showing that manganese induces the expression of c-Fos and

Fig. 5. Specificity of manganese treatment on DNA fragmentation. Other divalent cations failed to induce DNA fragmentation. PC12 cells were cultured in either in the absence (None) or in the presence of 1 mM of CaCl₂, Fe(NH₄)₂(SO₄)₂, MgCl₂, or MnCl₂ for 22 h or 0.1 mM CuCl₂, NiSO₄, Zn(CH₃COO)₂, or MnCl₂ for 46 h. Soluble DNA was analyzed using Southern blotting as described in the Materials and methods. Comparable results were obtained in several independent experiments.

binding to mitochondrial voltage-dependent anion channels [54]. These findings prompted us to test whether Bcl-2 is capable of preventing manganese-induced DNA fragmentation in PC12 cells. Overexpression of Bcl-2 protein in PC12 cells [53] completely prevented DNA fragmentation induced by manganese (Fig. 8A). Moreover, Bcl-2 protected the PC12 cells from manganese-induced cell death (Fig. 8B). In contrast, Bcl-2 did not block the manganese-induced accumulation of lactate in these PC12 cells (Fig. 8C), suggesting that cytochrome *c* release is independent of the inhibition of mitochondrial respiration.

3.3. Manganese-induced DNA fragmentation depends on transcription

One influential hypothesis posits that neural programmed cell death requires a genetic program [18]. Motivated by this, we next examined whether alterations in transcription are involved in manganese-induced apoptosis in PC12 cells. This was addressed by culturing the cells in the presence of both manganese and macromolecular synthesis inhibitors. Actinomycin D, an RNA synthesis inhibitor, effectively blocked manganese-induced DNA fragmentation when applied at concentrations of 5 and 10 μM. Cycloheximide, a protein synthesis inhibitor, also inhibited manganese-induced apoptosis when applied at low concentrations (0.1 μg/ml), but at high concentrations (0.5 and 1.0 μg/ml), cycloheximide itself was toxic, producing

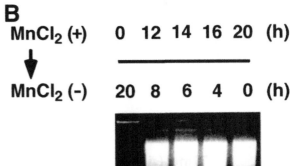

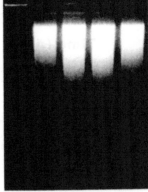

Fig. 6. Time of commitment to apoptosis following treatment with manganese. The fate of PC12 cells is apparently sealed within the first 12 h of exposure. (A) PC12 cells were first incubated in manganese-free medium [MnCl₂ (−)] for either 0, 4, 6, 8, or 20 h, and then incubated with 1 mM manganese [MnCl₂ (+)] for the indicated periods. (B) Reverse experiment to that in (A). At the 20-h time point, soluble DNA was isolated and analyzed using agarose gel electrophoresis and visualized under UV light after EtBr staining as described in Materials and methods. Comparable

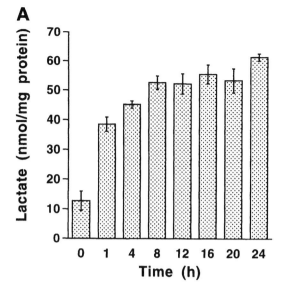

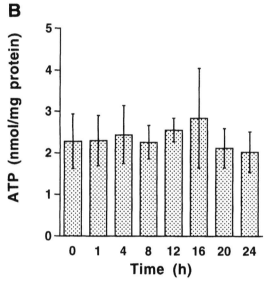

Fig. 7. Manganese impairs mitochondrial oxidative function but not ATP production. Effects were assessed by measuring intracellular lactate (A) and ATP (B) levels. PC12 cells were cultured in the presence of 1 mM $MnCl_2$ for the indicated periods. Lactate and ATP were extracted and measured as described in the Materials and methods. Data are means ± S.D. ($n = 3$).

c-Jun. c-Fos levels were increased 4 h after manganese treatment and thereafter remained elevated throughout the period examined (Fig. 9B). c-Jun was induced over a similar time course (Fig. 10A). Activation of either c-Fos or c-Jun expression appears to be a common element in many instances of mammalian programmed cell death. For example, an association between c-*fos* mRNA expression and genetically determined neurodegeneration has been shown [57], and recent studies using the microinjection of neutralizing antibodies or the expression of dominant negative mutants suggest a critical role for a transcription factor, c-Jun, in apoptosis induced by NGF withdrawal from sympathetic neurons [16,24].

3.4. Activation of various signaling pathways by manganese

3.4.1. JNK/p38 MAPK pathway

Previous studies have demonstrated that several apoptosis-inducing treatments activate JNKs/stress-activated protein kinases (SAPKs) and p38 MAPK. To determine whether these pathways are involved in manganese-induced apoptosis, studies were performed to establish whether manganese stimulates the phosphorylation of JNK/SAPK and p38 MAPK in PC12 cells. Western blot analysis revealed that manganese induces the expression of c-Jun protein as well as the concomitant phosphorylation of c-Jun at Ser63 and Ser73 (Fig. 10A). These results are consistent with the suggestion that manganese activates JNK/SAPK, which in turn phosphorylates Ser63 and Ser73 of the c-Jun transactivation domain [10,32]. The phosphorylated form of SEK1/MKK4 (JNK kinase) appeared over a similar time course as that of c-Jun (Fig. 10A), indicating that manganese activates the JNK pathway. While the phosphorylation of p38 MAPK was also induced by manganese (Fig. 10B), p38 protein levels in general were not significantly affected by manganese treatment. Interestingly, the time course of p38 MAPK activation, while much slower than that of the JNK pathway, was similar to the time course for the appearance of DNA ladders. This suggests that the JNK pathway is involved in early apoptotic events, whereas the p38 MAPK pathway is involved in later events concerned with carrying out the remaining apoptotic program. When PC12 cells are exposed to the p38 MAPK inhibitor, SB203580 (5–50 μM) in the presence of manganese (0.5 mM $MnCl_2$, 20 h), however, DNA fragmentation was not prevented. Rather, SB203580 potentiated the effects of manganese (data not shown). Further study is necessary to clarify the role of p38 MAPK in manganese-induced apoptosis.

3.4.2. ERK pathway

In addition to these death-related signaling cascades, manganese also induces the phosphorylation of both ERK1 and ERK2 (Fig. 10C), two ERKs involved in the activation of the MAPK kinase cascade. The phosphorylation of ERK1 and ERK2 was corroborated by the induction of phospho-CREB, a target molecule downstream from the ERKs in the MAPK kinase pathway. Western blot analyses further revealed that the ERKs phosphorylated ATF-1 (Fig. 10C), a CREB-related transcription factor that shares sequence identity surrounding Ser133, even more strongly than CREB. Manganese did not significantly affect CREB and ATF-1 protein levels in PC12 cells. These results indicate that the active form of ERKs (i.e., phosphorylated ERKs) increases the phosphorylation of CREB and ATF-1. It is surprising that the increased phosphorylation of CREB and ATF-1 lasts for at least 24 h. Taken together with previous findings, these results suggest that the induction of a variety of transcription factors, and persistent modifications by phosphorylation, allow the various combinations of these factors to form heterodimers that drive a program specific for apoptosis.

3.4.3. P70 S6 kinase

Manganese also induces the activation of p70 S6 kinase, a serine/threonine kinase involved in modulating cell cycles [33]. Western blot analysis showed that the mobility of p70

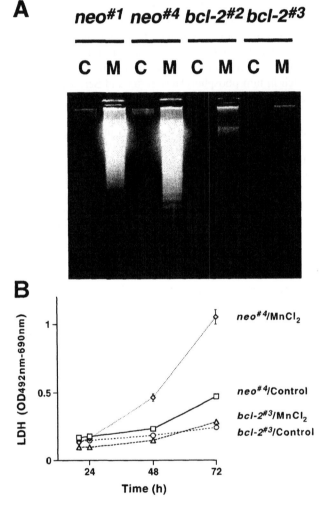

A

neo#1 neo#4 bcl-2#2 bcl-2#3

C M C M C M C M

B

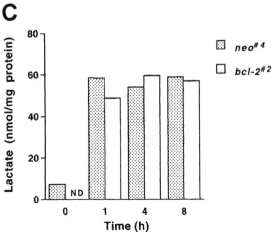

C

S6 kinase was reduced with manganese treatment, suggesting the enzyme is modified posttranslationally. The phosphorylation of p70 S6 kinase at Ser411 and/or Thr421/ Ser424 is shown in a Western blot stained with phospho-specific antibodies (Fig. 10D). Immunocomplex kinase assays revealed that the phosphorylation of p70 S6 kinase is accompanied by elevated enzyme activity [27]. However, rapamycin, a specific p70 S6 kinase inhibitor, inhibited the manganese-induced activation of p70 S6 kinase, but failed to prevent DNA fragmentation (data not shown). While these results indicate that the phosphorylation and subsequent activation of p70 S6 kinase is, to some extent, involved in the signaling pathway underlying manganese-induced apoptosis, the inability of rapamycin to prevent DNA fragmentation suggests that other pathways may also participate in manganese-induced apoptosis.

Next, we compared the apoptosis-inducing effects of manganese to those of various agents that act on different cell signaling pathways. All agents examined caused intensive DNA fragmentation under the conditions we used (Fig. 11). Among these agents, manganese is the most potent activator of c-Jun induction and phosphorylation (Fig. 12A). Although longer exposure to ECL detection system enabled the visualization of weak phospho-c-Jun signals in cell extracts treated with the other agents, only manganese induced the phosphorylation of SEK1/MKK4 (JNK kinase). It is not clear whether these agents induce c-Jun phosphorylation via different pathways, or simply, whether the sensitivity of the immunoblotting technique used is inadequate to detect the phosphorylation of SEK1/MKK4.

Manganese was also found to be a potent activator of p38 MAPK, ERK/CREB/ATF-1, and p70 S6 kinase (Fig. 12B,C,D). Among these, p38 MAPK was phosphorylated in a similar pattern as c-Jun phosphorylation. All of the other agents weakly induced the phosphorylation of p38 MAPK and ERK1 and ERK2, although, again, signals were

Fig. 8. Prophylaxis against manganese-induced toxicity by Bcl-2 proteins. (A) Prevention of manganese-induced DNA fragmentation by Bcl-2 proteins. The vector transfectants of PC12 cells (neo#1, neo#4) and human Bcl-2 overexpressed PC12 cells (bcl-2#2, bcl-2#3) were cultured in the absence (C) or presence (M) of 1 mM MnCl₂ for 24 h. Soluble DNA was analyzed using agarose gel electrophoresis and visualized under UV light after EtBr staining as described in Materials and methods. Comparable results were obtained in several independent experiments. (B) Prevention of manganese-induced cell death by Bcl-2 proteins. Cells (1×10^5/well/400 μl) were cultured in 48-well plates for one day and treated with 1 mM MnCl₂ (diamonds, neo#4/MnCl₂; triangles, bcl-2#3/MnCl₂) or no MnCl₂ (squares, neo#4/control; circles, bcl-2#3/control). At the times indicated, the culture medium (5 μl) was analyzed for LDH activity. Data are means ± S.D. ($n = 4$). LDH activity was measured with the cytotoxicity detection kit according to the manufacturer's protocol (Roche Molecular Biochemicals). (C) Overexpression of Bcl-2 proteins fails to prevent manganese-induced accumulation of lactate. PC12 cells were cultured in the presence of 1 mM MnCl₂ for the indicated periods. Lactate was extracted and measured as described in Materials and methods. Comparable results were obtained in several independent experiments using neo#1, neo#4, bcl-2#2, and bcl-2#3.

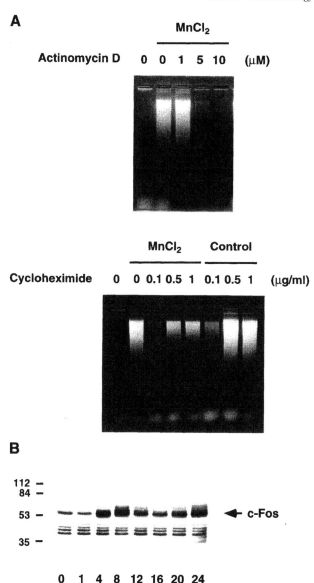

A

B

Fig. 9. Manganese-induced DNA fragmentation depends on transcription. (A) Actinomycin D and cycloheximide inhibit manganese-induced DNA fragmentation, as assessed by EtBr staining. PC12 cells were treated for 20 h under the indicated conditions. MnCl₂ was absent from the control medium. Soluble DNA was extracted and analyzed as described in Materials and methods. Comparable results were obtained in several independent experiments. (B) Manganese induced immediate early gene expression. Western blot analysis of manganese-treated PC12 cell extracts using c-Fos antibody (Santa Cruz). PC12 cells were treated with 1 mM MnCl₂ for the indicated periods. Whole-cell lysates were subjected to 10% SDS-PAGE and immunoblot as described in Materials and methods (from Hirata et al. [26]).

stronger with longer exposure to ECL detection. In contrast, manganese was the only agent to specifically induce the phosphorylation of CREB and ATF-1. In accordance with previous studies in which protein synthesis inhibitors were reported to induce S6 phosphorylation [15], we observed that cycloheximide and anisomycin reduced mobility and

stimulated phosphorylation of p70 S6 kinase at Ser411. Phosphorylation at Thr421/Ser424, however, was observed only in the cell cultures treated with manganese. The reason for this difference is unknown, but this observation suggests that the phosphorylation at Thr421/Ser424 is not always accompanied by a shift in p70 S6 kinase mobility.

3.4.4. Protein tyrosine phosphorylation

The tyrosine phosphorylation cascade plays an important role in cell proliferation and differentiation [30]. In addition, increased protein tyrosine phosphorylation is associated with cellular stress, such as changes in osmolarity [64]. To determine whether manganese affects tyrosine phosphorylation of protein in PC12 cells, we examined cell lysates after exposure to 1 mM MnCl₂. Several proteins were clearly activated only 1 h after exposure to manganese. Tyrosine phosphorylation peaks at 12 h and then begins to decline, so that at the 24-h point, the level of phosphorylation is equal to that at 1 h (Fig. 13). These results indicate that either tyrosine kinase receptors or nonreceptor tyrosine kinases are involved in manganese-induced apoptosis.

3.5. Involvement of the caspase family in manganese-induced apoptosis

Several lines of evidence indicate that the caspase family plays a major role in the execution of apoptosis. To determine whether caspases are involved in manganese-induced apoptosis of PC12 cells, we measured the formation of p17, a 17-kDa protein resulting from the activation of caspases using biotin-X-VAD-fmk, a biotinylated caspase inhibitor. The affinity labeling of p17 was previously demonstrated in the neuronal cell line GT1–7 undergoing apoptosis after treatment with staurosporine, C₂-ceramide, and serum free medium [59]. Fig. 14A shows that an affinity-labeled 17-kDa subunit of an activated member of the caspase family was detected in extracts from PC12 cells treated with either manganese or the kinase inhibitor, staurosporine, which served as a positive control. This indicates that the caspase family is indeed activated after exposure to manganese.

To determine whether caspases contribute to manganese-induced DNA fragmentation in PC12 cells, cells were treated with MnCl₂ and the caspase inhibitor, z-VAD-fmk. As expected, z-VAD-fmk (50 μM) completely prevented manganese-induced DNA fragmentation (Fig. 14B). z-VAD-fmk by itself does not induce DNA laddering in PC12 cells. Moreover, z-VAD-fmk protected the cells from manganese-induced cell death for up to 48 h (Fig. 14C). These results suggest that caspase-like activity is involved in manganese-induced apoptosis in PC12 cells.

4. Discussion

DNA fragmentation occurs in serum-starved PC12 cells, as already reported by Batistatou and Greene [1]. Here, we

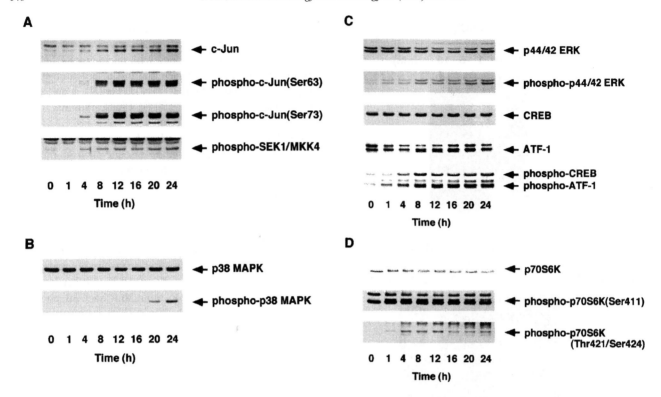

A — c-Jun, phospho-c-Jun(Ser63), phospho-c-Jun(Ser73), phospho-SEK1/MKK4; Time (h) 0 1 4 8 12 16 20 24

B — p38 MAPK, phospho-p38 MAPK; Time (h) 0 1 4 8 12 16 20 24

C — p44/42 ERK, phospho-p44/42 ERK, CREB, ATF-1, phospho-CREB, phospho-ATF-1; Time (h) 0 1 4 8 12 16 20 24

D — p70S6K, phospho-p70S6K(Ser411), phospho-p70S6K (Thr421/Ser424); Time (h) 0 1 4 8 12 16 20 24

Fig. 10. Manganese-induced activation of c-Jun, and kinase pathways associated. Western blot analysis of manganese-treated PC12 cell extracts using antibodies against the indicated proteins and their phosphorylated forms (New England BioLabs). PC12 cells were treated with 1 mM $MnCl_2$ for the indicated periods. Whole-cell lysates were subjected to SDS-PAGE and immunoblot as described in Materials and methods. (A) Activation and time course of c-Jun and JNK pathway products. Note that c-Jun and JNK (SEK1/MKK4) have similar activation time courses, and c-Jun and c-Fos have similar activation time courses (cf. Fig. 9B). (B) Activation and time course of p38 MAPK pathway products. Note that p38 MAPK activation was delayed compared to the JNK pathway (cf. A of this figure). (C) Activation and time course of ERK pathway products. Note that the phosphorylation of CREB and ATF-1, a CREB-related transcription factor, remains at an elevated level for at least 24 h. (D) Activation and time course of p70 S6 kinase pathway products.

document that manganese also induces DNA fragmentation in PC12 cells. There are, however, substantial differences between these two phenomena. First, the time course over which DNA laddering appears is much slower in manganese-treated PC12 cells compared to serum-starved PC12 cells. Internucleosomal cleavage of cellular DNA is detectable within 3 h of serum withdrawal [1], whereas DNA fragmentation is first observed 16 h following manganese treatment. Second, NGF and ATA, two substances that block DNA fragmentation induced by serum withdrawal, are less effective in blocking manganese-induced DNA fragmentation. Third, macromolecular synthesis is required in manganese-induced, but not in serum withdrawal-induced DNA fragmentation. These differences strongly indicate that the mechanisms underlying manganese-associated apoptotic events differ from those associated with serum withdrawal-associated apoptosis in nonneuronal PC12 cells. Moreover, the results of experiments conducted in our laboratory also indicate that manganese-induced apoptosis is transcription dependent, and is thus similar to apoptosis in NGF-deprived sympathetic neurons or neuronally differentiated PC12 cells [38].

Using the TUNEL technique and flow cytometry, Desole et al. [11,12] suggested an oxidative stress mechanism may in part be involved in manganese-induced apoptosis in PC12

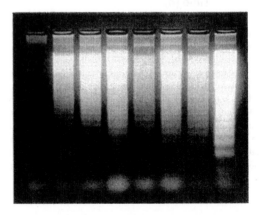

Fig. 11. Effects of various apoptosis-inducing agents on DNA fragmentation. PC12 cells were cultured in the absence (Control) or presence of 1 mM of $MnCl_2$, serum-free DMEM, 1 μM anisomycin, 1 μM camptothecin, 0.5 μg/ml cycloheximide, 100 μM cisplatin [cis-platinum (II) diammine dichloride (CDDP)], or 20 nM staurosporine for 20 h. Soluble DNA was analyzed by agarose gel electrophoresis and visualized under UV light after EtBr staining as described in Materials and methods (from Hirata et al. [26]).

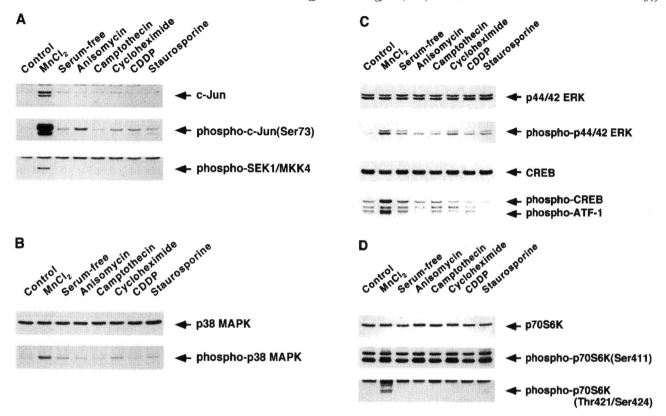

Fig. 12. Comparison of manganese with other established apoptosis-inducing agents on activating kinase pathways. Western blot analysis of PC12 cell extracts prepared from treatment with various apoptosis-inducing reagents. The conditions of cell culture were as in Fig. 11. Whole-cell lysates were subjected to SDS-PAGE and immunoblot as described in Materials and methods. (A) Manganese is the most potent of the tested agents in activating the JNK pathway. It also is a potent activator of (B) p38 MAPK, (C) ERK/CREB/ATF-1, and (D) p70 S6 kinases.

cells. In another study, however, manganese failed to stimulate lipid peroxidation, indicating that cell death may not be initiated by oxidative stress [51]. Our results suggest that manganese causes cell death by driving the caspase family-dependent mechanism of the apoptotic program.

As suggested previously, mitochondrial abnormalities may be involved in the pathogenesis of Parkinson's disease. Both manganese and MPTP are known to induce a parkinsonian-like syndrome, and the latter inhibits mitochondrial respiration [65]. In the present study, we showed that manganese also perturbed mitochondrial function. The persistent manganese-induced accumulation of lactic acid without concomitant ATP depletion is consistent with the observations of Denton and Howard [9] in which MPP^+, the active metabolite of MPTP, caused increased lactate production, but not a reduction of ATP, in PC12 cells. Inhibition of oxidative phosphorylation in PC12 cells does not readily result in depletion of ATP stores [50], and it has been suggested that both glycolysis and oxidative phosphorylation must be blocked for such depletion to occur [9]. The data presented here indicate that, similar to MPTP, manganese inhibits mitochondrial respiration. This conclusion is consistent with the hypothesis that the complete apoptotic program involves energy-requiring steps in some cells [35,42]. Our finding that manganese elicits apoptosis under conditions in which ATP levels

remain unchanged provides additional evidence that intracellular ATP may be required to execute apoptosis. One recent study showed, however, that manganese decreases

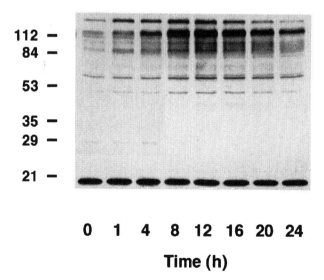

Fig. 13. Time-dependent activation of protein tyrosine phosphorylation by manganese in PC12 cells. Cells were treated with 1 mM MnCl$_2$ for the indicated periods. Whole-cell lysates were subjected to 10% SDS-PAGE and immunoblotting with an antiphosphotyrosine antibody, 4G10 (Upstate Biotechnology). Molecular mass markers (kDa) are shown on the left (from Hirata et al. [26]).

ATP levels in PC12 cells, and that necrosis contributes to manganese-induced cell death [51]. This discrepancy can be explained by considering that intracellular ATP production depends strongly on glucose present in the culture medium, thus the Roth et al. [51] findings are probably due to differences in experimental conditions, especially glucose concentration.

MTT assays demonstrated that the cleavage of MTT is significantly reduced (67% of control) after 24 h of incubation with 1 mM MnCl₂ (Fig. 3), even though at this point, cell death, as assessed by LDH release, is not evident (Fig. 2). Because MTT is reduced by mitochondrial dehydrogenases, these results also support the hypothesis that manganese

inhibits mitochondrial function. There is evidence that mitochondria actively concentrate Mn^{2+} [39]. As with MPP^+, therefore, it is likely that manganese also exerts its toxic effects by targeting mitochondria.

Our model system holds much promise as a useful paradigm for understanding how apoptosis is involved in neurodegenerative disease in general. The treatment of PC12 cells with 1 mM MPP^+ for 24 h induced slight DNA fragmentation (data not shown), and MPTP has also been reported to induce apoptosis in cerebellar granular and mesencephalic neurons [13,40]. The similarities between the effects of MPTP and manganese provide support for the hypothesis that exogenous agents or synthesized endogenous substances harm neurons via a universal mechanism relevant to Parkinson's disease. Our model system using PC12 cells treated with manganese could serve as a good tool for investigations aimed at elucidating the role of apoptosis in neurodegenerative disorders.

Activation of c-Fos or c-Jun expression has been suggested to be common to many situations in which apoptosis occurs [16,57]. We show here that activation of these two immediate early genes also occurs in manganese-treated PC12 cells. The expression of both c-Fos and c-Jun shows a similar time course. Interestingly, gene products of both c-Fos and c-Jun appear prior to DNA fragmentation, thereby providing the necessary components that comprise the form of a transcription activator protein-1 (AP-1) essential for cell death. In manganese-treated PC12 cells, c-Jun function appears to be regulated not only by the level of protein expression but also by protein phosphorylation. Western blot analysis clearly shows increased phosphorylation of c-Jun at Ser63 and Ser73, which is consistent with the finding that the phosphorylation of these two increases transcriptional activity [56]. The protein kinase involved in this phosphorylation reaction is SAPK/JNK [10,25]. JNKs are activated in response to inflammatory cytokines, a variety of genotoxic stresses (including treatment with alkylating

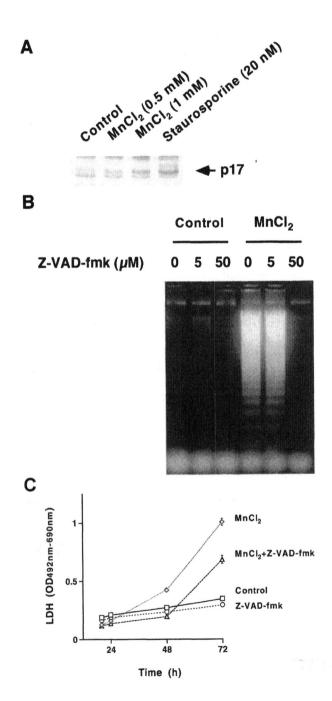

Fig. 14. Involvement of caspases in manganese-induced apoptosis. (A) Induction of the p17 subunit of an activated member of the caspase family by manganese. Western blots of biotinylated caspase inhibitor-labeled cell extracts from PC12 cells. Cells were treated with manganese or staurosporine (positive control) for 20 h, and cell extracts were prepared as described in Materials and methods. (B) Prevention of manganese-induced DNA fragmentation by z-VAD-fmk. PC12 cells were preincubated in the absence or presence of z-VAD-fmk at the indicated concentrations for 1 h. After the addition of MnCl₂, the incubation was continued for 20 h. Soluble DNA was analyzed by agarose gel electrophoresis and visualized under UV light after EtBr staining as described in Materials and methods. Comparable results were obtained in several independent experiments. (C) Prevention of manganese-induced cell death by z-VAD-fmk. Note that z-VAD-fmk protected the cells from manganese-induced cell death for up to 48 h. Cells (1×10^5/well/400 μl) were cultured in 48-well plates for one day and treated with 1 mM MnCl₂ and/or 50 μM z-VAD-fmk (squares, control; diamonds, 1 mM MnCl₂; circles, 50 μM z-VAD-fmk; triangles, MnCl₂ plus z-VAD-fmk). At the times indicated, culture media (5 μl) were analyzed for LDH activity. Data are means ± S.D. ($n = 4$). LDH activity was measured with the cytotoxicity detection kit according to the manufacturer's protocol (Roche Molecular Biochemicals).

agents, osmotic shock, ultraviolet light, and ionizing radiation), and NGF withdrawal [10,19,25,32,67]. In addition, SEK1/MKK4 (JNK kinase), which activates SAPK/JNK, is stimulated in manganese-treated PC12 cells. Thus, manganese is another activator of JNK pathways. Because c-Jun induction and phosphorylation by several apoptosis-inducing reagents were not correlated with DNA fragmentation, as shown in Fig. 11, signals generated by these agents appear to be transduced by multiple, independent pathways.

Similar to JNK/SAPK, p38 MAPK is activated by a variety of cellular stresses including osmotic shock, inflammatory cytokines, lipopolysaccharides (LPS), UV light, and withdrawal of growth factors through the dual phosphorylation of Thr180 and Tyr182 [49,67]. Manganese also strongly induces the phosphorylation of p38 MAPK at Thr180/Tyr182 in PC12 cells. These data are consistent with the findings of Roth et al. [51] that manganese is an efficient activator of p38 MAPK.

Activation of ERKs and phosphorylation of CREB and ATF-1 clearly occur in manganese-treated PC12 cells. Activation of ERK1 and ERK2 is required for manganese-induced neurite outgrowth in PC12 cells [66]. The time course of ERK phosphorylation is similar to that of CREB and ATF-1, suggesting that the ERKs could be the primary kinases responsible for the activation of CREB and ATF-1. Although p38 MAPK is an upstream signaling component of one pathway leading to the phosphorylation of CREB [62], it does not appear to be involved in the phosphorylation of CREB and ATF-1 in the present case. CREB is well recognized as a major mediator of neurotrophin signals [17]. The roles of phosphorylated CREB and ATF-1 in apoptosis is unclear. In cells treated with growth factors such as NGF, EGF, and FGF, peak phosphorylation of CREB and ATF-1 occurs within 30 min of exposure, and persistent, increased levels of phospho-CREB and phospho-ATF-1 are not usually observed [21,62]. It is possible, therefore, that the long-term phosphorylation of CREB and ATF-1 has adverse effects on cell survival.

The phosphorylation of S6 has been reported to alter the efficiency and specificity of protein synthesis [45]. In fact, p70 S6 kinase plays an important role in cell proliferation [15]. Upon stimulation with either PDGF, EGF, insulin, serum, or TPA, the activity of p70 S6 kinase rises rapidly. In addition to these mitogenic signals, NGF, a differentiation factor, also induces p70 S6 kinase activity [4]. We previously found that manganese also activates p70 S6 kinase by initiating phosphorylation of Ser411 and Thr421/Ser424, which is accompanied by increased enzyme activity [27]. Since the activation and phosphorylation of p70 S6 kinase occur prior to the time when cells are irreversibly committed to undergo apoptosis, the p70 S6 kinase signaling pathway may be important for determining the ultimate fate of the cells. Our results indicate that an apoptotic signal can stimulate p70 S6 kinase, which supports the idea that protein synthesis is required for manganese-induced apoptosis. Alternatively, manganese-activated p70 S6 kinase may be

involved in events underlying morphological changes in neurons similar to ERK1 and ERK2, two MAPKs involved in the manganese-induced neurite outgrowth in PC12 cells [66]. As illustrated by the data in Figs. 10 and 12, manganese can stimulate several different signal transduction pathways. However, it is difficult to ascertain how and to what extent each of these pathways contributes to manganese-induced apoptosis because manganese also elicits neurite extension in PC12 cells [36,66].

How does manganese trigger several signaling pathways in the cytoplasm? In the present study, manganese was found to increase the content of lactic acid. It is reasonable to speculate that acidification occurs in manganese-treated PC12 cells and that it is this stressful stimulus that activates the JNK/p38 MAPK pathway. This hypothesis is consistent with recent findings that low extracellular pH induced activation of JNK in A431 and Swiss 3T3 cells [68].

Our results demonstrate that protein tyrosine phosphorylation is greatly induced by manganese. This adds weight to the hypothesis that tyrosine kinases are involved in the stress response. For example, the c-Abl nonreceptor tyrosine kinase is activated by DNA-damaging agents, but not by tumor necrosis factor [31]. Further experiments are necessary to determine the potential role of c-Abl in manganese-induced apoptosis. While it is possible that tyrosine-phosphorylated molecules are potential regulators of upstream components in JNK signal transduction pathways, this does not mean that this is the only system mediating manganese-induced apoptosis. Although tyrosine phosphorylation preceded DNA fragmentation in our PC12 cell model, manganese-associated tyrosine phosphorylation was slower compared to when the cells are exposed to growth factors or cytokines of receptor tyrosine kinase. Further study is required to determine the precise role of protein tyrosine phosphorylation in manganese-induced apoptosis. Such studies would shed light on the exact signal transduction pathways uniquely involved in manganese-induced apoptosis.

The apoptotic death of manganese-treated PC12 cells appears to be regulated by pathways that contain many of the elements discussed above. Consistent with apoptosis in other cell systems, overexpression of Bcl-2 and inhibition of the caspase family proteases attenuate PC12 death caused by manganese. Similar results are obtained with human B cells [52]. Manganese-induced apoptosis of different Burkitt lymphoma cell lines is also completely abolished by the overexpression of Bcl-2 and by treatment with z-VAD-fmk, a broad-spectrum caspase inhibitor. At variance with these findings is the failure of caspase family inhibitors to prevent manganese toxicity in PC12 cells that displayed caspase-3 activation [51]. The different experimental conditions under which each of these studies was carried out could explain these inconsistencies. In the Roth et al. study [51], necrosis was involved because ATP levels in the PC12 cells were decreased. In contrast, apoptosis was involved in our experiments because ATP levels in our PC12 cell preparations remained unchanged. This again supports the idea that intracellular ATP acts as a switch that either initiates events leading to either apoptosis or necrosis [35,42].

In conclusion, our data and that of others suggest that manganese can stimulate several signal transduction pathways known to be responsible for apoptosis. Furthermore, manganese also induces caspase-dependent apoptosis that is blocked by Bcl-2.

Acknowledgments

Bcl-2 overexpressed PC12 cells were generously provided by Drs. Yutaka Eguchi and Yoshihide Tsujimoto. This work was supported by the Frontier Research Program, RIKEN.

References

[1] A. Batistatou, L.A. Greene, Aurintricarboxylic acid rescues PC12 cells and sympathetic neurons from cell death caused by nerve growth factor deprivation: Correlation with suppression of endonuclease activity, J. Cell Biol. 115 (1991) 461–471.

[2] A. Batistatou, L.A. Greene, Internucleosomal DNA cleavage and neuronal cell survival/death, J. Cell Biol. 122 (1993) 523–532.

[3] A. Batistatou, D.E. Merry, S.J. Korsmeyer, L.A. Greene, Bcl-2 affects survival but not neuronal differentiation of PC12 cells, J. Neurosci. 13 (1993) 4422–4428.

[4] J. Blenis, R.L. Erikson, Regulation of protein kinase activities in PC12 pheochromocytoma cells, EMBO J. 5 (1986) 3441–3447.

[5] E.P. Brouillet, L. Shinobu, U. McGarvey, F. Hochberg, M.F. Beal, Manganese injection into the rat striatum produces excitotoxic lesions by impairing energy metabolism, Exp. Neurol. 120 (1993) 89–94.

[6] D.B. Calne, N.-S. Chu, C.-C. Lu, C.-S. Huang, W. Olanow, Manganism and idiopathic parkinsonism: Similarities and differences, Neurology 44 (1994) 1583–1586.

[7] D.G. Cook, S. Fahn, K.A. Brait, Chronic manganese intoxication, Arch. Neurol. 30 (1974) 59–64.

[8] G.C. Cotzias, Manganese in health and disease, Physiol. Rev. 38 (1958) 502–532.

[9] T. Denton, B.D. Howard, A dopaminergic cell line variant resistant to the neurotoxin 1-methyl-4-phenyl-1,2,3,6-tetrahydropyridine, J. Neurochem. 49 (1987) 622–630.

[10] B. Derijard, M. Hibi, I.-H. Wu, T. Barrett, B. Su, T. Deng, M. Karin, R.J. Davis, JNK1: A protein kinase stimulated by UV light and Ha-Ras that binds and phosphorylates the c-Jun activation domain, Cell 76 (1994) 1025–1037.

[11] M.S. Desole, L. Sciola, M.R. Delogu, S. Sircana, R. Migheli, Manganese and 1-methyl-4-(2′-ethylphenyl)-1,2,3,6-tetrahydropyridine induce apoptosis in PC12 cells, Neurosci. Lett. 209 (1996) 193–196.

[12] M.S. Desole, L. Sciola, M.R. Delogu, S. Sircana, R. Migheli, E. Miele, Role of oxidative stress in the manganese and 1-methyl-4-(2′-ethylphenyl)-1,2,3,6-tetrahydropyridine-induced apoptosis in PC12 cells, Neurochem. Int. 31 (1997) 169–176.

[13] B. Dipasquale, A.M. Marini, R.J. Youle, Apoptosis and DNA degradation induced by 1-methyl-4-phenylpyridinium in neurons, Biochem. Biophys. Res. Commun. 181 (1991) 1442–1448.

[14] H. Ehringer, O. Hornykiewicz, Verteilung von Noradrenalin und Dopamin (3-Hydroxytyramin) im Gehirn des Menschen und ihr Verhalten bei Erkrankungen des extrapyramidalen Systems, Klin. Wochenschr. 38 (1960) 1236–1240.

[15] R.L. Erikson, Structure, expression, and regulation of protein kinases involved in the phosphorylation of ribosomal protein S6, J. Biol. Chem. 266 (1991) 6007–6010.

[16] S. Estus, W.J. Zaks, R.S. Freeman, M. Gruda, R. Bravo, E.M. Johnson Jr., Altered gene expression in neurons during programmed cell death: Identification of c-jun as necessary for neuronal apoptosis, J. Cell Biol. 127 (1994) 1717–1727.

[17] S. Finkbeiner, CREB couples neurotrophin signals to survival messages, Neuron 25 (2000) 11–14.

[18] R.S. Freeman, S. Estus, K. Horigome, E.M. Johnson Jr., Cell death genes in invertebrates and (maybe) vertebrates, Curr. Opin. Neurobiol. 3 (1993) 25–31.

[19] Z. Galcheva-Gargova, B. Derijard, I.-H. Wu, R.J. Davis, An osmosensing signal transduction pathway in mammalian cells, Science 265 (1994) 806–808.

[20] Y. Gavrieli, Y. Sherman, S.A. Ben-Sasson, Identification of programmed cell death in situ via specific labeling of nuclear DNA fragmentation, J. Cell Biol. 119 (1992) 493–501.

[21] D.D. Ginty, A. Bonni, M.E. Greenberg, Nerve growth factor activates a Ras-dependent protein kinase that stimulates c-fos transcription via phosphorylation of CREB, Cell 77 (1994) 713–725.

[22] L.A. Greene, A.S. Tischler, Establishment of a noradrenergic clonal line of rat adrenal pheochromocytoma cells which respond to nerve growth factor, Proc. Natl. Acad. Sci. U. S. A. 73 (1976) 2424–2428.

[23] L.A. Greene, G. Rein, Release storage and uptake of catecholamines by a clonal cell line of nerve growth factor (NGF) responsive pheochromocytoma cells, Brain Res. 129 (1977) 247–263.

[24] J. Ham, C. Babij, J. Whitfield, C.M. Pfarr, D. Lallemand, M. Yaniv, L.L. Rubin, A c-Jun dominant negative mutant protects sympathetic neurons against programmed cell death, Neuron 14 (1995) 927–939.

[25] M. Hibi, A. Lin, T. Smeal, A. Minden, M. Karin, Identification of an oncoprotein- and UV-responsive protein kinase that binds and potentiates the c-Jun activation domain, Genes Dev. 7 (1993) 2135–2148.

[26] Y. Hirata, K. Adachi, K. Kiuchi, Activation of JNK pathway and induction of apoptosis by manganese in PC12 cells, J. Neurochem. 71 (1998) 1607–1615.

[27] Y. Hirata, K. Adachi, K. Kiuchi, Phosphorylation and activation of p70 S6 kinase by manganese in PC12 cells, NeuroReport 9 (1998) 3037–3040.

[28] D. Hockenbery, G. Nunez, C. Milliman, R.D. Schreiber, S.J. Korsmeyer, Bcl-2 is an inner mitochondrial membrane protein that blocks programmed cell death, Nature 348 (1990) 334–336.

[29] C.-C. Huang, N.-S. Chu, C.-S. Lu, J.-D. Wang, J.-L. Tsai, J.-L. Tzeng, E.C. Wolters, D.B. Calne, Chronic manganese intoxication, Arch. Neurol. 46 (1989) 1104–1106.

[30] T. Hunter, The Croonian Lecture 1997. The phosphorylation of proteins on tyrosine: Its role in cell growth and disease, Philos. Trans. R. Soc. London, Ser. B 353 (1998) 583–605.

[31] S. Kharbanda, R. Ren, P. Pandey, T.D. Shafman, S.M. Feller, R.R. Weichselbaum, D.W. Kufe, Activation of the c-Abl tyrosine kinase in the stress response to DNA damaging agents, Nature 376 (1995) 785–788.

[32] J.M. Kyriakis, P. Banerjee, E. Nikolakaki, T. Dai, E.A. Rubie, M.F. Ahmad, J. Avruch, J.R. Woodgett, The stress-activated protein kinase subfamily of c-Jun kinases, Nature 369 (1994) 156–160.

[33] H.A. Lane, A. Fernandez, N.J.C. Lamb, G. Thomas, p70^{s6k} function is essential for G1 progression, Nature 363 (1993) 170–172.

[34] S.C. Lane, R.D. Jolly, D.E. Schmechel, J. Alroy, R.-M. Boustany, Apoptosis as the mechanism of neurodegeneration in Batten's disease, J. Neurochem. 67 (1996) 677–683.

[35] M. Leist, B. Single, A.F. Castoldi, S. Kuhnle, P. Nicotera, Intracellular adenosine triphosphate (ATP) concentration: A switch in the decision between apoptosis and necrosis, J. Exp. Med. 185 (1997) 1481–1486.

[36] W.H. Lin, D. Higgins, M. Pacheco, J. Aletta, S. Perini, K.A. Marcucci, J.A. Roth, Manganese induces spreading and process outgrowth in rat pheochromocytoma (PC12) cells, J. Neurosci. Res. 34 (1993) 546–561.

[37] S.P. Mah, L.T. Zhong, Y. Liu, A. Roghani, R.H. Edwards, D.E. Bredesen, The protooncogene bcl-2 inhibits apoptosis in PC12 cells, J. Neurochem. 60 (1993) 1183–1186.

[38] D.P. Martin, R.E. Schmidt, P.S. DiStefano, O.H. Lowry, J.G. Carter, E.M. Johnson Jr., Inhibitors of protein synthesis and RNA synthesis prevent neuronal death caused by nerve growth factor deprivation, J. Cell Biol. 106 (1988) 829–844.

[39] L.S. Maynard, G.C. Cotzias, Partition of manganese among organs and intracellular organelles in the rat, J. Biol. Chem. 214 (1955) 489–495.

[40] H. Mochizuki, N. Nakamura, K. Nishi, Y. Mizuno, Apoptosis is induced by 1- methyl-4-phenylpyridinium ion (MPP$^+$) in ventral mesencephalic–striatal coculture in rat, Neurosci. Lett. 170 (1994) 191–194.

[41] N.H. Neff, R.E. Barrett, E. Costa, Selective depletion of caudate nucleus dopamine and serotonin during chronic manganese dioxide administration to squirrel monkeys, Experientia 25 (1969) 1140–1141.

[42] P. Nicotera, M. Leist, E. Ferrando-May, Intracellular ATP, a switch in the decision between apoptosis and necrosis, Toxicol. Lett. 102–103 (1998) 139–142.

[43] C.W. Olanow, W.G. Tatton, Etiology and pathogenesis of Parkinson's disease, Annu. Rev. Neurosci. 22 (1999) 123–144.

[44] R.W. Oppenheim, Cell death during development of the nervous system, Annu. Rev. Neurosci. 14 (1991) 453–501.

[45] E. Palen, J.A. Traugh, Phosphorylation of ribosomal protein S6 by cAMP-dependent protein kinase and mitogen-stimulated S6 kinase differentially alters translation of globin mRNA, J. Biol. Chem. 262 (1987) 3518–3523.

[46] M. Patel, B.J. Day, J.D. Crapo, I. Fridovich, J.O. McNamara, Requirement for superoxide in excitotoxic cell death, Neuron 16 (1996) 345–355.

[47] C. Portera-Cailliau, J.C. Hedreen, D.L. Price, V.E. Koliatsos, Evidence for apoptotic cell death in Huntington disease and excitotoxic animal models, J. Neurosci. 15 (1995) 3775–3787.

[48] O. Rabin, L. Hegedus, J.-M. Bourre, Q.R. Smith, Rapid brain uptake of manganese(II) across the blood–brain barrier, J. Neurochem. 61 (1993) 509–517.

[49] J. Raingeaud, S. Gupta, J.S. Rogers, M. Dickens, J. Han, R.J. Ulevitch, R.J. Davis, Pro-inflammatory cytokines and environmental stress cause p38 mitogen-activated protein kinase activation by dual phosphorylation on tyrosine and threonine, J. Biol. Chem. 270 (1995) 7420–7426.

[50] E.E. Reynolds, W.P. Melega, B.D. Howard, Adenosine 5′triphosphate independent secretion from PC12 pheochromocytoma cells, Biochemistry 21 (1982) 4795–4799.

[51] J.A. Roth, L. Feng, J. Walowitz, R.W. Browne, Manganese-induced rat pheochromocytoma (PC12) cell death is independent of caspase activation, J. Neurosci. Res. 61 (2000) 162–171.

[52] N. Schrantz, D.A. Blanchard, F. Mitenne, M.-T. Auffredou, A. Vazquez, G. Leca, Manganese induces apoptosis of human B cells: Caspase-dependent cell death blocked by Bcl-2, Cell Death Differ. 6 (1999) 445–453.

[53] S. Shimizu, Y. Eguchi, W. Kamiike, S. Waguri, Y. Uchiyama, H. Matsuda, Y. Tsujimoto, Retardation of chemical hypoxia-induced necrotic cell death by Bcl-2 and ICE inhibitors: Possible involvement of common mediators in apoptotic and necrotic signal transductions, Oncogene 12 (1996) 2045–2050.

[54] S. Shimizu, M. Narita, Y. Tsujimoto, Bcl-2 family proteins regulate the release of apoptogenic cytochrome *c* by the mitochondrial channel VDAC, Nature 399 (1999) 483–487.

[55] W.N. Sloot, A.J. van der Sluijs-Gelling, J.B.P. Gramsbergen, Selective lesions by manganese and extensive damage by iron after injection into rat striatum or hippocampus, J. Neurochem. 62 (1994) 205–216.

[56] T. Smeal, B. Binetruy, D. Mercola, M. Birrer, M. Karin, Oncogenic and transcriptional cooperation with Ha-Ras requires phosphorylation of c-Jun on serines 63 and 73, Nature 354 (1991) 494–496.

[57] R.J. Smeyne, M. Vendrell, M. Hayward, S.J. Baker, G.G. Miao, K. Schilling, L.M. Robertson, T. Curran, J.I. Morgan, Continuous c-*fos* expression precedes programmed cell death in vivo, Nature 363 (1993) 166–169.

[58] W.P.J.M. Spooren, C. Gentsch, C. Wiessner, Tunel-positive cells in the substantia nigra of C57BL/6 mice after a single bolus of 1-methyl-4-phenyl-1,2,3,6-tetrahydropyridine, Neuroscience 85 (1998) 649–651.

[59] A. Srinivasan, L.M. Foster, M.-P. Testa, T. Ord, R.W. Keane, D.E. Bredesen, C. Kayalar, Bcl-2 expression in neural cells blocks activation of ICE/CED-3 family proteases during apoptosis, J. Neurosci. 16 (1996) 5654–5660.

[60] B. Storrie, E.A. Madden, Isolation of subcellular organelles, Methods Enzymol. 182 (1990) 203–225.

[61] B.L. Strehler, J.K. Totter, Determination of ATP and related compounds: Firefly luminescence and other methods, Methods Biochem. Anal. 1 (1954) 341–346.

[62] Y. Tan, J. Rouse, A. Zhang, S. Cariati, P. Cohen, M.J. Comb, FGF and stress regulate CREB and ATF-1 via a pathway involving p38 MAP kinase and MAPKAP kinase-2, EMBO J. 15 (1996) 4629–4642.

[63] N.A. Tatton, S.J. Kish, In situ detection of apoptotic nuclei in the substantia nigra compacta of 1-methyl-4-phenyl-1,2,3,6-tetrahydropyridine-treated mice using terminal deoxynucleotidyl transferase labelling and acridine orange staining, Neuroscience 77 (1997) 1037–1048.

[64] B.C. Tilly, N. van den Berghe, L.G.J. Tertoolen, M.J. Edixhoven, H.R. de Jonge, Protein tyrosine phosphorylation is involved in osmoregulation of ionic conductances, J. Biol. Chem. 268 (1993) 19919–19922.

[65] I. Vyas, R.E. Heikkila, W.J. Nicklas, Studies on the neurotoxicity of 1-methyl-4-phenyl-1,2,3,6-tetrahydropyridine: Inhibition of NAD-linked substrate oxidation by its metabolite, 1-methyl-4-phenylpyridinium, J. Neurochem. 46 (1986) 1501–1507.

[66] J.L. Walowitz, J.A. Roth, Activation of ERK1 and ERK2 is required for manganese-induced neurite outgrowth in rat pheochromocytoma (PC12) cells, J. Neurosci. Res. 57 (1999) 847–854.

[67] Z. Xia, M. Dickens, J. Raingeaud, R.J. Davis, M.E. Greenberg, Opposing effects of ERK and JNK-p38 MAP kinases on apoptosis, Science 270 (1995) 1326–1331.

[68] L. Xue, J.M. Lucocq, Low extracellular pH induces activation of ERK2, JNK, and p38 in A431 and Swiss 3T3 cells, Biochem. Biophys. Res. Commun. 241 (1997) 236–242.

ELSEVIER

Neurotoxicology and Teratology 24 (2002) 655–666

NEUROTOXICOLOGY
AND
TERATOLOGY

www.elsevier.com/locate/neutera

Review article

Physiological role of salsolinol:
Its hypophysiotrophic function in the regulation of pituitary prolactin secretion

Béla E. Tóth[a,1], Ibolya Bodnár[a,1], Krisztián G. Homicskó[a,1], Ferenc Fülöp[b],
Márton I.K. Fekete[c], György M. Nagy[a,*]

[a]*Neuroendocrine Research Laboratory, Department of Human Morphology and Developmental Biology,*
Semmelweis University, Tuzoltó u. 58, Budapest H-1094, Hungary
[b]*Institute of Pharmaceutical Chemistry, Faculty of Pharmacy, University of Szeged, Szeged, Hungary*
[c]*Institute of Experimental Medicine, Hungarian Academy of Sciences, Budapest, Hungary*

Received 9 January 2002; accepted 25 January 2002

Abstract

We have recently observed that 1-methyl-6,7-dihydroxy-1,2,3,4-tetrahydroisoquinoline (salsolinol) produced by hypothalamic neurons can selectively release prolactin from the anterior lobe (AL) of the pituitary gland. Moreover, high affinity binding sites for SAL have been detected in areas, like median eminence (ME) and the neuro-intermediate lobe (NIL) that are known terminal fields of the tuberoinfundibular DAergic (TIDA) and tuberohypophysial (THDA)/periventricular (PHDA) DAergic systems of the hypothalamus, respectively. However, the *in situ* biosynthesis and the mechanism of action of SAL are still enigmatic, these observations clearly suggest that sites other than the AL might be targets of SAL action. Based on our recent observations it may be relevant to postulate that an "autosynaptocrine" regulatory mechanism functioning at the level of the DAergic terminals localized in both the ME and NIL, may play a role in the hypophyseotrophic regulation of PRL secretion. Furthermore, SAL may be a key player in these processes. The complete and precise mapping of these intra-terminal mechanisms should help us to understand the tonic DAerg regulation of PRL secretion. Moreover, it may also give insight into the role of pre-synaptic processes that most likely have distinct and significant functional as well as pathological roles in other brain areas using DAergic neurotransmission, like striatonigral and mesolimbic systems. © 2002 Elsevier Science Inc. All rights reserved.

Keywords: Dopamine; Salsolinol; Prolactin; Aromatic amino acid decarboxylase; Dopamine transporter; Synaptocrine regulation

1. Introduction

Prolactin (PRL) is a polypeptide hormone that is synthesized in and secreted from one of the specific cell types of the anterior lobe (AL) of the pituitary gland called mammotropes. This hormone not only subserves multiple roles during reproduction but it also plays an essential role in the general homeostasis of the organism [80,95]. The secretion of PRL from mammotropes is under a dominant inhibitory control that is exercised by the medial–basal hypothalamus. Moreover, in vivo, this hypothalamic influence tonically restrains PRL secretion. Experiments evid-

ence that drugs affecting catecholamine metabolism can also alter PRL secretion [24,69] and that dopamine (DA) is present in high concentration in both the median eminence [43] and the hypophysial stalk plasma [12,48,100], and several investigators have concluded that DA is the hypothalamic PRL-inhibiting factor (PIF). Experimental evidence provided by MacLeod [80] that DA inhibits PRL release from pituitary mammotropes in vitro provided a strong support to this conclusion [13,80,81]. Subsequently, receptors for DA have been detected on pituitary membranes [19,21,25,51,84]. Now, it is also known that this receptor belongs to the D_2 subclass of the DA receptor family [20]. Hence, all of the required conditions seem to be present for considering DA as a major physiological hypothalamic PIF.

Although much progress has been made in identifying hypothalamic neuroendocrine dopaminergic (NEDA) neurons (vide infra) involved in the regulation of PRL secretion,

---

* Corresponding author. Tel.: +36-1-215-6920; fax: +36-1-215-3064.
E-mail address: nagy-gm@ana2.sote.hu (G.M. Nagy).

[1] B.E. Tóth, I. Bodnár, and K.G. Homicskó have equally contributed in the work presented in this article.

several elements of their function still remain elusive. One of them is the mechanism of the tonic DA release of NEDA neurons. Since DA level in the long portal vessels (LPV) should determine the baseline level of D_2 receptor stimulation present on the surface of mammotropes, changes in the tonic DA release at the terminal sites of NEDA neurons must elicit homeostatic compensatory act to restore the original DA level. This statement is supported by observations that permanent separation of pituitary cells from the hypothalamic regulatory influences, like (a) surgical disconnection of the AL and the hypothalamus (e.g., pituitary stalk section) [70], (b) transplantation of the AL under the kidney capsule [37], or (c) culturing AL cells in vitro [96], all result in a gradual and permanent elevation of PRL secretion from mammotropes due to the lack of compensatory hypothalamic events.

The question whether DA is the sole PIF mediating tonic hypothalamic inhibition in vivo arises from time to time. In early studies of this issue, investigators reported that the amount of DA in stalk blood was sufficient to account for only about two-thirds of the PRL inhibition normally observed [48,55,100]. This conclusion was based on quantitative studies in which DA was infused into rats depleted of endogenous DA with the rate-limiting enzyme, tyrosine hydroxylase (TH) inhibitor, α-methyl-p-tyrosine (α-MpT), and the rate of DA infusion was set to mimic the levels measured in stalk blood of intact animals [48]. The established influence of DA on pituitary PRL secretion, however, is more complex and still holds controversy when one considers the facts that DA in pituitary stalk blood largely reflects a preferential release of DA from a relatively small, newly synthesized pool of this catecholamine [58]. Close to the 75–80% reduction of DA level in stalk plasma but only a slight (25–30%) decrease of DA concentration in the median eminence 1 h after intravenous administration of α-MpT have clearly suggested that the newly synthesized DA is preferentially and spontaneously released from dopaminergic terminals of the median eminence [57,68].

It is well established that DA produced by neurons is packed into vesicles at the terminals and released through a calcium-mediated vesicle docking process [6]. Once released from neurons, DA either acts at its target (i.e., post- and presynaptic DA receptors) or it is taken up into the presynaptic neurons by DA transporters (DATs) and oxidized by mitochondrial monoamine oxidase (MAO) to produce 3,4-dihydroxyphenylacetic acid (DOPAC) [49]. According to the general view, DA not packed in the secretory granules in NEDA terminals is also susceptible to the enzymatic metabolism by MAO [77]. Based upon the previously mentioned observation that the newly synthesized DA is preferentially released to the LPV, another mechanism of release can coexist with the classical exocytosis. It is the reverse transport of the cytosolic DA from terminals by the DA uptake carrier molecule, DAT, as it has been already established in the midbrain [39,75]. A slow and continuous nonvesicular release of DA to the perivas-

cular space may well be responsible to the persistent diffusion of DA into the portal blood, which is then transported to the pituitary gland and can exercise the established tonic inhibitory regulation on mammotropes. This view is supported by the fact that DA terminals at the median eminence do not form conventional synapses. Furthermore, elimination of the majority of released dopamine is thought to be due to the clearing effect of the continuous blood flow in the capillary loops of the median eminence. A possible consequence of this microanatomical environment may well be fitted to a carrier-mediated release of DA that can bypass the classical transmission through conventional synapses.

At the same time, an inverse relationship does not always exist between hypothalamic secretion of DA and pituitary secretion of PRL. For example, DA levels in hypophysial stalk plasma are five to seven times lower in males than in females [12,30,59,100], but plasma levels of PRL are not much different. Moreover, several exteroceptive stimuli such as suckling of the nipples of the mothers by their litters sharply elevate plasma PRL [41,95], and a lack of mirror image relationship between DA concentrations of the median eminence and plasma PRL has also been demonstrated in lactating rats during simulated suckling stimulus [56,101,102]. According to the classical view, these stimuli may act by decreasing the inhibitory influence of hypothalamic DA and/or enhancing the activity of a still unknown hypothalamic neuronal system that secretes a PRL-releasing factor (PRF). Isolation of such regulatory factor(s) has(have) been sought for over two decades [7,112,116]. However, difference in the proportion or balance of the cytosolic DA compartments (among vesicular storage, reverse transport, or enzymatic catabolism) may well explain these controversies.

Salsolinol (SAL) belongs to a family of compounds referred to as tetrahydroisoquinolines (TIQ). TIQs have mainly been regarded as potential neurotoxins [47]. It is well demonstrated that several TIQs are endogenously produced in the central nervous system (CNS), however, research on DA-derived TIQs has mainly focused on the involvement of the N-methylated metabolite of SAL in the progressive degeneration of dopaminergic neurons in the substantia nigra leading to Parkinson's disease [36,83, 89,93]. Moreover, recent experimental findings of Antkiewicz-Michaluk et al. [4,5] and Vetulani et al. [117] have provided evidence about a possible role of SAL as an endogenous regulator of the dopaminergic system in the striatum. According to their conclusions, the neurotoxic properties of N-methyl-SAL may be a side effect that is not eliminated during evolution. At the same time, very little has been known about the effects of TIQ on the hypothalamo-pituitary axis until now [18,103]. Based on our recent findings, it is fairly obvious that SAL can be produced in the hypothalamo-hypophysial system in situ and it is supposedly plays a pivotal role in the regulation of PRL secretion. This finding strongly supports the view that SAL indeed has certain physiological role.

Based on all of this, our present review will focus on the development of our knowledge in the physiological hypophysiotrophic function of SAL in the regulation of pituitary PRL secretion. At the same time, it is tempting to raise a hypothesis to explain the "tonic" function of NEDA neurons as well as the relative insensitiveness of their terminal sites i.e., median eminence and neuro-intermediate lobe (NIL) to known neurotoxins and suggest a crucial role of SAL in both of these processes.

2. Hypothalamic NEDA neurons and their regulatory properties

2.1. Anatomy of the hypophysiotrophic NEDA neurons

In the early 1960s, Falk et al. [38] discovered that biogenic amines in histological sections treated with vapor of paraformaldehyde produced fluorescence upon exposure to ultraviolet light [38]. Using this amine fluorescence method Dahlström and Fuxe [27] presented the full picture of the anatomy of the brain catecholaminergic system. They classified it from A1 to A15 according to the location of the neurons in different brain nuclei.

Three populations of hypothalamic NEDA neurons, which contribute to the regulation of the secretion of PRL, can be distinguished (Fig. 1) [14,42,52,53]. According to the rostro-caudal location of the neuronal cells located in the periventricular–arcuate nucleus (PeV–ARC) region (also termed A14 and A12), which is clearly reflected in their terminal sites at different lobes of the pituitary gland, they can be divided into three different neuronal systems. Tubero-infundibular dopaminergic neurons (TIDA), located in the middle and posterior portion of PeV–ARC that project to the median eminence where DA diffuses to the LPVs and transported to the AL, are well accepted as a major and physiological regulator of PRL secretion [74]. The intermediate (IL) and neural lobes (NL) of the pituitary gland are innervated by two virtually independent groups of hypothalamic dopaminergic neurons [10,53,54]. The periventriculo-hypophysial dopaminergic system (PHDA) is located at the most rostral subdivision of the PeV–ARC compared to TIDA neurons and terminates only in the IL [54]. Tubero-hypophysial (THDA) dopaminergic neurons sit in the PeV–ARC between the two abovementioned cell groups but they project to both the IL and the NL of the pituitary gland [63,64]. Holzbauer and Racké [64] have described first some aspects of the microanatomy and biochemistry as well as the independent physiological functions of these two groups of dopaminergic neurons. While DA released from TIDA neurons has been considered the primary source of DA inhibiting the secretion of PRL [11,41,81,88,95,96], it is now evident that DA released to the NIL also plays a critical role in the regulation of the secretion of PRL [41]. Moreover, several significant characteristics and differences in their activity have only been recently demonstrated [52,60,78,82,99].

The rate-limiting step of DA synthesis is the formation of L-3,4-dihydroxyphenylalanine (L-DOPA) from tyrosine by TH. L-DOPA is further metabolized by L-amino acid decarboxylase (AADC) into DA. At the beginning of the introduction of TH immunocytochemistry [62], it was already noted that TH-immunoreactive neurons are distributed in a bit of wider area than catecholamine-fluorescent neurons [23]. The clarification of the obvious question whether it is only due to the different sensitivities of these two methods or it really reflects a different distribution became possible when antibodies against the amines themselves had been produced for immunocytochemistry [46]. Using this technique, it was demonstrated that the immunoreactions of DA and its immediate biosynthetic precursor, L-DOPA, are heterogeneous suggesting the existence of a "nonclassical" amine producing neuronal system containing only TH but lacking the enzyme AADC. These neurons must have produced only L-DOPA but not DA [66,87,98]. As far as NEDA neurons are concerned, using the abovementioned catecholamine markers, it has been clearly shown that the PHDA, THDA, and dorso-medial part of the TIDA system (DM-TIDA) produce DA as an end-product because they can be immunostained with TH, AADC, and DA antibodies. However, TH-immunostained neurons (Fig. 1) could be detected not only in the DM-TIDA, but also in the ventro-lateral part of TIDA (VL-TIDA) just dorsal to the ventral surface of the brain [15,97,98]. They were negative when AADC or DA antibody has been used [85,97,98,15]. In contrast, a strong L-DOPA immunoreactivity could be detected in this area [87,97]. Therefore, TH immunopositive VL-TIDA has the capacity to produce L-DOPA that may not be converted to DA, since no AADC exists in these neurons. These findings suggest that L-DOPA itself may be transported to the median eminence and secreted into the LPV, or the "end-product" of one of these hypophysiotrophic neurons may be a substrate of further enzymatic processes resulting to a different hypothalamic hypophysiotrophic factor.

2.2. Contribution of different subpopulations of NEDA neurons in the regulation of PRL secretion

The relative contribution of the hypothalamic NEDA and the supposedly existing L-DOPAergic systems in the control of PRL secretion is a very important but still an open question. TIDA neurons are well accepted as a main regulator of PRL secretion of pituitary mammotropes. The role of the DA released at NIL has attracted the most attention during the past few years. For example, both PHDA and THDA neurons, unlike TIDA system, are not dependent upon the presence of circulating gonadal steroids [82,88] and there are no differences between male and female rats in the activity of the THDA neurons [60]. In addition, TIDA and THDA neurons, but not PHDA neurons, regulate the control of the secretion of PRL in response to suckling stimulus [31,33,90,91,92]. Surgical removal of the NIL results in a three- to fourfold increases in basal

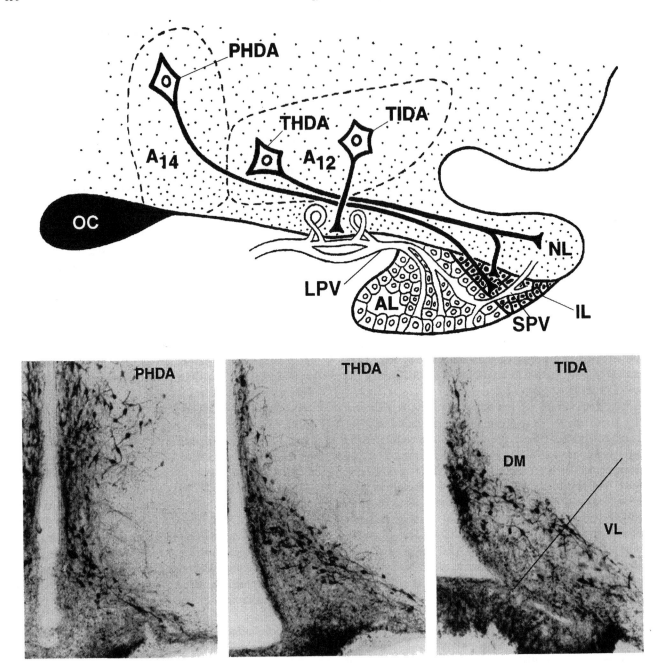

Fig. 1. Frontal sections of the medial basal hypothalamus where the rostro-caudal and the medio-lateral organization of the hypothalamic dopaminergic/
L-DOPAergic neuronal systems can be seen. Frontal sections represent all three NEDA systems (PHDA, THDA, and TIDA) with TH immunostaining. DM:
dorso-medial TIDA; VM: ventro-lateral TIDA; OC: optic chiasma; NL: neural lobe; IL: intermediate lobe; AL: anterior lobe; LPV: long portal vessels; SPV:
short portal vessels.

plasma PRL levels in male, cycling, and lactating female
rats [31,41,116]. Consistent with this finding is the report
that electrochemically detectable DA in the AL is reduced
after surgical removal of the posterior lobe [90]. Moreover,
when lactating rats have been exposed to relatively short-
term (24–48 h) dehydration, basal- and suckling-induced
pituitary PRL secretion were immediately depressed [90].
Since THDA system can be selectively activated by dehydration [90,91,41], one can conclude that DA released by
nerve terminals in the NIL of the pituitary gland may travel

to the AL to affect PRL secretion. Thus, a reduction or an
elevation of DA level in blood carried by the short portal
vessels (SPVs) is able to provide an important signal
(directly or indirectly) by which PRL secretion of mammotropes can change.

It is well known that DA, released from terminals of
PHDA neurons in the IL [111], tonically inhibits the secretion
of α-melanocyte-stimulating hormone (α-MSH) from mela-
notropes of the IL [72,114]. Therefore, we can expect parallel
changes in plasma levels of PRL and α-MSH during an acute

stimulus like suckling. Recent data have clearly shown that there is no change in plasma α-MSH in response to nursing [91]. Therefore, an acute diminution in the activity of the PHDA system does not occur during suckling stimulus. However, the observations that DA concentration of NIL is lower and the basal level plasma α-MSH is higher in lactating than in cycling female rats [64] strongly support a permissive role of PHDA neurons in the regulation of PRL secretion during lactation [31,33,34].

2.3. DATs and NEDA neurons

It is generally accepted that the plasma membrane DAT and the vesicular monoamine transporter (VMAT2) are essential for normal DA neurotransmission. DAT, a 12-transmembrane domain transporter belonging to a larger family of sodium/chloride-dependent transporters [1], is regulated

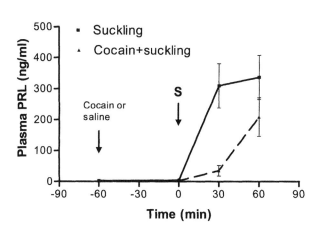

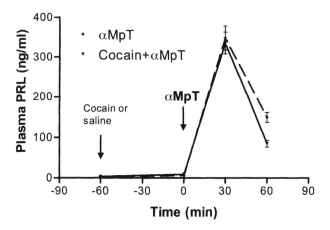

Fig. 2. Effect of DAT inhibition by cocaine (15 mg/kg bw sc at − 60 min) on the suckling- (at 0 time) and αMpT- (8 mg/kg bw iv at 0 time) induced plasma PRL responses.

at the mRNA level by ovarian steroids [17] and at the protein level by phosphorylation [113]. DAT mRNA has been localized in several regions within the rat brain, including the ARC [22,61,79] and periventricular nuclei of the hypothalamus [40,79]. Moreover, DAT protein has been localized in the median eminence [104]. These observations are in sharp contrast to previous reports that both TIDA and THDA neurons lack functional DAT systems [2,3,29]. However, administration of DAT blockers was shown to disrupt the estrous cycle of the rat [73].

It is clear that the DA uptake mechanism in the median eminence, IL, and NL is less robust than in other brain regions [2,3,29]. However, the fact that the administration of 6-hydroxydopamine (6-OHDA, a neurotoxin requiring uptake through transporters) decreases the concentration of DA and the presence of dopaminergic nerve terminals in the median eminence, IL, and NL also suggests the presence of DAT in these terminal areas [26,76]. Moreover, DAT knockout mice display AL hypoplasia and hyperdopaminosis in the pituitary gland [16]. Recently, it has been demonstrated that administration of DAT blockers significantly decreased the turnover of DA at the terminal areas of TIDA (median eminence), THDA (NL and IL), and PHDA (IL) neurons in ovariectomized (OVX) rats during acute estradiol treatment [9,32]. Similarly, cocaine pretreatment of lactating rats can attenuate suckling- but not αMpT-induced plasma PRL elevations (Fig. 2). Therefore, blockade of DAT inhibits the DA uptake mechanism on NEDA neurons thus increasing the concentration of DA in the perivascular space and thus enhanced diffusion into the portal vessels to diminish pituitary PRL secretion [9,16,32,45,76].

3. Role of SAL in the regulation of PRL secretion

We have recently made a serendipitous observation [115] using high-performance liquid chromatography coupled with electrochemical detection (HPLC-EC) running perchloric acid extract of NIL. (R)-SAL, a DA-related TIQ [35,36,94], has been found to be present in NIL as well as in median eminence extracts of male and in intact and OVX as well as lactating female rats [115]. Moreover, analysis of SAL concentrations in NIL revealed parallel increases with plasma PRL in lactating rats exposed to a brief (10 min) suckling stimulus following 4-h separation. SAL is sufficiently potent in vivo to account for the massive discharge of PRL that occurs after physiological stimuli (i.e., suckling). At the same time, it has shown no effect on the secretion of other pituitary hormones [116]. SAL, therefore, appears to be a selective stimulator of PRL secretion in vivo (Fig. 3). We have also found that SAL can elevate PRL release, though to a lesser extent, in pituitary cell cultures as well as in hypophysiotomized rats bearing AL transplants under the kidney capsule [115]. Lack of interference of SAL with [³H]-spiperone binding to AL homogenates indicates that SAL does not act at the D_2-DA receptor [115]. Moreover, SAL has a specific

A.

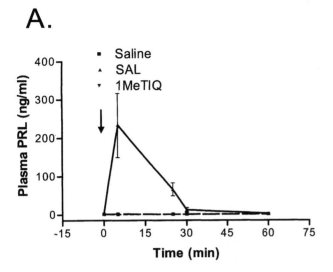

B.

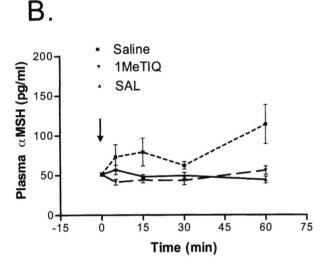

Fig. 3. Effect of intravenous injection of SAL (5 mg/kg bw iv) or 1-methyl-1,2,3,4-TIQ (1MeTIQ; 5 mg/kg bw iv) on plasma PRL and α-MSH secretion.

and saturable binding site in the hypothalamus, median eminence, and in the NIL of the pituitary gland [65,71, 115]. K_D values of these bindings are in the nanomolar range in all tissue tested. It has been also shown that SAL could not displace the D_1 receptor-specific ligand, ^{3}H-SCH23390, from the striatal synaptosome [4,5] or the D_2 receptor-specific ligand, ^{3}H-spiperone, from both striatal [4,5] and pituitary gland homogenates [115]. Consonant with these data is that none of the tested agonists and/or antagonists of different DA receptors (D_1–D_5) were able to displace ^{3}H-SAL except DA itself and in a lesser extent apomorphine (Table 1). Our results are in agreement with the data obtained by Antkiewicz-Michaluk et al. [5] using ^{3}H-apomorphine binding on striatal synaptosomes. All these together suggest that the binding of SAL is closely related to a site that can also recognize and may use DA as a signaling molecule, but its property differs from any known DA receptors [50]. Taken together, these data clearly suggest that SAL is synthesized in situ in all three

subpopulations of NEDA system. Based on all of these data, it is clear that the endogenous SAL could play a role in the regulation of pituitary PRL secretion.

One of the most interesting and unexpected findings of our recent studies [65,115] is that several inhibitors (carbidopa, benserazide, and α-methyl-DOPA) and one of the substrates (L-DOPA) of the known enzyme of DA biosynthesis, called AADC, could effectively displace ^{3}H-SAL (Table 1). This enzyme(s) catalyzes the synthesis of DA and serotonin from L-DOPA and 5-hydroxtryptophane (5-HTP), respectively [106]. One can conclude that a possible site of binding for SAL is this enzyme. A major concern against this possibility is that AADC predominantly locates in the soluble fraction of rat brain homogenate [119]. However, there are reports that a substantial proportion (between 35% and 50%) of AADC (using L-DOPA and 5-HTP as a substrate, respectively) is associated with the synaptic membranes [44,119]. Therefore, the particulate fraction of AADC may serve as (an AADC substrate/inhibitor displaceable) binding site for ^{3}H-SAL. As far as the obligatory functional consequence of an interaction between AADC and SAL is concerned (induction of PRL release), one can only conceive an inhibitory influence on AADC located in the median eminence and/or the NIL. It will result in a decrease of DA released into the bloodstream of LPV and/or SPVs. Consequently, a reduction of dopaminergic influence on AL can lead to an increase of PRL release. The results of our experiments, when carbidopa and SAL were injected alone or in combination and changes in plasma PRL was measured, clearly indicate a positive cooperative interaction between SAL and the AADC inhibitor inducing PRL release [65]. On the other hand, the lack of displacing activity of NSD-1015 (one of the most selective AADC inhibitor at the CNS) and 5-HTP (the other substrate of the AADC) on ^{3}H-SAL binding argues against this possibility. Moreover, our preliminary data indicate no direct influence of SAL on the enzyme activity of AADC (unpublished data). A possible solution to explain these controversial data is that SAL may change the distribution of AADC between the soluble and particulate fraction rather than the activity of the enzyme. The site and the mechanism of action of SAL influencing pituitary PRL secretion is still enigmatic, therefore, to learn more about the abovementioned possibilities needs further experiments.

The in vivo relevance of our binding studies is strengthened by the following pharmacological evidence (Table 2). In our present experiments, an inhibitor of VMAT, reserpine

Table 1
The capacity of selected compounds that can (+) or cannot (−) influence the binding of ^{3}H-SAL

SAL	++	NSD-1015	−
L-DOPA	+++	5-HTP	−
DA	++	Haloperidol	−
Apomorphine	++	Domperidone	−
Carbidopa	++	Quinelorane	−
Benserazide	++	Papaverine	−
α-Methyl-DOPA	++	Mazindol	−

Table 2
The effect of reserpine (2.5 mg/kg bw ip at −45 min) and pargyline (75 mg/kg bw ip at −120 min) alone or in combination on SAL- (10 mg/kg bw iv at 0 min) induced PRL release

Treatment	Time (min)					
	−120	−45	0	10	30	60
Saline + reserpine + saline	13.6 ± 2.0	7.7 ± 0.8	575.3 ± 91.9	637.4 ± 97.9	711.4 ± 38.1	727.9 ± 50.4
Saline + reserpine + SAL	10.6 ± 2.0	8.7 ± 0.6	582.5 ± 84.9	595.4 ± 51.6	646.4 ± 97.5	873.4 ± 211.6
Pargyline + saline + SAL	23.9 ± 3.7	5.8 ± 0.2	9.2 ± 1.2	359.3 ± 45.0	154.1 ± 13.4	68.3 ± 12.2
Pargyline + reserpine + SAL	23.6 ± 1.8	2.3 ± 0.6	89.7 ± 24.7	923.9 ± 130.2	924.1 ± 178.2	491.9 ± 173.0

(as well as tetrabenazine; data not shown), completely inhibits the PRL releasing effect of SAL. It is well known that reserpine decreases the level of DA in the LPV [7,8,58]. Furthermore, the decreasing level of DA causes a compensatory increase in AADC activity (i.e., DA synthesis) through a positive feedback mechanism [118]. Blockage of MAO by pretreatment of the animals with pargyline prior to reserpine brings back SAL-induced PRL response (Table 2). In a parallel experiment, the maximal binding capacity of ^{3}H-SAL was increased, while K_D value was just slightly effected following reserpine treatment [65]. The involvement and requirement of the presence of the intraterminal dopamine in SAL's action clearly suggests that one possible target of SAL action is the dopaminergic terminals.

4. Influence of neurotoxins on hypophysiotrophic NEDA neurons

4.1. Sensitivity of hypophysiotrophic NEDA neurons to DAT inhibitor and to 6-OHDA

As it has been already stated (vide supra), uptake mechanism for DA in the median eminence, IL, and NL is less robust than in other brain regions [2,3,29]. However, the fact that intravenous administration of 6-OHDA (a neurotoxin requiring uptake through DAT) decreases the concentration of DA in the IL and NL suggests the presence of DAT in these terminal areas [28,76,105,108,109,110].

It is interesting that the nerve terminals in the external layer of the median eminence show a higher resistance to treatment with 6-OHDA compared with other DA nerve terminals [68]. Injection of 6-OHDA into the medio-basal hypothalamus had no effect upon basal PRL secretion of male rats. Examination of catecholamine fluorescence indicated that 6-OHDA neither destroyed TIDA neurons of the ARC nucleus nor the majority of the catecholamine terminals in the median eminence [28]. The resistance of TIDA neurons to the toxic effect of 6-OHDA has been also noted by others [26] and was attributed in the conclusion that there is a lack of a high-affinity uptake system for catecholamine in the median eminence [29]. Learning more about the reverse transport function of DAT, which is responsible for the basal (tonic) release of DA at the terminal sites of the NEDA neurons, may also help to solve these controversies.

4.2. Sensitivity of NEDA neurons to 1-methyl-4-phenyl-1,2,3,6-tetrahydropyridine (MPTP)

It has been shown that the ARC nucleus and the PeV hypothalamic region of the rat brain possess very high densities of MPTP binding sites [67]. Acute microinfusion of MPTP bilaterally into the ARC−ventromedial area of the hypothalamus produced a dose-dependent facilitation of lordosis behavior in oestrogen- and oestrogen−progestorone-treated OVX rats. This facilitation could be blocked by pretreatment of this site of the brain with pargyline [107]. Using systemic route of administration, MPTP (10 mg/kg bw/day, ip) treatments of rats for 7 or 14 days produced a significant decrease of DOPAC levels in the median eminence and the PeV region, and an increase in serum PRL levels. This suggests that systemic administration of MPTP can influence terminals of the TIDA neurons [76].

Recent experimental data have clearly suggested that perturbation of a tightly regulated balance between DAT and VMAT2 predisposes dopaminergic neurons to a damaging influence of a variety of toxins [86]. Although VMAT2 appears to provide a certain degree of protection, the best known neurotoxins appear to gain access to the DA neurons by DAT. The difference or similarity between the SAL binding site and the transporter, which is responsible for taking up neurotoxins, have been tested in the ^{3}H-SAL binding assay using median eminence and NIL homogenates. As shown on Fig. 4, 6-OHDA could displace ^{3}H-SAL from its sites in the NIL, but MPTP could not influence it.

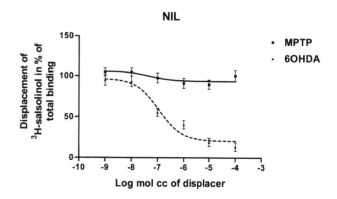

Fig. 4. Displacement of ^{3}H-SAL by different dopaminergic compounds. 6-OHDA could, but MPTP could not, displace ^{3}H-SAL. Representative data from an experiment repeated two times with three parallels in each.

None of these compounds influenced ³H-SAL in the median eminence (data not shown).

5. Conclusions

A consensus view developed over the last two decades holds that the hypophysiotropic signal mediating tonic as well as stimulus-induced release of PRL travels from the median eminence to the AL of the pituitary gland via the LPV of the hypothalamo-hypophysial portal system. All of these messages are believed to involve a transient change in the secretion of hypothalamic DA. The most direct studies, in this respect, have been performed in the laboratory of Jimmy D. Neill [56,101,102], in which the mammary nerve was electrically stimulated to simulate suckling and DA release was detected in hypophysial stalk blood [56,101] or in the median eminence with an electrochemical probe [102]. Only a brief (3–5 min) 60–70% decline in DA release was observed [56,101,102]. This decline was followed by series of rapid pulses of DA above the baseline, which lasted for the duration of mammary nerve stimulation [101]. These results have led to the conclusion that change in DA outflow from the hypothalamus itself may be sufficient for initiating a series of events, but insufficient to account for the prolonged elevation of PRL release induced by suckling. Although much progress has been made in identifying neuronal systems and events involved in this process, the precise nature of the terminal signal initiating PRL release has remained elusive.

Based on our recent findings, it may be relevant to postulate that an "auto-synaptocrine" regulatory mechanism functions in the dopaminergic terminals (Fig. 5) at the level of both the median eminence and NIL. Furthermore, SAL may be a key player in these regulatory processes. The complete and precise mapping of the complexity of these "auto-synaptocrine" mechanisms will help to understand the regulation of PRL secretion. Fig. 5 summarizes all of the possible sites, where SAL may influence DA transmission. SAL is most likely synthesized in DA terminals by SAL synthase [94]. One possible way to eliminate SAL from the cytosol is its sequestration into the secretory vesicles. Consequently, together with DA, SAL will be released to the LPV through the exocytotic process of the vesicles. Following its release, it might affect pre- and/or postsynaptic receptor sites (D_{2S} and D_{2L}, respectively). Parallel with the release of DA through the exocytotic process, another mechanism of release can also exist, namely the reverse transport. Newly synthesized cytosolic DA uses DAT, ordinarily responsible for uptake function, for this reverse mode of transport. This "atypical" mode of release can be influenced directly at the preterminal level by endogenous factors. One possible candidate for this regulatory role is SAL. However, it is possible that reverse transport of DA uses a completely different carrier protein. The third possible way of SAL to have an influence on DA release is an also "unconventional" inhibition of DA synthesis at the level of the AADC.

Our discussion has focused on the possible mechanism and site of action of SAL participating in the hypophysiotrophic regulation of PRL secretion. However, our data may

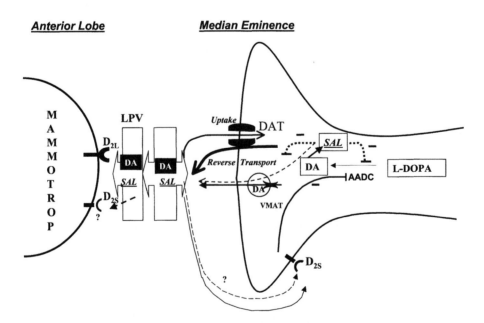

Fig. 5. Summary of proposed "synaptocrine" regulatory processes. Cytoplasmic DA is produced by synthesis, carrier-mediated uptake, and displacement from vesicular stores. DA can be catabolized by MAO, synthesis of SAL, exocytosis following vesicularization, and release directly through DAT. DA released to the LPVs is transported to the pituitary gland where it binds to its postsynaptic receptors (D_{2L} or D_{2S}). DA can also influence presynaptic D_{2S} receptor. SAL can also be sequestered into synaptic vesicles and released to LPV. According to our proposal, SAL can influence (inhibit) tonic reverse transport through DAT and/or DA synthesis by inhibiting aromatic AADC.

also give insight into the functional role of presynaptic processes that most likely have distinct and significant roles in other brain areas like nigrostriatal (NSDA) and mesolimbic (MSDA) systems also using dopaminergic transmission. Previous studies have already shown that SAL itself has no action on NSDA activity but it could effectively antagonize behavioral and biochemical changes accompanied with dopaminergic hyperactivity induced by lesion or drug treatment of the rats [5,117]. All together, these results clearly support the hypothesis that SAL can alter dopaminergic transmission and/or DA synthesis metabolism at the presynaptic level.

Acknowledgments

This work was sponsored by the Hungarian National Research Grant (OTKA 030748 to G.M.N.) and by the Research Grant of the Hungarian Ministry of Public Health (ETT 277/2001 to G.M.N.). The PhD Program of the Semmelweis University has also supported this work (to B.E.T., I.B., and K.G.H.). K.G. Homicskó was a recipient of a 2000 Summer Research Fellowship of the US Endocrine Society.

References

[1] S.G. Amara, M.J. Kuhar, Neurotransmitter transporters: Recent progress, Annu. Rev. Neurosci. 16 (1993) 73–93.

[2] L. Annunziato, P. Leblanc, C. Kordon, R.L. Weiner, Differences in the kinetics of dopamine uptake in synaptosome preparations of the median eminence relative to other dopaminergically innervated brain regions, Neuroendocrinology 31 (1980) 316–320.

[3] L. Annunziato, R.I. Weiner, Characteristics of dopamine uptake 3,4-dihydroxyphenylacetic acid (DOPAC) formation in the dopaminergic terminals of the neurointermediate lobe of the pituitary gland, Neuroendocrinology 31 (1980) 8–12.

[4] L. Antkiewicz-Michaluk, I. Romanska, I. Papla, J. Michaluk, M. Bakalarz, J. Vetulani, A. Krygowska-Wajs, A. Szcudlik, Neurochemical changes induced by acute and chronic administration of 1,2,3,4-tetrahydroisoquinoline and salsolinol in dopaminergic structures of rat brain, Neuroscience 96 (2000) 59–64.

[5] L. Antkiewicz-Michaluk, J. Michaluk, I. Romanska, I. Papla, J. Vetulani, Antidopaminergic effects of 1,2,3,4-tetrahydroisoquinoline and salsolinol, J. Neural Transm. 107 (2000) 1009–1019.

[6] G.W. Arbuthnott, I.S. Fairbrother, S.P. Butcher, Brain microdialysis studies on the control of dopamine release and metabolism in vivo, J. Neurosci. Methods 34 (1990) 73–81.

[7] R.L. Averill, D.R. Grattan, S.K. Norris, Posterior pituitary lobectomy chronically attenuates the nocturnal surge of PRL in early pregnancy, Endocrinology 128 (1991) 705–709.

[8] C.A. Barraclough, C.H. Sawyer, Induction of pseudopregnancy in the rat by reserpine and chlorpromazine, Endocrinology 65 (1959) 563.

[9] M.H. Baumann, R.B. Rothman, Effects of acute and chronic cocaine on the activity of tuberoinfundibular dopamine neurons in the rat, Brain Res. 608 (1993) 175–179.

[10] H.G. Baumgarten, A. Björklund, A.F. Holstein, A. Nobin, Organization and ultrastructural identification of the catecholamine nerve terminals in the neural lobe and pars intermedia of the rat pituitary, Z. Zellforsch. 126 (1972) 483–517.

[11] N. Ben-Jonathan, Dopamine: A prolactin-inhibiting hormone, Endocrinol. Rev. 6 (1985) 564–589.

[12] N. Ben-Jonathan, C. Oliver, H.J. Weiner, R.S. Mical, J.C. Porter, Dopamine in hypophysial portal plasma of the rat during the estrous cycle and throughout pregnancy, Endocrinology 100 (1977) 452–480.

[13] C.A. Birge, L.S. Jacobs, C.T. Hammeo, W.H. Daughaday, Catecholamine inhibition of prolactin secretion by isolated rat adenohypophyses, Endocrinology 86 (1970) 120.

[14] A. Bjorklund, R.Y. Moore, A. Nobin, U. Stenevi, The organization of tubero-hypophyseal and reticulo-infundibular catecholamine neuron systems in the rat brain, Brain Res. 51 (1973) 171–191.

[15] I. Bodnár, P. Gooz, H. Okamura, B.E. Tóth, M. Vecsernyés, B. Halász, G.M. Nagy, Effect of neonatal treatment with monosodium glutamate on dopaminergic and L-DOPAergic neurons of the medial basal hypothalamus and on prolactin and MSH secretion of rats, Brain Res. Bull. 55 (2001) 767–774.

[16] R. Bossé, F. Fumagalli, M. Jaber, B. Giros, R.R. Gainetdinov, W.C. Wetsel, C. Missale, M.G. Caron, Anterior pituitary hypoplasia and dwarfism in mice lacking the dopamine transporter, Neuron 19 (1997) 127–138.

[17] R. Bossé, R. Rivest, T. Di Paolo, Ovariectomy and estradiol treatment affect the dopamine transporter and its gene expression in the rat brain, Mol. Brain Res. 46 (1997) 343–346.

[18] R.D. Britton, C. Rivier, T. Shier, F. Bloom, W. Vale, In vivo and in vitro effects of tetrahydroisoquinolines and other alkaloids on rat pituitary function, Biochem. Pharmacol. 31 (1982) 1205–1211.

[19] G.M. Brown, P. Seeman, T. Lee, Dopamine/neuroleptic receptors in basal hypothalamus and pituitary, Endocrinology 99 (1976) 1407–1410.

[20] J.R. Bunzow, H.H.M. Van Tol, D.K. Grandy, P. Albert, J. Salon, M. Christie, C.A. Machida, K.A. Neve, O. Civelli, Cloning and expression of a rat D_2 dopamine receptor cDNA, Nature 336 (1988) 783–787.

[21] M.G. Caron, N. Amlaiky, B.F. Kilpatrick, D_2-dopamine receptor: Biochemical characterization, in: W.F. Ganong, L. Martini (Eds.), Frontiers in Neuroendocrinology, Raven Press, New York, 1986, pp. 205–224.

[22] C. Cerruti, D. Walther, M.J. Kuhar, G.R. Uhl, Dopamine transporter mRNA expression is intense in rat brain neurons and modest outside midbrain, Mol. Pharmacol. 18 (1993) 181–186.

[23] V. Chan-Palay, L. Zaborszky, C. Kohler, M. Goldstein, S.L. Palay, Distribution of tyrosine-hydroxylase-immunoreactive neurons in the hypothalamus of rats, J. Comp. Neurol. 227 (1984) 467–496.

[24] J.A. Coppola, R.G. Leonardi, W. Lippman, J.W. Perrine, I. Ringler, Induction of pseudopregnancy in rats by depletors of endogenous catecholamines, Endocrinology 77 (1965) 485.

[25] I. Creese, R. Schneider, S.H. Snyder, [3H]-Spiroperidol labels dopamine receptors in pituitary and brain, Eur. J. Pharmacol. 46 (1977) 377.

[26] A.C. Cuello, W.J. Shoemaker, W.F. Ganong, Effect of 6-hydroxydopamine on hypothalamic norepinephrine and dopamine content, ultrastructure of the median eminence, and plasma corticosterone, Brain Res. 78 (1974) 57–69.

[27] K. Dahlström, K. Fuxe, Evidence for the existence of monoaminergic containing neurons in the central nervous system. I. Demonstration of monoamines in cell bodies of brain stem neurons, Acta Physiol. Scand. 62 (Suppl. 232) (1964) 1–54.

[28] T.A. Day, P.M. Jervois, M.F. Menadue, J.O. Willoughby, Catecholamine mechanisms in medio-basal hypothalamus influence prolactin but not growth hormone secretion, Brain Res. 253 (1982) 213–219.

[29] K.T. Demarest, K.E. Moore, Lack of a high affinity transport system for dopamine in the median eminence and posterior pituitary, Brain Res. 171 (1979) 545–551.

[30] K.T. Demarest, D.W. McKay, G.D. Riegle, K.E. Moore, Sexual differences in tuberoinfundibular dopamine nerve activity induced by neonatal androgen exposure, Neuroendocrinology 32 (1981) 108.

[31] J.E. DeMaria, D. Zelena, M. Vecsernyés, G.M. Nagy, M.E. Freeman,

The effect of neurointermediate lobe denervation on hypothalamic neuroendocrine dopaminergic neurons, Brain Res. 806 (1998) 89–94.

[32] J.E. DeMaria, G.M. Nagy, A. Lerant, M.I.K. Fekete, C.W. Levenson, M.E. Freeman, Dopamine transporters participate in the physiological regulation of prolactin, Endocrinology 141 (2000) 366–374.

[33] J.E. DeMaria, A. Lerant, M.E. Freeman, Prolactin activates all three populations of hypothalamic neuroendocrine dopaminergic neurons in ovariectomized rats, Brain Res. 837 (1999) 236–241.

[34] J.E. DeMaria, J.D. Livingstone, M.E. Freeman, Characterization of the dopaminergic input to the pituitary gland throughout the estrous cycle of the rat, Neuroendocrinology 67 (1998) 377–383.

[35] Y. Deng, W. Maruyama, P. Dostert, T. Takahashi, M. Kawai, M. Naoi, Determination of the (R)- and (S)-enantiomers of salsolinol and N-methylsalsolinol by use of a chiral high-performance liquid chromatographic column, J. Chromatogr., Biomed. Appl. 650 (1995) 47–54.

[36] P. Dostert, M. Strolin Benedetti, V. Bellotti, C. Allievi, G. Dordain, Biosynthesis of salsolinol, a tetrahydroisoquinoline alkaloid, in healthy subjects, J. Neural Transm. 81 (1990) 215–223.

[37] J.W. Everett, Functional corpora lutea maintained for months by autografts of rat hypophysis, Endocrinology 58 (1956) 786–796.

[38] B. Falk, N.A. Hillarp, G. Thieme, A. Torp, Fluorescence of catecholamines and related compounds condensed with formaldehyde, J. Histochem. Cytochem. 10 (1962) 348–354.

[39] B.H. Falkenburger, K.L. Barstow, I.M. Mintz, Dendrodendritic inhibition through reversal of dopamine transport, Science 293 (2001) 2465–2470.

[40] C. Freed, R. Revay, R.A. Vaughan, E. Kriek, S. Grant, G.R. Uhl, M.J. Kuhar, Dopamine transporter immunoreactivity in rat brain, J. Comp. Neurol. 359 (1995) 340–349.

[41] M.E. Freeman, B. Kanyicska, A. Lerant, G.M. Nagy, Prolactin: Structure, function, and regulation of secretion, Physiol. Rev. 80 (2000) 1523–1631.

[42] K. Fuxe, Cellular localization of monoamines in the median eminence and in the infundibular stem of some mammals, Acta Physiol. Scand. 58 (1964) 383–384.

[43] K. Fuxe, Evidence for the existence of monoamine neurons in the central nervous system, Acta Physiol. Scand., Suppl. (1965) 239–247.

[44] R. Gardner, M.H. Richards, Use of D,L-α-monofluoromethyldopa to distinguish subcellular pools of aromatic amino acid decarboxylase in mouse brain, Brain Res. 216 (1981) 291–298.

[45] P.A. Garris, N. Ben-Jonathan, Effects of reuptake inhibitors on dopamine release from the stalk–median eminence and posterior pituitary in vitro, Brain Res. 556 (1991) 123–129.

[46] M. Geffard, R.M. Buijs, P. Seguela, C.W. Pool, M. Le Moal, First demonstration of highly specific and sensitive antibodies against dopamine, Brain Res. 294 (1984) 161–165.

[47] M. Gerlach, E. Koutsilieri, P. Riederer, N-Methyl-(R)-salsolinol and its relevance to Parkinson's disease, Lancet 351 (1998) 850–851.

[48] D.M. Gibbs, J.D. Neill, Dopamine levels in hypophysial stalk blood in the rat are sufficient to inhibit prolactin secretion in vivo, Endocrinology 102 (1978) 1895–1900.

[49] B. Giros, M.G. Caron, Molecular characterization of the dopamine transporter, Trends Pharmacol. Sci. 14 (1993) 43–49.

[50] J. Glowinski, L.L. Iversen, Regional studies of catecholamines in the rat brain: I. The disposition of [³H]norepinephrine, [³H]dopamine and [³H]dopa in various regions of the brain, J. Neurochem. 13 (1966) 655–669.

[51] P.C. Goldsmith, M.J. Cronin, R.I. Weiner, Dopamine receptor sites in the anterior pituitary, J. Histochem. Cytochem. 27 (1979) 1205.

[52] J.L. Goudreau, K.J. Lookingland, K.E. Moor, 5-Hydroxytryptamine 2 receptor-mediated regulation of periventricular–hypophysial dopaminergic neuronal activity and the secretion of alpha-melanocyte-stimulating hormone, J. Pharmacol. Exp. Ther. 268 (1994) 175–179.

[53] J.L. Goudreau, S.E. Lindley, K.J. Lookingland, K.E. Moor, Evidence that hypothalamic periventricular dopamine neurons innervate the

[54] J.L. Goudreau, W.M. Falls, K.J. Lookingland, K.E. Moore, Periventricular–hypophysial dopaminergic neurons innervate the intermediate but not the neural lobe of the rat pituitary gland, Neuroendocrinology 62 (1995) 147–154.

[55] J.W. Greef, J.D. Neill, Dopamine levels in hypophysial stalk plasma of the rat during surges of prolactin secretion induced by cervical stimulation, Endocrinology 105 (1979) 1093.

[56] J.W. Greef, P.M. Plotsky, J.D. Neill, Dopamine levels in hypophysial stalk plasma and prolactin levels in peripheral plasma of the lactating rat: Effect of a simulated suckling stimulus, Neuroendocrinology 32 (1981) 229.

[57] G.A. Gudelsky, H.Y. Meltzer, Function of tuberoinfundibular dopamine neurons in pargyline- and reserpine-treated rats, Neuroendocrinology 38 (1984) 51–55.

[58] G.A. Gudelsky, J.C. Porter, Release of newly synthesized dopamine into the hypophysial portal vasculature of the rat, Endocrinology 104 (1979) 583–587.

[59] G.A. Gudelsky, J.C. Porter, Sex-related difference in the release of dopamine into hypophysial portal blood, Endocrinology 109 (1981) 1394–1398.

[60] T. Higuchi, K. Honda, S. Takano, H. Negoro, Estrogen fails to reduce tuberoinfundibular dopaminergic neuronal activity and to cause a prolactin surge in lactating, ovariectomized rats, Brain Res. 576 (1992) 143–146.

[61] B.J. Hoffman, S.R. Hansson, E. Mezey, M. Palkovits, Localization and dynamic regulation of biogenic amine transporters in the mammalian central nervous system, Front. Neuroendocrinol. 19 (1998) 187–231.

[62] T. Hökfelt, O. Johansson, K. Fuxe, M. Goldstein, D. Park, Immunohistological studies on the localization and distribution of monoamine neuron systems in the rat brain: I. Tyrosine hydroxylase in the mes-, and diencephalon, Med. Biol. 54 (1976) 427–453.

[63] M. Holzbauer, D.F. Sharman, U. Godden, Observations on the function of the dopaminergic nerves innervating the pituitary gland, Neuroscience 3 (1978) 1251–1262.

[64] M. Holzbauer, K. Racké, The dopaminergic innervation of the intermediate lobe and of the neural lobe of the pituitary gland, Med. Biol. 63 (1985) 97–116.

[65] K.G. Homicsko, I. Kertész, B. Radnai, B.E. Toth, G. Toth, F. Fülöp, M.I.K. Fekete, G.M. Nagy, Binding site of salsolinol: Its properties in different regions of the brain and pituitary gland of the rat, Neurochem. Int., 2002 (submitted for publication).

[66] C.B. Jaeger, D.A. Ruggiero, V.R. Albert, V.R. Park, T.H. Joh, D.J. Reis, Aromatic L-amino acid decarboxylase in the rat brain: Immunocytochemical localization in neurons of the brain stem, Neuroscience 11 (1984) 691–713.

[67] J.A. Javitch, G.R. Uhl, S.H. Snyder, Parkinsonism-inducing neurotoxin, N-methyl-4-phenyl-1,2,3,6-tetrahydropyridine: Characterisation and localization of receptor binding sites in rat and human brain, Proc. Natl. Acad. Sci. U. S. A. 81 (1984) 4591–4595.

[68] G. Jonnson, K. Fuxe, T. Hökfelt, On the catecholamine innervation of the hypothalamus, with special reference to the median eminence, Brain Res. 40 (1972) 271–281.

[69] S. Kanematsu, J. Hillard, C.H. Sawyer, Effect of reserpine on pituitary prolactin content and its hypothalamic site of action in the rabbit, Acta Endocrinol. 44 (1963) 467.

[70] S. Kanematsu, C.H. Sawyer, Elevation of plasma prolactin after hypophysial stalk section in the rat, Endocrinology 93 (1973) 238–241.

[71] I. Kertész, F. Fülöp, G. Tóth, G.M. Nagy, Synthesis, tritiation and separation of R,S-salsolinol, in: U. Pleiss, R. Voges (Eds.), Synthesis and Applications of Isotopically Labelled Compounds, vol. 7, Wiley, New York, 2001, pp. 151–154.

[72] O. Khorram, J.C. Bedran deCastro, S.M. McCann, The influence of suckling on the hypothalamic and pituitary secretion of immuno-

reactive α-melanocyte stimulating hormone, Brain Res. 398 (1986) 361–365.

[73] T.S. King, R.S. Schenken, I.S. Kang, M.A. Javors, R.M. Riehl, Cocaine disrupts estrous cyclicity and alters the reproductive neuroendocrine axis in the rat, Neuroendocrinology 51 (1990) 15–22.

[74] D.A. Leong, L.S. Frawley, J.D. Neill, Neuroendocrine control of PRL secretion, Annu. Rev. Physiol. 45 (1983) 109–127.

[75] V. Leviel, The reverse transport of DA, what physiological significance? Neurochem. Int. 38 (2001) 83–106.

[76] J.Y. Lin, L.M. Mai, J.T. Pan, Effects of systemic administration of 6-hydroxydopamine, 6-hydroxydopa and 1-methyl-4-phenyl-1,2,3,6-tetrahydroxypyridine (MPTP) on tuberoinfundibular dopaminergic neurons in the rat, Brain Res. 624 (1993) 126–130.

[77] K.J. Lookingland, H.D. Jarry, K.E. Moore, The metabolism of dopamine in the median eminence reflects the activity of tuberinfundibular neurons, Brain Res. 419 (1987) 303–310.

[78] K.J. Lookingland, J.W. Gunnet, K.E. Moor, Stress-induced secretion of alpha-melanocyte-stimulating hormone is accompanied by a decrease in the activity of tuberohypophysial dopaminergic neurons, Neuroendocrinology 53 (1991) 91–96.

[79] D. Lorang, S.G. Amara, R.B. Simerly, Cell-type specific expression of catecholamine transporters in the rat brain, J. Neurosci. 14 (1994) 4903–4914.

[80] R.M. MacLeod, Influence of norepinephrine and catecholamine-depleting agents on the synthesis and release of prolactin and growth hormone, Endocrinology 85 (1969) 916–923.

[81] R.M. MacLeod, Regulation of prolactin secretion, in: L. Martini, W.F. Ganong (Eds.), Frontiers in Neuroendocrinology, Raven Press, New York, 1976, pp. 169–194.

[82] J. Manzanares, T.W. Toney, Y. Tian, M.J. Eaton, K.E. Moor, K.J. Lookingland, Sexual differences in the activity of periventricular–hypophysial dopaminergic neurons in the rats, Life Sci. 51 (1992) 995–1001.

[83] W. Maruyama, T. Abe, H. Tohgi, P. Dostert, M. Naoi, A dopaminergic neurotoxin, (R)-N-methylsalsolinol, increases in parkinsonian cerebrospinal fluid, Ann. Neurol. 40 (1996) 119–122.

[84] J.H. Meador-Woodruff, A. Mansour, J.R. Bunzow, H.H.M. Van Tol, S.J. Watson Jr., O. Civelli, Distribution of D_2 dopamine receptor mRNA in rat brain, Proc. Natl. Acad. Sci. U. S. A. 86 (1989) 7625–7628.

[85] B. Meister, T. Hokfelt, H.W. Steinbusch, G. Skagerberg, O. Lindvall, M. Geffard, T.H. Joh, A.C. Cuello, M. Goldstein, Do tyrosine hydroxylase-immunoreactive neurons in the ventrolateral arcuate nucleus produce dopamine or only L-dopa? J. Chem. Neuroanat. 1 (1988) 59–64.

[86] G.W. Miller, R.R. Gainetdinov, A.I. Levey, M.G. Caron, Dopamine transporters and neuronal injury, Trends Pharmacol. Sci. 20 (1999) 424–429.

[87] Y. Misu, Y. Goshima, H. Ueda, H. Okamura, Neurobiology of L-DOPAergic systems, Prog. Neurobiol. 49 (1996) 415–454.

[88] K.E. Moore, K.T. Demarest, Tuberoinfundibular and tuberohypophyseal dopaminergic neurons, in: W.F. Ganong, L. Martini (Eds.), Frontiers in Neuroendocrinology, Raven Press, New York, 1982, pp. 161–190.

[89] F. Musshoff, P. Schmidt, R. Dettmeyer, F. Priemer, K. Jachau, B. Madea, Determination of dopamine and dopamine-derived (R)-/(S)-salsolinol and norsalsolinol in various human brain areas using solid-phase extraction and gas chromatography/mass spectrometry, Forensic Sci. Int. 113 (2000) 359–366.

[90] G.M. Nagy, A. Arendt, Zs. Banky, B. Halasz, Dehydration attenuates plasma prolactin response to suckling through a dopaminergic mechanism, Endocrinology 130 (1992) 819–824.

[91] G.M. Nagy, M. Vecsernyés, I. Barna, Dehydration decreases plasma level of α-melanocyte stimulating hormone (α-MSH) and attenuates suckling-induces β-endorphin (β-END) but not ACTH response in lactating rats, Neuroendocrinol. Lett. 16 (1994) 275–284.

[92] G.M. Nagy, J.E. DeMaria, M.E. Freeman, Changes in the local metabolism of dopamine in the anterior and neural lobes but not in the intermediate lobe of the pituitary gland during nursing, Brain Res. 790 (1998) 315–317.

[93] M. Naoi, W. Maruyama, P. Dostert, Y. Hashizume, N-Methyl-(R)-salsolinol as a dopaminergic neurotoxin: From an animal model to an early marker of Parkinson's disease, J. Neural. Transm., Suppl. 50 (1997) 89–105.

[94] M. Naoi, W. Maruyama, P. Dostert, K. Kohda, T. Kaiya, A novel enzyme enantio-selectively synthesize (R)salsolinol, a precursor of a dopaminergic neurotoxin, N-methyl(R)salsolinol, Neurosci. Lett. 212 (1996) 183–186.

[95] J.D. Neill, G.M. Nagy, Prolactin secretion and its control, in: E. Knobil, J.D. Neill (Eds.), The Physiology of Reproduction, Raven Press, New York, 1994, pp. 1833–1860.

[96] J.D. Neill, Neuroendocrine regulation of prolactin secretion, in: L. Martini, W.F. Ganong (Eds.), Frontiers in Neuroendocrinology, Raven Press, New York, 1980, pp. 129–155.

[97] H. Okamura, K. Kitahama, L. Nagatsu, M. Geffard, Comparative topography of dopamine- and tyrosine hydroxylase-immunoreactive neurons in the rat arcuate nucleus, Neurosci. Lett. 95 (1988) 347–353.

[98] H. Okamura, K. Kitahama, B. Raynaud, I. Nagatsu, C. Borri-Volttatorni, M. Weber, Aromatic L-amino acid decarboxylase (AADC)-immunoreactive cells in the tuberal region of the rat hypothalamus, Biomed. Res. 9 (1988) 261–267.

[99] J.T. Pan, Y. Tian, K.J. Lookingland, K.E. Moor, Neurotensin-induced activation of hypothalamic dopaminergic neurons is accompanied by a decrease in pituitary secretion of prolactin and alpha-melanocyte-stimulating hormone, Life Sci. 50 (1992) 2011–2017.

[100] P.M. Plotsky, D.M. Gibbs, J.D. Neill, Liquid chromatographic–electrochemical measurement of dopamine in hypophysial stalk blood of rats, Endocrinology 102 (1978) 1887–1894.

[101] P.M. Plotsky, W.J. de Greef, J.D. Neill, In situ voltametric microelectrodes: Application to the measurement of the median eminence catecholamine release during simulated suckling, Brain Res. 250 (1982) 251.

[102] P.M. Plotsky, J.D. Neill, The decrease in hypothalamic dopamine secretion induced by suckling: Comparison of voltametric and radioisotopic methods of measurement, Endocrinology 110 (1982) 691–696.

[103] I. Putscher, H. Haber, A. Winkler, J. Fickel, M.F. Melzing, Effect of S(−)- and R(+)-salsolinol on the POMC gene expression and ACTH release of an anterior pituitary cell line, Alcohol 12 (1995) 447–452.

[104] R. Revay, R. Vaughan, S. Grant, M.J. Kuhar, Dopamine transporter immunohistochemistry in median eminence, amygdala, and other areas of the rat brain, Synapse 22 (1996) 93–99.

[105] W.J. Sheward, A.G. Watts, G. Fink, G.C. Smith, Effects of intravenously administered 6-hydroxydopamine on the content of monoamines in the median eminence and neurointermediate lobe of the rat, Neurosci. Lett. 55 (1985) 141–144.

[106] K.L. Sims, G.A. Davis, F.E. Bloom, Activities of DOPA and 5-HTP decarboxylases in rat brain: Assay characteristics and distribution, J. Neurochemistry 20 (1973) 449–464.

[107] D.J. Sirinathsinghji, Intrahypothalamic infusions of a synthetic heroin substitute, N-methyl-4-phenyl-1,2,3,6-tetrahydropyridine, potentiate mating behaviour in the female rat, Brain Res. 346 (1985) 130–135.

[108] G.C. Smith, P.G. Courtney, N.G. Wreford, M.M. Walker, Further studies on the effects of intravenously administered 6-hydroxydopamine on the median eminence of the rat, Brain Res. 234 (1982) 101–110.

[109] G.C. Smith, W.J. Sheward, G. Fink, Effect of 6-hydroxydopamine lesions of the median eminence and neurointermediate lobe on the secretion of pituitary hormones in the male rat, Brain Res. 246 (1982) 330–333.

[110] G.C. Smith, R.D. Helme, Ultrastructural and fluorescence histochemical studies on the effects of 6-hydroxydopamine on the rat median eminence, Cell Tissue Res. 152 (1974) 493–512.

[111] S. Taleisnik, Control of melanocyte-stimulating hormone (MSH) secretion, in: S.L. Jeffcoate, J.S.M. Hutchinson (Eds.), The Endocrine Hypothalamus, Academic Press, London, 1978, pp. 421–439.

[112] G.B. Thomas, J.T. Cummins, B.J. Canny, S.E. Rundle, N. Griffin, S. Katasahambas, I.J. Clarke, The posterior pituitary regulates prolactin, but not adrenocorticotropin or gonadotropin, secretion in the sheep, Endocrinology 125 (1989) 2204–2211.

[113] Y. Tian, G. Kapatos, J.G. Granneman, M.J. Bannon, Dopamine and γ-aminobutyric acid transporters: Differential regulation by agents that promote phosphorylation, Neurosci. Lett. 173 (1994) 143–146.

[114] F.J. Tilders, H.F. Berkenbosch, S. Melik, Control of secretion of peptides related to adrenocorticotropin, melanocyte-stimulating hormone and endorphin, in: T.J.B. Wimersma Griedanus (Ed.), Frontiers in Hormone Research, Karger, Basel, 1985, pp. 161–196.

[115] B.E. Toth, K.G. Homicsko, B. Radnai, W. Maruyama, J.E. DeMaria, M. Vecsernyés, M.I.K. Fekete, F. Fülöp, M. Naoi, M.E. Freeman, G.M. Nagy, Salsolinol is a putative endogenous neuro-

[116] M. Vecsernyés, K. Krempels, B.E. Tóth, J. Julesz, G.B. Makara, G.M. Nagy, Effect of posterior pituitary denervation (PPD) on prolactin (PRL) and α-melanocyte-stimulating hormone (α-MSH) secretion of lactating rats, Brain Res. Bull. 43 (1998) 313–319.

[117] J. Vetulani, I. Nalepa, L. Antkiewicz-Michaluk, M. Sansone, Opposite effect of simple tetrahydroisoquinolines on amphetamine- and morphine-stimulated locomotor activity in mice, J. Neural. Transm. 108 (2001) 513–526.

[118] T.C. Wessel, T.H. Joh, Parallel upregulation of catecholamine-synthesizing enzymes in rat brain and adrenal gland: Effects of reserpine and correlation with immediate early genes, Mol. Brain Res. 15 (1992) 349–360.

[119] M.-Y. Zhu, A.V. Juorio, Aromatic L-amino acid decarboxylase: Biological characterization and functional role, Gen. Pharmacol. 26 (1995) 681–696.

intermediate lobe prolactin-releasing factor, J. Neuroendocrinol. 13 (2001) 1042–1050.

ELSEVIER

Neurotoxicology and Teratology 24 (2002) 667–673

NEUROTOXICOLOGY

AND

TERATOLOGY

www.elsevier.com/locate/neutera

Review article

Aliphatic propargylamines as symptomatic and neuroprotective treatments for neurodegenerative diseases ☆

M.D. Berry, A.A. Boulton*

ALviva Biopharmaceuticals Inc., 218-111 Research Drive, Saskatoon, SK, Canada S7N 3R2

Received 2 April 2001; accepted 24 January 2002

Abstract

Over the past several years, we have developed a number of novel aliphatic propargylamine-related compounds. These can be divided into 14 main chemical families. These families have been shown to possess members that selectively and stereochemically (i.e. R-enantiomer) rescue neurons from p53-dependent apoptosis in vitro. In contrast, no rescue has been observed by the enantiomers of the opposite configuration or in p53-independent apoptosis. In vivo, several compounds have been shown to possess neural rescue properties in models of unilateral hypoxia/ischaemia, focal ischaemia, facial nerve axotomy, pmn mice, 1-methyl-4-phenyl-1,2,3,6-tetrahydropyridine (MPTP) mouse and MPTP non-human primate. Our prototype compound, R-2HMP, has been shown to be metabolised in a manner analogous to that of R-deprenyl but devoid of amphetaminergic metabolites. These compounds have been shown to be active through an interaction with the same binding site as R-deprenyl and CGP 3466. This site is suggested to be the glycolytic enzyme glyceraldehyde-3-phosphate dehydrogenase (GAPDH). © 2002 Elsevier Science Inc. All rights reserved.

Keywords: Antiapoptotic; Neural rescue; Aliphatic propargylamine; Neurodegeneration

1. Introduction

Over the past several years, the potential role of apoptosis in a number of neurodegenerative conditions has been suggested (see Table 1). This has led to much interest in the development of antiapoptotic agents as potential therapies to either block or reduce the progression of these disorders. We have developed a number of novel, aliphatic propargylamine-related compounds that can be divided into 14 broad chemical families, examples of which are shown in Fig. 1. Several of these compounds have been tested in a variety of cellular models of apoptosis (see Table 2). Routinely, we have tested compounds in the cultured cerebellar granule cell model system that we have previously described in detail [6,26]. This model system has been shown to be amenable to the induction of multiple apoptotic pathways

[15,16], with cytosine arabinoside (Ara-c) inducing a p53-dependent pathway, and low K media inducing a p53-independent apoptosis. Using this model system, we have shown that a number of compounds stereoselectively inhibit apoptosis induced by Ara-c treatment. In contrast, we have yet to see activity against low K-induced apoptosis. These results suggest a level of specificity in the antiapoptotic action of the compounds, with activity only being demonstrated against p53-dependent forms of apoptosis. This is an important distinction, since, in general, p53-dependent apoptosis appears to be triggered by pathological/toxic insults, whereas p53-independent apoptosis appears to be triggered in a more physiological fashion. In agreement with the p53 dependence of the rescue observed, in every instance where we have observed a neural rescue effect (see Table 2), the toxin used to initiate apoptosis has previously been demonstrated to result in a p53-dependent apoptosis. This activity of our compounds has previously been reviewed elsewhere [7] and will not be discussed further here. The subsequent sections of this paper will review the activity of our compounds in disease-specific in vivo animal models. A list of in vivo models in which the aliphatic propargylamines have been shown to be active is provided in Table 3.

☆ Paper presented at the 9th International Catecholamine Symposium and 5th International Joint Congresses on Progress in Alzheimer's and Parkinson's Disease, Kyoto, Japan, March 31–April 5, 2001.
* Corresponding author. Tel.: +1-306-956-6880; fax: +1-306-956-6877.
E-mail address: aab@alviva.ca (A.A. Boulton).

2. Stroke models

2.1. Unilateral hypoxia/ischaemia

We have tested a number of our compounds in a modified Levine preparation of unilateral global cerebral ischaemia in rats. We have found this model to be more reproducible than the gerbil model, avoids the complications associated with anaesthesia, and the unilateral nature of the ischaemia provides an internal control. The basic protocol for this procedure is outlined below. On Day-1, the left common carotid artery is ligated under isoflurane/nitrous oxide anaesthesia. Following recovery in a warm cage, animals are returned to their home cage for 24 h. On Day 0, cerebral ischaemia is induced by placing the animals in a 0.2% CO atmosphere for 25 min at an environmental temperature of 22–25 °C. Animals are allowed to recover for at least 1 h at the same temperature prior to the onset of drug treatment. This drug treatment is maintained from Day 1 to Day 6. Twenty-four hours after the last drug treatment, animals are deeply anaesthetised and killed by transcardiac

Table 1
Human diseases showing increased apoptosis

Disease	References
Parkinson's disease	[2,9,24,34,36]
Glaucoma	[18]
HIV encephalitis	[27]
Amyotrophic lateral sclerosis	[37,40]
Huntington's disease	[10,12,28,35]
Alzheimer's disease	[1,20,21,32,33]
Ebola virus (non recoverers)	[3]
Diabetic retinopathy	[5]
Retinitis pigmentosum	[39]
Brain ischaemia (stroke)	[22]

perfusion with FAM (40% formaldehyde; glacial acetic acid; methanol, 1:1:8, v/v/v). Neuronal death is assessed in the CA1, CA3 and CA4 regions of the hippocampus following H&E staining.

Using this protocol, we have shown R-2HMP and R-2HPA to be considerably more potent rescue agents than is R-deprenyl (Fig. 2). Such an effect is seen following both subcutaneous and oral administration, with R-deprenyl

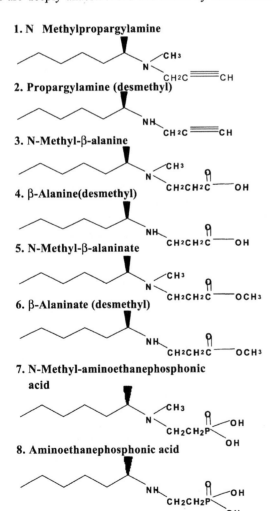

Fig. 1. Chemical families of aliphatic propargylamine-related compounds. Aliphatic propargylamine-related compounds have been divided into 14 broad chemical classes. Shown are heptyl examples of each family. The example shown for Families 1, 2 and 3 are abbreviated as R-2HMP, R-2HPA and R-2HPcAc.

Table 2
In vitro models of apoptosis in which aliphatic propargylamines have been tested

Model	Response
Cgc (ara-C)	p53-Dependent apoptosis, rescue by aliphatic propargylamines (R-isomers only)
Cgc (β-amyloid)	Rescue by R-2HMP, R-2HPA and R-2HPcAc in a dose-dependent manner
Cgc (low K⁺)	p53-Independent apoptosis, no rescue by aliphatic propargylamines
PC12 cells trophic withdrawal	Rescue by R-2HMP
Thymocytes (Dexamethasone)	A p53-independent nonneuronal model. No rescue by aliphatic propargylamines
Mesencephalic neurons (MPP+, aging)	Rescue by R-2HMP

Table 3
In vivo animal models of apoptosis in which aliphatic propargylamines have been tested

Model	Disease indication
Unilateral hypoxia/ischaemia	Stroke
Focal ischaemia (pial artery)	Stroke
Facial nerve axotomy	Amyotrophic lateral sclerosis
pmn mouse	Amyotrophic lateral sclerosis
DSP-4	–
Kainate	Seizures
MK 801	–
MPTP mouse	Parkinson's disease
MPTP primate	Parkinson's disease

unable to elicit any protection when given orally. In all instances, at least a doubling of neuronal survival, as compared to untreated controls was noted (Fig. 2). It is important to note that in all cases, drug treatment was not started until 1 h following lesion. Pretreatment failed to

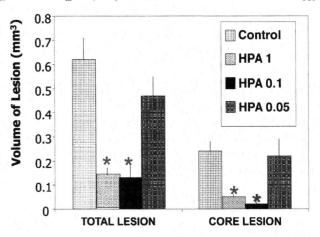

Fig. 3. Effect of intraperitoneal R-2HPA on lesion volume when administered 3 h after lesion. Total lesion is the volume of lesion including the penumbral region. Core lesion is the volume of lesion excluding the penumbral region. Lesion volumes were assessed following H&E staining. *$P < .05$, Newman–Keuls with respect to controls.

produce any rescue effect (data not shown). A number of other compounds also showed varying degrees of rescue when administered subcutaneously (data not shown).

2.2. Permanent focal ischaemia

We have extended these studies to determine the effects of certain selected compounds in a permanent focal ischaemia model. We have chosen to utilise the pial artery disruption model. Here, anaesthetised animals are placed in a stereotaxic frame and a 5-mm diameter hole placed in the skull 1 mm lateral and 1 mm rostral to bregma. The dura are carefully torn, and all terminal blood vessels burst under a dissecting microscope. The area is flushed with sterile, isotonic saline to remove blood, and the wound was closed.

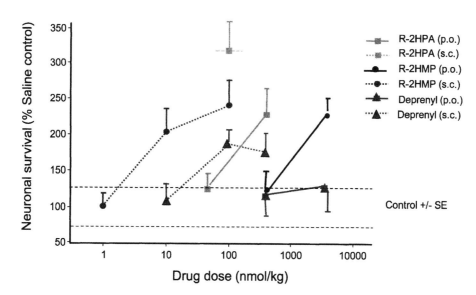

Fig. 2. Efficacy of R-2HMP, R-2HPA and R-deprenyl in the unilateral hypoxia/ischaemia model. Neuron counts were performed in the CA1 region of the hippocampus following staining with H&E. Values are expressed as percent of saline controls. All values in excess of 150% are significantly different from saline controls ($P < .01$).

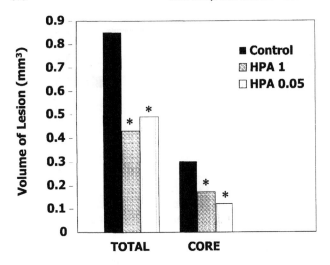

Fig. 4. Effect of oral *R*-2HPA on lesion volume when administered 3 h after lesion. *R*-2HPA was administered orally via gavage using a feeding tube, in a volume of 1 ml/kg. Total lesion is the volume of lesion including the penumbral region. Core lesion is the volume of lesion excluding the penumbral region. Lesion volumes were assessed following H&E staining. * $P < .05$, Newman–Keuls with respect to controls.

We have found that blood flow from the wound rarely continues beyond 45 s. At all times during surgery and recovery, body temperature is monitored and maintained. At various times postlesion, drug is administered. A second administration occurs 24 h after the first. Lesion size is determined 24 h after the last dose of compound as described above. Using this model, we have found that control animals suffer an approximately conical lesion with distinct core and penumbral regions. We have therefore estimated the size of the lesion by measuring the maximal width and depth. The width is used to determine the radius, the depth as the height, and the volume of lesion was calculated according to the formula

$$\text{Volume} = \pi r^2 h / 3$$

As can be seen in Figs. 3 and 4, *R*-2HPA, when administered initially 3 h after lesion, either orally or intraperitoneally, dose-dependently decreases both core and total lesion size. Following intraperitoneal administration, this results in an approximate 80% reduction in lesion size. Under the same conditions, oral administration reduced lesion size by approximately 50%. In contrast, *R*-2HMP was less efficacious when administered 3 h following lesion. Efficacy returned, however, when *R*-2HMP was initially administered 1 h following lesion (data not shown).

3. Parkinson's disease

MPTP-induced dopaminergic neurotoxicity is still regarded as the gold standard animal model for Parkinson's disease [29]. We have tested *R*-2HMP in two MPTP animal models. The first involved administration of MPTP (30 mg/kg

subcutaneous) to male C57BI/6J mice twice per day on Days 1 and 4. *R*-2HMP treatment began on Day 6 and continued twice daily until Day 23. Twenty-four hours following the last drug treatment, animals received an anaesthetic overdose and were transcardially perfused with 4% paraformaldehyde. Sections containing the substantia nigra were paraffin embedded and 5-μm sections were cut. Lesion size was determined by staining every tenth section for tyrosine hydroxylase, and the number of tyrosine hydroxylase-positive neurons was determined. As can be seen in Fig. 5, *R*-2HMP, at a dose of 0.1 mg/kg twice daily, resulted in an approximate halving of the loss of tyrosine hydroxylase-positive neurons. Since *R*-2HMP treatment was not started until 2 days after the final MPTP administration, this represents a true neural rescue effect. The degree of rescue obtained compares favourably with that reported for CGP 3466, a putative novel antiapoptotic agent [38].

Although many compounds have been tested in the MPTP mouse model, this is a less than ideal model [29]. We have therefore also tested, in a pilot study, *R*-2HMP in the more relevant non-human primate MPTP model. For this study, we adopted the intracarotid administration model of Bankiewicz et al. [4]. Briefly, young adult female cynamolgus monkeys (*Macaca fascicularis*) were anaesthetized with ketamine HCl (10 mg/kg im) and maintained with isoflurane. The internal carotid artery was isolated, the external and superior carotid arteries were clamped, and MPTP (2 mg) was infused in a total volume of 60 ml over a 15-min period. This method results in a unilateral Parkinsonian

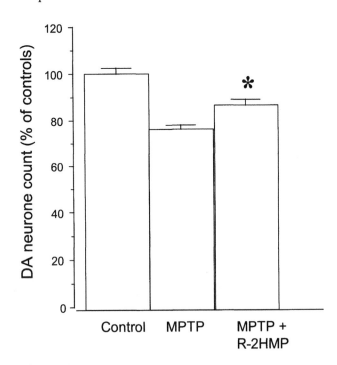

Fig. 5. Rescue of dopaminergic neurons by *R*-2HMP following MPTP administration. Mice were treated with MPTP as described in the text. Tyrosine hydroxylase-positive neurons in the substantia nigra were counted and expressed as percent of control. * $P < .05$, Mann–Whitney *U*-test, with respect to MPTP treated animals.

syndrome [4,30], which allows animals to maintain normal nutrition without the need for L-DOPA therapy. As such, this model is particularly amenable to the demonstration of neural rescue effects, without the possible confounding factor of L-DOPA symptomatic relief.

Two hours following MPTP administration, *R*-2HMP treatment commenced. *R*-2HMP was administered at a dose of 0.1 mg/kg im twice per day for 28 days. Twenty-four hours after the last drug administration, animals received an overdose of anaesthetic, brains were removed, and an 18-mm block containing the substantia nigra was immersed and fixed in 10% formalin. The degree of lesion was determined by measuring tyrosine hydroxylase immunoreactive neurons in 30-μm thick slices. This has previously been shown to provide an accurate reflection of lesion size. It can be clearly seen in Fig. 6 that twice daily administration of *R*-2HMP

Table 4
Effect of chronic *R*-2HMP administration on tyrosine hydroxylase-positive neuron counts

	Ipsilateral	Contralateral
MPTP[a] alone	27[b]	24
MPTP alone	N/A[c]	68
MPTP + *R*-2HMP[d]	44	124
MPTP + *R*-2HMP	82	153

[a] MPTP was administered by intracarotid infusion, followed 2 h later by the first dose of *R*-2HMP or vehicle.

[b] Tyrosine hydroxylase-positive neuronal cell counts in the substantia nigra pars compacta were obtained stereologically as previously described [14].

[c] N/A neuronal counts were not available at the same level of the substantia nigra pars compacta in this animal.

[d] *R*-2HMP was administered for a total of 28 days at a dose of 0.1 mg/kg im twice daily. Tyrosine hydroxylase-positive neuron counts are expressed as neuron/mm^2.

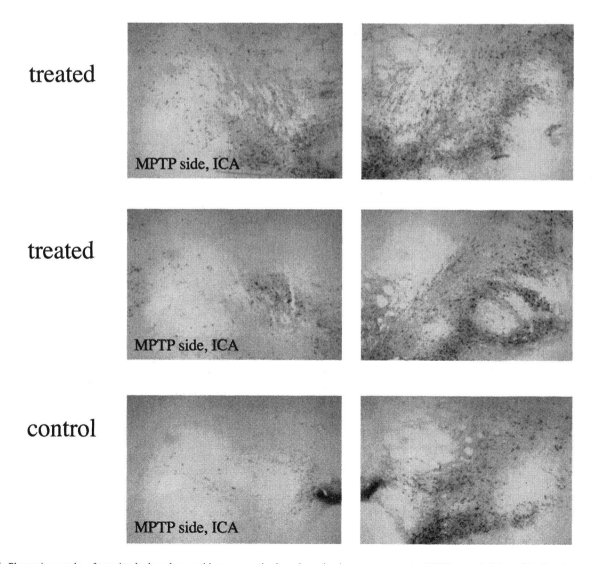

Fig. 6. Photomicrographs of tyrosine hydroxylase-positive neurons in the substantia nigra pars compacta. MPTP was administered to female cynamolgus primates via the right carotid artery. All images are from the same level of the substantia nigra. Right side = contralateral; left side = ipsilateral. Control = MPTP alone; treated = MPTP + *R*-2HMP.

resulted in a marked increase in surviving tyrosine hydroxylase-positive neurons in the substantia nigra in comparison to animals treated with MPTP alone. The actual neuron counts from these slices are shown in Table 4.

Although R-2HMP is known to be a selective MAO-B inhibitor [7], with a potency similar to that of R-deprenyl [41], this is not the mechanism by which the above effects occur. Thus, while MPTP requires toxification to MPP^+ for activity [17,23], R-2HMP was not administered until 2 h following MPTP intracarotid administration. This is a time point at which MPTP has been completely cleared from the brain, and conversion to MPP^+ has already occurred (Bankiewicz, personal communication). Therefore, once again, this effect of R-2HMP represents a true neural rescue effect.

4. Conclusions

From the studies described above, it is readily apparent that the aliphatic propargylamines and their derivatives represent a novel class of antiapoptotic compounds with a clear clinical potential in a number of disease states. R-2HPA, shows excellent neural rescue properties when administered 3 h following a permanent focal ischaemic insult. We are currently investigating whether R-2HPA maintains activity when administered at later time points in a reversible global ischaemia model. R-2HMP, which is also active in ischaemia models but not at later time points, shows excellent rescue properties in the definitive Parkinson's disease model. The activity of aliphatic propargylamines and their derivatives in other neurodegenerative disease model systems are currently under investigation.

The mechanism by which these compounds are active has yet to be definitively demonstrated, but what has been established to date is summarized in Table 5. We have previously shown that in cell culture model systems, these compounds are active through an interaction with the same binding site as R-deprenyl and CGP 3466 [26]. This site has recently been suggested to be glyceraldehyde-3-phosphate dehydrogenase (GAPDH) [11,19]. During the past 5 years, it has become increasingly apparent that GAPDH is a multifunctional enzyme, with numerous activities other than its well-documented role in glycolysis [31]. One of these novel roles appears to be as an initiator of at least some forms of apoptosis [8], although the mechanism by which this occurs remains to be determined. We are currently investigating the ability of the aliphatic propargylamines to interact with GAPDH.

In summary, the aliphatic propargylamines represent a novel class of antiapoptotic compounds that show excellent in vivo efficacy in a number of disease-specific model systems. While these compounds appear to share a common site of action with R-deprenyl and have analogous metabolism [13], unlike R-deprenyl, these metabolites are not amphetaminergic in nature. This offers

Table 5
Temporal events and proposed mechanism(s) of action of aliphatic propargylamines

Mitochondrial membrane potential is decreased 4–6 h after induction of p53-dependent apoptosis.
GAPDH, in nuclear and mitochondrial fractions, is increased 6–12 h after induction of p53-dependent apoptosis.
GAPDH mRNA is increased 1 h after induction of p53-dependent apoptosis.
Antisense GAPDH mRNA blocks p53-dependent apoptosis.

The aliphatic propargylamines prevent:
  p53-dependent apoptosis
  increased expression of GAPDH
  reduction in mitochondrial membrane potential
  increase in GAPDH mRNA.

R-2HMP (like L-deprenyl and CGP 3466) binds to GAPDH.

a considerable advantage over R-deprenyl, since the amphetaminergic metabolites appear to limit the in vivo efficacy of R-deprenyl [25] and may explain the poor in vivo efficacy of R-deprenyl following oral administration (see Fig. 3).

References

[1] A.J. Anderson, J.H. Su, C.W. Cotman, DNA damage and apoptosis in Alzheimer's disease: Colocalization with c-Jun immunoreactivity, relationship to brain area, and effect of postmortem delay, J. Neurosci. 16 (1996) 1710–1719.

[2] P. Anglade, S. Vyas, F. Javoy-Agid, M.T. Herrero, P.P. Michel, J. Marquez, A. Mouatt-Prigent, M. Ruberg, E.C. Hirsch, Y. Agid, Apoptosis and autophagy in nigral neurons of patients with Parkinson's disease, Histol. Histopathol. 12 (1997) 25–31.

[3] S. Baize, E.M. Leroy, M.C. Georges-Courbot, M. Capron, J. Lansoud-Soukate, P. Debre, S.P. Fisher-Hoch, J.B. McCormick, A.J. Georges, Defective humoral responses and extensive intravascular apoptosis associated with fatal outcome in Ebola virus-infected patients, Nat. Med. 5 (1999) 423–426.

[4] K.S. Bankiewicz, E.H. Oldfield, C.C. Chiueh, J.L. Doppman, D.M. Jacobowitz, I.J. Kopin, Hemiparkinsonism in monkeys after unilateral internal carotid artery infusion of 1-methyl-4-phenyl-1,2,3,6-tetrahydropyridine (MPTP), Life Sci. 39 (1986) 7–16.

[5] A.J. Barber, E. Lieth, S.A. Khin, D.A. Antonetti, A.G. Buchanan, T.W. Gardner, Neural apoptosis in the retina during experimental and human diabetes. Early onset and effect of insulin, J. Clin. Invest. 102 (1998) 783–791.

[6] M.D. Berry, N^8-Acetyl spermidine protects rat cerebellar granule cells from low K^+-induced apoptosis, J. Neurosci. Res. 55 (1999) 341–351.

[7] M.D. Berry, R-2HMP: An orally active agent combining independent antiapoptotic and MAO-B-inhibitory activities, CNS Drug Rev. 5 (1999) 105–124.

[8] M.D. Berry, A.A. Boulton, Glyceraldehyde-3-phosphate dehydrogenase and apoptosis, J. Neurosci. Res. 60 (2000) 150–154.

[9] R.E. Burke, N.G. Kholodilov, Programmed cell death: Does it play a role in Parkinson's disease? Ann. Neurol. 44 (Suppl. 1) (1998) S126–S133.

[10] N.G. Butterworth, L. Williams, J.Y. Bullock, D.R. Love, R.L. Faull, M. Dragunow, Trinucleotide (CAG) repeat length is positively correlated with the degree of DNA fragmentation in Huntington's disease striatum, Neuroscience 87 (1998) 49–53.

[11] G.W. Carlile, R.M.E. Chalmers-Redman, N.A. Tatton, A. Pong, W.G. Tatton, Reduced apoptosis after NGF and serum withdrawal: Conversion of tetrameric glyceraldehyde-3-phosphate dehydrogenase to a dimer, Mol. Pharmacol. 57 (2000) 2–12.

[12] M. Dragunow, R.L.M. Faull, P. Lawlor, E.J. Beilhartz, K. Singleton, E.B. Walker, E. Mee, In situ evidence for DNA fragmentation in Huntington's disease striatum and Alzheimer's disease temporal lobes, NeuroReport 6 (1995) 1053–1057.

[13] D.A. Durden, L.E. Dyck, B.A. Davis, Y.D. Liu, A.A. Boulton, Metabolism and pharmacokinetics, in the rat, of (R)-N-(2-heptyl)-methyl-propargylamine (R-2HMP), a new potent monoamine oxidase inhibitor and antiapoptotic agent, Drug Metab. Dispos. 28 (2000) 147–154.

[14] J.L. Eberling, K.S. Bankiewicz, S. Jordan, H.F. vanBrocklin, W.J. Jagust, PET studies of functional compensation in a primate model of Parkinson's disease, NeuroReport 8 (1997) 2727–2733.

[15] Y. Enokido, T. Araki, S. Aizawa, H. Hatanaka, p53 Involves cytosine arabinoside-induced apoptosis in cultured cerebellar granule neurons, Neurosci. Lett. 203 (1996) 1–4.

[16] Y. Enokido, T. Araki, K. Tanaka, S. Aizawa, H. Hatanaka, Involvement of p53 in DNA strand break-induced apoptosis in postmitotic CNS neurons, Eur. J. Neurosci. 8 (1996) 1812–1821.

[17] R.E. Heikkila, L. Manzino, F.S. Cabbat, R.C. Duvoisin, Protection against the dopaminergic neurotoxicity of 1-methyl-4-phenyl-1,2,3,6-tetrahydropyridine by monoamine oxidase inhibitors, Nature 311 (1984) 467–469.

[18] L.A. Kerrigan, D.J. Zack, H.A. Quigley, S.D. Smith, M.E. Pease, TUNEL-positive ganglion cells in human primary open-angle glaucoma, Arch. Opthalmol. 115 (1997) 1031–1035.

[19] E. Kragten, I. Lalande, K. Zimmermann, S. Roggo, P. Schindler, D. Muller, J. van Oostrum, P. Waldmeier, P. Furst, Glyceraldehyde-3-phosphate dehydrogenase, the putative target of the antiapoptotic compounds CGP 3466 and R-(−)-deprenyl, J. Biol. Chem. 273 (1998) 5821–5828.

[20] J.W. Kusiak, J.A. Izzo, B. Zhao, Neurodegeneration in Alzheimer disease. Is apoptosis involved? Mol. Chem. Neuropathol. 28 (1996) 153–162.

[21] H. Lassmann, C. Bancher, H. Breitschopf, J. Wegiel, M. Bobinski, K. Jellinger, H.M. Wisniewski, Cell death in Alzheimer's disease evaluated by DNA fragmentation in situ, Acta Neuropathol. 89 (1995) 35–41.

[22] S. Love, R. Barber, G.K. Wilcock, Apoptosis and expression of DNA repair proteins in ischaemic brain injury in man, NeuroReport 9 (1998) 955–959.

[23] S.P. Markey, J.N. Johannessen, C.C. Chiueh, R.S. Burns, M.A. Herkenham, Intraneuronal generation of a pyridinium metabolite may cause drug-induced parkinsonism, Nature 311 (1984) 464–467.

[24] H. Mochizuki, K. Goto, H. Mori, Y. Mizuno, Histochemical detection of apoptosis in Parkinson's disease, J. Neurol. Sci. 137 (1996) 120–123.

[25] C. Oh, B. Murray, N. Bhattacharya, D. Holland, W.G. Tatton, (−)-Deprenyl alters the survival of adult murine facial motoneurons after axotomy: Increases in vulnerable C57BL strain but decreases in motor neuron degeneration mutants, J. Neurosci. Res. 38 (1994) 64–74.

[26] I.A. Paterson, D. Zhang, R.C. Warrington, A.A. Boulton, R-Deprenyl and R-2-heptyl-N-methylpropargylamine prevent apoptosis in cerebel-

lar granule neurons induced by cytosine arabinoside but not low extracellular potassium, J. Neurochem. 70 (1998) 515–523.

[27] C.K. Petito, B. Roberts, Evidence of apoptotic cell death in HIV encephalitis, Am. J. Pathol. 146 (1995) 1121–1130.

[28] C. Portera-Cailliau, J.C. Hedreen, D.L. Price, V.E. Koliatsos, Evidence for apoptotic cell death in Huntington's disease and excitotoxic animal models, J. Neurosci. 15 (1995) 3775–3787.

[29] S. Przedborski, V. Jackson-Lewis, A.B. Naini, M. Jakowec, G. Petzinger, R. Miller, M. Akram, The parkinsonian toxin 1-methyl-4-phenyl-1,2,3,6-tetrahydropyridine (MPTP): A technical review of its utility and safety, J. Neurochem. 76 (2001) 1265–1274.

[30] S. Przedborski, V. Jackson-Lewis, S. Popilski, V. Kostic, M. Levivier, S. Fahn, J.L. Cadet, Unilateral MPTP-induced parkinsonism in monkeys. A quantitative autoradiographic study of dopamine D1 and D2 receptors and re-uptake sites, Neurochirurgie 37 (1991) 377–382.

[31] M.A. Sirover, New insights into an old protein: The functional diversity of mammalian glyceraldehyde-3-phosphate dehydrogenase, Biochim. Biophys. Acta 1432 (1999) 159–184.

[32] G. Smale, N.R. Nichols, D.R. Brady, C.E. Finch, W.E. Horton Jr., Evidence for apoptotic cell death in Alzheimer's disease, Exp. Neurol. 133 (1995) 225–230.

[33] J.H. Su, A.J. Anderson, B.J. Cummings, C.W. Cotman, Immunohistochemical evidence for apoptosis in Alzheimer's disease, NeuroReport 5 (1994) 2529–2533.

[34] N.A. Tatton, A. Maclean-Fraser, W.G. Tatton, D.P. Perl, C.W. Olanow, A fluorescent double-labeling method to detect and confirm apoptotic nuclei in Parkinson's disease, Ann. Neurol. 44 (Suppl. 1) (1998) S142–S148.

[35] L.B. Thomas, D.J. Gates, E.K. Richfield, T.F. O'Brien, J.B. Schweitzer, D.A. Steindler, DNA end labelling (TUNEL) in Huntington's disease and other neuropathological conditions, Exp. Neurol. 133 (1995) 265–272.

[36] M.M. Tompkins, E.J. Basgall, E. Zamrini, W.D. Hill, Apoptotic-like changes in Lewy-body-associated disorders and normal aging in substantia nigral neurons, Am. J. Pathol. 150 (1997) 119–131.

[37] D. Troost, J. Aten, F. Morsink, J.M. de Jong, Apoptosis in amyotrophic lateral sclerosis is not restricted to motor neurons. Bcl-2 expression is increased in unaffected post-central gyrus, Neuropathol. Appl. Neurobiol. 21 (1995) 498–504.

[38] P.C. Waldmeier, A.A. Boulton, A.R. Cools, A.C. Kato, W.G. Tatton, Neurorescuing effects of the GAPDH ligand CGP 3466B, J. Neural. Transm., Suppl. 60 (2000) S215–S226.

[39] P. Wong, Apoptosis, retinitis pigmentosa, and degeneration, Biochem. Cell Biol. 72 (1994) 489–498.

[40] Y. Yoshiyama, T. Yamada, K. Asanuma, T. Asahi, Apoptosis related antigen, Le(Y) and nick-end labelling are positive in spinal motor neurons in amyotrophic lateral sclerosis, Acta Neuropathol. 88 (1994) 207–211.

[41] P.H. Yu, B.A. Davis, A.A. Boulton, Aliphatic propargylamines: Potent, selective, irreversible monoamine oxidase B inhibitors, J. Med. Chem. 35 (1992) 3705–3713.

ELSEVIER

Neurotoxicology and Teratology 24 (2002) 675–682

NEUROTOXICOLOGY
AND
TERATOLOGY

www.elsevier.com/locate/neutera

Review article

Neuroprotection by propargylamines in Parkinson's disease Suppression of apoptosis and induction of prosurvival genes

Wakako Maruyama[a],*, Yukihiro Akao[b], Maria Chrisina Carrillo[c], Ken-ichi Kitani[a],
Moussa B.H. Youdium[d], Makoto Naoi[e]

[a]Laboratory of Biochemistry and Metabolism, Department of Basic Gerontology, National Institute for Longevity Sciences, Obu, Aichi 474-8522, Japan
[b]Gifu International Institute of Biotechnology, Mitake, Gifu, Japan
[c]National University of Rosario, Suipacha, Rosario, Argentina
[d]Eve Topf and NPF Center, Department of Pharmacology, Faculty of Medicine, Technion, Haifa, Israel
[e]Institute of Applied Biochemistry, Mitake, Gifu, Japan

Received 18 February 2002; accepted 19 February 2002

Abstract

In Parkinson's disease (PD), therapies to delay or suppress the progression of cell death in nigrostriatal dopamine neurons have been proposed by use of various agents. An inhibitor of type B monoamine oxidase (MAO-B), (−)deprenyl (selegiline), was reported to have neuroprotective activity, but clinical trials failed to confirm it. However, the animal and cellular models of PD proved that selegiline protects neurons from cell death. Among selegiline-related propargylamines, $(R)(+)$-N-propargyl-1-aminoindan (rasagiline) was the most effective to suppress the cell death in in vivo and in vitro experiments. In this paper, the mechanism of the neuroprotection by rasagiline was examined using human dopaminergic SH-SY5Y cells against cell death induced by an endogenous dopaminergic neurotoxin N-methyl(R)salsolinol $(N$M(R)Sal). NM(R)Sal induced apoptosis (but not necrosis) in SH-SY5Y cells, and the apoptotic cascade was initiated by mitochondrial permeability transition (PT) and activated by stepwise reactions. Rasagiline prevented the PT in mitochondria directly and also indirectly through induction of antiapoptotic Bcl-2 and a neurotrophic factor, glial cell line-derived neurotrophic factor (GDNF). Long-term administration of propargylamines to rats increased the activities of antioxidative enzymes superoxide dismutase (SOD) and catalase in the brain regions containing dopamine neurons. Rasagiline and related propargylamines may rescue degenerating dopamine neurons through inhibiting death signal transduction initiated by mitochondria PT. © 2002 Elsevier Science Inc. All rights reserved.

Keywords: Rasagiline; Apoptosis; N-methyl(R)salsolinol; Parkinson's disease; Mitochondrial permeability transition; Dopamine neuron

1. Introduction

Strategies to rescue or protect declining neurons in Parkinson's disease (PD) and other neurodegenerative disorders involve interfering with cell death process or promoting growth and function of dopamine neurons in the nigrostriatum. There are three major mechanisms to induce neuronal cell death: metabolic compromise, excitotoxicity and oxidative stress [8]. Metabolic compromise of neurons is caused, in most cases, by inhibitors of mitochondria respiratory chains: cyanide, azide and toxins, such as 1-methyl-4-phenyl-1,2,3,6-tetrahydropyridine (MPTP), 3-nitropropionic acid and rotenone. Mitochondrial dysfunction results in Ca^{2+} release and increased generation of reactive oxygen and nitrogen species (ROS and RNS, respectively), leading to oxidative stress. Excitotoxicity occurs by glutamate and other excitatory amino acids, but only few indirect evidences support the involvement in PD. Oxidative stress is due to the increased production of ROS and RNS (especially hydroxyl radicals and peroxynitrite) by dopamine, Ca^{2+} and nitric oxide synthase (NOS) or decreased activity of antioxidative system. In PD, the involvement of oxidative stress was suggested by increase in the oxidative stress products [44]. To protect dopamine neurons from cell death induced by these mechanisms, various kinds of agents have been proposed, as summarized in Table 1.

* Corresponding author. Tel.: +81-562-46-2311x822; fax: +81-562-46-3157.
E-mail address: maruyama@nils.go.jp (W. Maruyama).

Table 1
Neuroprotective agents for rescuing dopamine neurons in PD

Antioxidants and ions chelators: Desferrioxamine, Glutathione
Free radical scavengers: α-Tocopherol, spin-trap agents, Nitric oxide
    synthase inhibitors (7-NI, L-NAME), SOD
Neurotrophic factors: GDNF, BDNF, aFGF, EDF, Immunophilins,
    AMG-474
Excitatory amino acid antagonists: MK 801, KW-602, Glutamate-release
    inhibitors, Riluzole, Poly(ADP-ribose) polymerase inhibitors
Bioenergetic supplement: Coenzyme Q_{10}, *Gingo bioloba*, Nicotinamide
Immunosuppressants: Cyclosporin A, FK506, GPI-1046, V-10 367,
    rapamycin, AMG-474
Agents preventing protein aggregation and accumulation:
    Proteasome inhibitors
Others: Lithium, Nicotine, Adenosine

Recently, apoptosis, a death process with quite different characteristics than necrosis, was proposed to cause cell death in PD [3], and findings suggesting the activation of apoptotic cascade were detected in dopamine neurons of parkinsonian substantia nigra [15,35,48]. However, the final conclusion on the death mode in PD (apoptosis or necrosis) remains to be clarified [19]. Apoptosis is induced by various insults, but the death process is mediated by stepwise activation of common apoptotic cascades. In neurons, the most decisive site to initiate apoptotic process is the mitochondria [24,52]. Apoptosis is induced by opening a large conductance channel, so-called mitochondrial permeability transition (PT) pore. The PT pore is associated with adenine nucleotide translocator (ANT), voltage-dependent anion channel (VDAC), peripheral benzodiazepine receptor and cyclophilin. Mitochondrial PT is a sudden increase of the inner membrane permeability to proteins with molecular mass below 1500 Da and induces decline in mitochondrial membrane potential ($\Delta\Psi$m). Apoptogenic proteins cytochrome c and apoptosis inducing factor are released through the outer membrane into cytosol and activate caspases (which are effectors of apoptosis). PT is regulated by antiapoptotic oncogenes (Bcl-2 and Bcl-x_L stabilizing mitochondrial membrane) and by proapoptotic Bax analogues disrupting $\Delta\Psi$m [46,52]. A new strategy for neuroprotection is now proposed by amelioration of well-controlled apoptotic process [2,51]. As summarized in Table 2, antiapoptotic agents include stabilizers of PT pore, *bcl-2*, cyclosporin A [59], inhibitors of caspases, *N*-benzyloxycarbonyl-Val-Ala-Asp-fluoromethylketone (z-VAD.FMK) [14] and inhibitors of nuclear translocation of glyceraldehyde-3-phosphate dehydrogenase (GAPDH), CGB 3466 [54] and (*S*)-1-[*N*-(4-chlorobenzyl)succinamonyl]pyrrolidine-2carbaldehyde (ONO-1603) [20].

Propargylamine derivatives of type B monoamine oxidase (MAO-B) inhibitors are gathering attention as neuroprotective agents. (−)Deprenyl (selegiline), an MAO-B inhibitor, is used as adjuncts of L-DOPA therapy and the beneficial effects are considered to inhibit oxidation of dopamine and reduce the generation of ROS and cytotoxic

dopaquinone [9], prevent the oxidation of protoxicants (MPTP) into true neurotoxins (MPP$^+$) and directly scavenge ROS. The neuroprotective activity of selegiline was examined by a controlled clinical trial, "Deprenyl and Tocopherol Antioxidative Therapy of Parkinsonism (DATA-TOP)" study [41,42]. However, this study failed to present concrete evidences of neuroprotection with selegiline. On the other hand, in vivo and in vitro studies on the animal and cellular models of PD demonstrated the protection of dopamine neurons by a series of propargylamines structurally related to selegiline [16,22,29,36,58]. More recently, we screened the structure–activity relationship for protecting neurons from apoptosis induced by neurotoxins and peroxynitrite, and (*R*)(+)-*N*-propargyl-1-aminoindan (rasagiline) was found to be the most potent among propargylamines examined [29,33].

On the other hand, a dopamine-derived isoquinoline, *N*-methyl(*R*)salsolinol (*N*M(*R*)Sal), was found to be a dopaminergic neurotoxin and induce apoptosis in dopamine neurons [27,31,38], suggesting that it may be involved in the pathogenesis of PD [37,39]. The mechanism underlying apoptosis induced by *N*M(*R*)Sal was clarified using dopaminergic SH-SY5Y cells and isolated mitochondria. The neurotoxin induces reduction of $\Delta\Psi$m, activation of caspase 3, nuclear translocation of GAPDH and fragmentation of nucleosomal DNA [1,31,32]. By use of Bcl-2-overexpressed SH-SY5Y cells, Bcl-2 was found to suppress the apoptosis induced by this neurotoxin [32].

In this review, the mechanism underlying neuroprotection by propargylamines was examined to clarify how they interfere the apoptotic pathway induced by *N*M(*R*)Sal

Table 2
Candidates of antiapoptotic agents

Antiapoptotic agent	Mechanism of function
Propargylamines	
Rasagiline	Suppression of PT
	Induction of Bcl-2 expression
	Increased neurotrophic factors
Selegiline, Desmethylselegiline Aliphatic *N*-methylpropargylamines	
CGB 3466	Inhibition of GAPDH translocation
N-propargyl-(3*R*)-aminoindan-5-yl-ethyl,methylcarbamate (TV 3326)	
ONO-1603	Inhibition of GAPDH translocation
Caspase inhibitors	Inhibition of caspase 1,3
N-Benzyloxycarbonyl-Val-Ala-Asp-fluoromethylketone (z-VAD.FMK)	
Acetyl-Tyr-Val-Ala-Asp-chloromethylketone (Ac-YVAD)	
Bcl-2, Bcl-x_L	Stabilization of $\Delta\Psi$m
Agents preventing PT	
Cyclosporin A	Inhibition of ANT

in SH-SY5Y cells. The involvement of Bcl-2 protein family in regulation of mitochondrial apoptotic cascade was studied in SH-SY5Y cells and mitochondria prepared from liver of *bcl-2* transgenic mice. In addition, to intervene the abovementioned three cell death mechanisms, it was examined whether rasagiline can induce prosurvival genes *bcl-2* and *bcl-x_L*, a neurotrophic factor specific for dopamine neurons, glial cell line-derived neurotrophic factor (GDNF) and antioxidant enzymes superoxide dismutase (SOD) and catalase. The results are discussed in relation to the possible protection or rescue of deteriorating neurons by rasagiline and related compounds in PD and other neurodegenerative diseases.

2. Materials and methods

2.1. Chemicals

Rasagiline, prepared as reported [55], was kindly donated by Teva Pharmaceutical (Netanya, Israel) and aliphatic propargylamines was supplied by Professor A. Boulton (Saskatoon, Canada). 5,5′,6,6′tetrachloro-1,1′,3,3′-tetraethyl-benzimidazoyl-carbocyanine iodide (JC-1) and Hoechst 33342 were purchased from Molecular Probes (Eugene, OR). Propidium iodide (PI) and rhodamine 123 were from Sigma (St. Louis, MO). Minimum essential medium (MEM) and other reagents were purchased from Nacalai Tesque (Kyoto, Japan).

2.2. Detection of apoptotic cell death by NM(R)Sal

To confirm apoptosis by morphological observation, SH-SY5Y cells cultured in a six-well poly-D-lysine-coated flask were preincubated with 1 μM of rasagiline for 30 min then cultured with NM(R)Sal for 18 h. To the culture medium, PI or Hoechst 33342 solution was added to be 20 μg/ml and the cells were observed by fluorescence microscopy. Cells ($n = 100$) were examined for four independent experiments, and apoptotic and necrotic dead cells were determined as reported previously [31].

Nuclear translocation of GAPDH in the apoptotic process induced by NM(R)Sal was followed by histochemical observation after staining with anti-GAPDH antibody, as reported previously [32].

2.3. Measurement of $\Delta\Psi m$

The cells cultured in a six-well poly-D-lysine-coated flask were incubated with JC-1 (4 μg/ml) in MEM for 30 min and treated with 1 μM of rasagiline for 30 min. Then, NM(R)Sal was added in the medium to be 250 μM at the final concentration. JC-1 fluorescence was monitored either as green fluorescent monomers at depolarized membrane potentials or as red fluorescent J-aggregates at hyperpolarized membrane. The fluorescent intensity was quantified by

computer-assisted image analysis system using NIH imaging software, as reported previously [31]. Four fields containing about 200 cells were analyzed and the change in the relative intensity of red to green fluorescence was determined as an indicator of $\Delta\Psi m$ in the cells treated with NM(R)Sal with or without pretreatment of rasagiline.

The changes in $\Delta\Psi m$ were also quantitatively analyzed from reduction in the fluorescence of rhodamine 123 preloaded in SH-SY5Y cells, as reported previously [43].

2.4. Effects of NM(R)Sal and rasagiline on $\Delta\Psi m$ in isolated mitochondria

The change in $\Delta\Psi m$ was followed by measurement of $\Delta\Psi m$-dependent uptake of rhodamine 123 into the mitochondria prepared from male Donryu rat liver [40]. Rhodamine 123 (the final concentration of 10 μM) was added to the mitochondria suspension, and the fluorescence at 535 nm was measured with excitation at 505 nm. The effects of NM(R)Sal and rasagiline on $\Delta\Psi m$ were quantitatively assessed from the reduction in rhodamine fluorescence.

2.5. In vitro effects of rasagiline on the level of prosurvival genes

SH-SY5Y cells were cultured in the presence or absence of 10 μM–10 pM of rasagiline for 24 h and the level of Bcl-2 was analyzed by the Western blot analysis [31] and that of *bcl-2* mRNA by reverse transcription (RT)-PCR method. GDNF was quantified by the enzyme-linked immunosorbent assay (ELISA) (Nitta et al., in preparation).

2.6. In vivo effects of rasagiline on the activity of antioxidative enzymes

Male F-344 rats were purchased from Harlan–Sprague–Dawley (Indianapolis, ID) and raised in the National Institute for Longevity Sciences (Obu, Aichi, Japan) under the contact between the company and the National Institute on Aging [7]. At the time of sacrifice, animals were at the age of approximately 8.5 months. Rats were implanted subcutaneously (sc) in the back with an osmotic minipump (Alzet, Alza, Palo, CA). The experimental group received continuous subcutaneous infusion of saline containing rasagiline at doses of 0.5 mg/kg/day. Control animals received continuous subcutaneous infusion of saline. Each group consisted of four animals. About 24–26 days after the start of infusions, animals were sacrificed by decapitation and selected regions in the brain (the substantia nigra, striatum, hippocampus and frontal cerebral cortex) were removed. The samples were homogenized and sonicated in the distilled water and centrifuged for 2 min at $10,000 \times g$, and the supernatant was used for the analysis.

The activity of catalase was assayed by the method of Beers and Sizer [4] and that of SOD was determined according to McCord and Fridovich [34].

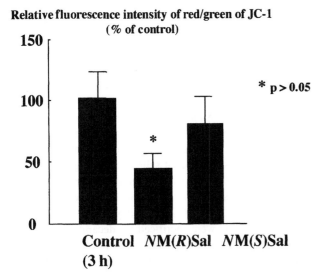

Fig. 1. Chemical structure of the neuroprotective propargylamines selegiline, rasagiline and R-2HMP.

2.7. Statistics

Statistical significance was assessed by analysis of variance (ANOVA). All differences with $P < .05$ were considered to be statistically significant.

3. Results

3.1. Rasagiline protected SH-SY5Y cells against apoptosis by NM(R)Sal

Fig. 1 shows chemical structure of selegiline, rasagiline and (R)-N-(2-heptyl)-N-methyl-propargylamine [(R)-2HMP] used in this paper. These compounds have a propargylamine residue and a hydrophobic structure, such as a phenethyl, cyclic benzyl and aliphatic chain, in the structure.

The (R)-enantiomer of N-methylsalsolinol induced apoptosis in SH-SY5Y cells, but the (S)-enantiomer did not. As a preapoptotic signal, mitochondria PT was in-

duced by the toxin, resulting in the decline in membrane potential, $\Delta\Psi$m, as visualized by a fluorescence indicator, JC-1. In cells treated with 250 μM $NM(R)$Sal, the red fluorescence of J-aggregates declined in a time-dependent way, whereas in control and $NM(S)$Sal treated cells it did not change for 3 h, as shown in Fig. 2. These results suggest that a protein molecule in the mitochondria may distinguish the enantiomeric configuration of N-methylsalsolinol and initiate the death signal transduction. Following the decline in $\Delta\Psi$m, caspase 3 was activated, then GAPDH was translocated from cytosol to nuclei, and nucleosomal DNA was fragmented and condensed, sequentially. Tranfection-induced overexpression of Bcl-2 in SH-SY5Y cells completely prevented the activation of apoptotic cascade, such as $\Delta\Psi$m decline, caspase activation, nuclear GAPDH translocation, and cell death by $NM(R)$Sal [32].

The preincubation with 1 μM of rasagiline suppressed the reduction of $\Delta\Psi$m in the cells treated with $NM(R)$Sal. In rasagiline-treated cells, the fluorescence intensities of red JC-1 aggregates and of rhodamine 123 were almost the same as before $NM(R)$Sal treatment or in control cells. The results on the quantitation of the change in rhodamine fluorescence are summarized in Table 3. The activation of following apoptotic cascade and the cell death were completely suppressed by pretreatment with propargylamines.

Relative fluorescence intensity of red/green of JC-1
(% of control)

* p > 0.05

Fig. 2. Reduction in $\Delta\Psi$m by $NM(R)$Sal but not $NM(S)$Sal. SH-SY5Y cells were stained with a mitochondrial fluorescence dye JC-1 and then incubated with 250 μM of $NM(R)$Sal or $NM(S)$Sal for 3 h. Red and green fluorescence of JC-1 were digitalized by CCD camera and quantified by NIH image. Relative intensity of red/green of JC-1 represents the level of $\Delta\Psi$m in the cells. $*P < .05$ by ANOVA.

Table 3
Effects of $NM(R)$Sal and pretreatment with rasagiline on $\Delta\Psi$m

	Relative fluorescence of rhodamine 123 (percent of control)
Control cells treated with	100
$NM(R)$Sal (100 μM)	$46.9 \pm 10.1*$
$NM(R)$Sal (250 μM)	$35.5 \pm 9.3*$
Rasagiline + $NM(R)$Sal (100 μM)	
Rasagiline (10 μM)	$136.9 \pm 12.7**$
Rasagiline (1 μM)	$103.4 \pm 9.5**$
Rasagiline (100 nM)	$124.0 \pm 13.6**$

SH-SY5Y cells were treated with rasagiline and then with $NM(R)$Sal before subjected to measure rhodamine fluorescence, as described in Materials and methods. The results are expressed as the mean ± S.D. of three samples from three independent experiments.

  * $P < .01$ vs. control.
  ** $P < .01$ vs. $NM(R)$Sal (100 μM)-treated cells.

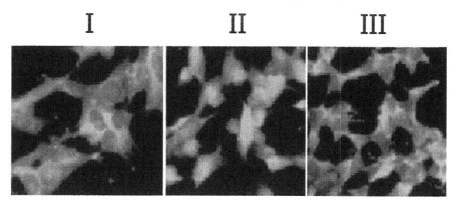

Fig. 3. Inhibition of *NM(R)*Sal-induced nuclear translocation of GAPDH by rasagiline. SH-SY5Y cells were treated with 1 μM of rasagiline for 30 min and then with 250 μM of *NM(R)*Sal for 12 h. The nuclear translocation of GAPDH was detected immunohistochemically using anti-GAPDH antibody, as described in Materials and methods. I: Control cells, II: Cells treated with 250 μM of *NM(R)*Sal for 12 h, III: Cells treated with 1 μM of rasagiline for 30 min and then with *NM(R)*Sal for 12 h.

Fig. 3 shows that rasagiline completely prevents the nuclear translocation GAPDH induced by *NM(R)*Sal. These results demonstrate that *NM(R)*Sal induces apoptosis by the opening mitochondria PT pore, which rasagiline antagonizes, as in the case of Bcl-2 overexpression.

3.2. Rasagiline suppressed $\Delta\Psi m$ decline in isolated mitochondria

The effects of *NM(R)*Sal on $\Delta\Psi m$ were examined using mitochondria isolated from rat liver to confirm whether the toxin induced PT by direct interaction on mitochondrial PT pore. After addition of 500, 250 or 100 μM of *NM(R)*Sal, rhodamine 123 fluorescence reduced in a dose-dependent manner, whereas the fluorescence remained unchanged even after 35-min incubation in control mitochondria. An inhibitor of ANT, cyclosporin A, completely prevented the $\Delta\Psi m$ decline, indicating that the toxin induced opening of

PT pore in mitochondria, as in the case with the whole cells. The pretreatment of mitochondria with 1 μM of rasagiline prevented the *NM(R)*Sal-induced reduction in $\Delta\Psi m$, as in the case of SH-SY5Y cells. These results demonstrate that *NM(R)*Sal induces the PT by directly acting on the mitochondria and rasagiline inhibits the effect of *NM(R)*Sal.

3.3. Rasagiline increased prosurvival Bcl-2 and GDNF in SH-SY5Y cells

The effects of rasagiline on the levels of Bcl-2 mRNA and protein were examined in SH-SY5Y cells. By the Western blot analysis, Bcl-2 protein level increased significantly in the cells treated with 0.1–1 μM of rasagiline from 3 h and the level continued to increase until 24 h. The mRNA level of *bcl-2* and *bcl-xL* increased after treatment with rasagiline, as shown by RT-PCR analyses (Akao et al.,

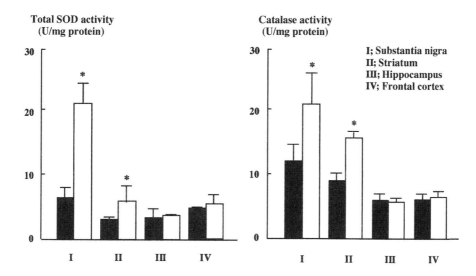

Fig. 4. Increase in the activity of SOD and catalase in the rat brain by rasagiline. Male F-344 rats were continuously injected with 0.5 mg/kg of rasagiline for 4 weeks and the activities of antioxidative enzymes were analyzed in the dissected brain regions. Each bar and column represents mean and S.D. of the data obtained in four rats. Filled column: rats injected with vehicle, open column: rats injected with rasagiline. *$P < .05$ by ANOVA.

submitted for publication). As shown by ELISA, GDNF increased by the incubation with rasagiline at 0.1 μM from 3 to 24 h.

3.4. Rasagiline increased catalase and SOD activities in rats

The infusion of rasagiline at the doses used for this study did not induce any behavioral changes. Fig. 4 shows the effects of rasagiline on the activity of catalase and SOD in the rat brain. The activity of catalase increased about twofolds in the substantia nigra and the striatum but not in the hippocampus or the frontal cortex. By infusion of rasagiline (0.5 mg/kg/day), SOD activity increased two- to fourfolds of the control in the substantia nigra and the striatum. The effects of rasagiline were more manifest in the rat treated with 0.5 mg/kg/day than those with a higher dose (1 mg/kg/day).

4. Discussion

This paper presents our recent results on the prosurvival effects of rasagiline and related propargylamines. As motioned above, selegiline was examined whether it can protect neurons in neurodegenerative disorders, but selegiline is metabolized into amphetamine and methamphetamine [26], which may cause adverse side effects in PD patients [10]. Among structurally related compounds, N-propargyl-1-aminoindan (AGN-1135) is a potent irreversible selective inhibitor of MAO-B but is not metabolized into amphetamine derivatives [11]. The (R)-enantiomer of AGN-1135, rasagiline, is a more selective and potent MAO-B inhibitor than the (S)-enantiomer [12,25] Rasagiline is now under phase 3 clinical trial as a monotherapy drug for PD [21]. It increased the survival of dopaminergic cells in a primary culture of rat fetal mesencephalic cells [12] and protected dopamine neurons and PC12 cells from cell death induced by serum withdrawal [13]. The neuroprotective activity of rasagiline was further indicated by in vivo experiment, where rasagiline reversed or corrected parkinsonism induced by a neurotoxin, α-methyl-p-tyrosine [47].

Recently, we found that, in addition to selegiline and rasagiline, a series of propargylamines protected SH-SY5Y cells from apoptosis induced by a peroxynitrite donor, N-morpholino sydnonimine (SIN-1) and NM(R)Sal [28–33,39]. Now, four groups of propargylamines have similar neuroprotective function: rasagiline (a secondary cyclic benzylamine) [56], selegiline (a tertiary aralkylamine), N-(2-hexyl)-N-methylpropargylamine (a tertiary branched alkylamine) [58] and dibenzo[b,f]oxepin-10-ylmethyl-methyl-prop-2-ynyl-amine (CGP 3466, a tricyclic amine) [54]. All these propargylamines, except CGP 3466, are the inhibitors of MAO with specific affinity to type B, as reported by Yu et al. [57]. It suggests that the propargy-

lamine residue may bind to a selective point of the catalyzing site in MAO-B. However, SH-SY5Y cells contain only type A MAO [27], indicating that the binding protein should not be MAO-B and that inhibition of MAO-B activity is not involved in the neuroprotection. In addition, among the enantiomers of propargylamines, the (R)-enantiomers were more potent than the (S)-enantiomers, suggesting that in mitochondria there should be a protein detecting the enantiomeric configuration of propargylamines [21,29,33]. At present, we have not determined the target protein of rasagiline. Previously, GAPDH was considered to be a possible binding site of selegiline [23], and the involvement of GAPDH to apoptotic cascade was proposed [18]. However, as mentioned above, GAPDH nuclear translocation is a downstream signal of mitochondrial PT [32].

The induction of PT, detected as the decline in ΔΨm, is considered to play a key role to initiate apoptosis [24]. Bcl-2 protein family in the mitochondria regulates the PT, even though the detailed mechanism remains as an enigma. Using SH-SY5Y cells overexpressing Bcl-2 and mitochondria isolated from Bcl-2 transgenic mice, we proved also that the PT induced by NM(R)Sal could be completely prevented by Bcl-2. Pretreatment with propargylamines prevented the collapse of ΔΨm [31], which was further confirmed using isolated rat mitochondria (Akao et al., J. Neurochem., in press). These results suggest that the inhibition of the PT by propargylamines may be mediated by Bcl-2 in mitochondria.

We proved that rasagiline increases the protein and mRNA levels of Bcl-2 and GDNF. Previously, the induction of new synthesis of neuroprotective proteins and expression of immediate early genes was proposed to be involved in the antiapoptotic process of selegiline [49,50,53]. However, the concrete data had not been presented until our reports on the induction of antiapoptotic Bcl-2 protein family by rasagiline (Akao et al., J. Neurochem., in press).

As another mechanism for neuroprotection, selegiline was reported to increase the synthesis of neurotrophic substances, such as nerve growth factor [45], and antioxidant enzymes, such as SOD and catalase [6]. Our data clearly present the evidences that rasagiline increases the expression of GDNF, a selective nutrition factor of dopamine neurons, and the activities of antioxidant enzymes, suggesting that the increase can suppress the death process and also promote survival of dopamine neurons. The signal transduction to increase the prosurvival gene expression will be clarified in near future.

A series of N-propargylamine derivatives were reported to protect not only dopaminergic neurons but also noradrenergic [60] and cholinergic neurons [5] against cell deaths induced by various stimuli. Rasagiline was reported to protect rat brain against ischemic damage [47] and mouse brain against closed head injury [17]. These results suggest that rasagiline may protect or rescue any type of neurons against apoptosis elicited during the process of neurodege-nerative diseases and also in aging.

Acknowledgments

This work was supported by grants from the following programs: Grants-in-Aid for Scientific Research (Y.A.) (no. 09670859) and for Scientific Research (C) (W.M.) from the Ministry of Education, Culture, Sports and Science and Technology, Japan; Grant-in-Aid for Scientific Research on Priority Areas (C) from Japan Society for the Promotion of Science (W.M.); Grants for Longevity Sciences (W.M.), Grants for Comprehensive Research on Aging and Health (W.M. and M.N.), and Medical Frontier Strategy Research (W.M., Y.A. and M.N.) from the Ministry of Health, Labour and Welfare, Japan. Rasagiline was kindly supplied by TEVA Pharmaceutical (Netanya, Israel) and aliphatic N-propargylamines was supplied by Professor A. Boulton (Saskatoon, Canada).

References

[1] Y. Akao, Y. Nakagawa, W. Maruyama, T. Takahashi, M. Naoi, Apoptosis induced by endogenous neurotoxin, N-methyl(R)salsolinol, is mediated by activation of caspase 3, Neurosci. Lett. 267 (1999) 153–156.

[2] T. Alexi, C. Borlongan, R. Faull, C. Williams, R. Clark, P. Gluckman, P. Hughes, Neuroprotective strategies for basal ganglia degeneration: Parkinson's and Huntington's diseases, Prog. Neurobiol. 60 (2000) 409–470.

[3] P. Anglade, S. Vyas, F. Javoy-Agid, M.T. Herrero, P.P. Nichel, J. Marquez, A. Mouratt-Prigent, E.C. Hirsh, Y. Agid, Apoptosis and autophagy in nigral neurons of patients with Parkinson's disease, Histol. Histopathol. 12 (1997) 25–31.

[4] R.F. Beers, I.W. Sizer, A spectrophotometric method for measuring the breakdown of hydrogen peroxide by catalase, J. Biol. Chem. 195 (1952) 133–140.

[5] E. Bronzetti, L. Felici, F. Ferrante, B. Valsecchi, Effect of ethylcholine mustard aziridinium (AF64A) and of the monoamine oxidase-B-inhibitor L-deprenyl on the morphology of the rat hippocampus, Int. J. Tissue React. XIV (1992) 175–182.

[6] M.-C. Carrillo, S. Kanai, M. Nokubo, G.O. Ivy, Y. Sato, K. Kitani, (−)Deprenyl increases activities of superoxide dismutase and catalase in striatum but not in hippocampus: The sex and age-related differences in the optimal dose in the rat, Exp. Neurol. 116 (1992) 286–294.

[7] M.C. Carrilo, C. Minami, K. Kitani, W. Maruyama, K. Ohashi, T. Yamamoto, M. Naoi, S. Kanai, M.B.H. Youdim, Enhancing effect of rasagiline on superoxide dismutase and catalase activities in the dopaminergic system in the rat, Life Sci. 67 (2000) 577–585.

[8] D.S. Cassarino, J.P. Bennett Jr., An evaluation of the role of mitochondria in neurodegenerative diseases: Mitochondrial mutations and oxidative pathology, protective nuclear responses, and cell death in neurodegeneration, Brain Res. Rev. 29 (1999) 1–25.

[9] G. Cohen, M.B. Spina, L-Deprenyl suppresses the oxidant stress associated with increased dopamine turnover, Ann. Neurol. 26 (1989) 689–690.

[10] G. Engberg, T. Elebring, H. Nissbrand, Deprenyl (selegiline), a selective MAO-B inhibitor with active metabolites, J. Pharmacol. Exp. Ther. 259 (1991) 841–847.

[11] J.P.M. Finberg, M. Tenne, M.B.H. Youdim, Tyramine antagonistic properties of AGN-1135—an irreversible inhibitor of monoamine oxidase type B, Br. J. Pharmacol. 73 (1981) 65–74.

[12] J.P. Finberg, L. Lamensdorf, J.W. Commissiong, M.B.H. Youdim, Pharmacology and neuroprotective properties of rasagiline, J. Neural Transm., Suppl. 48 (1996) 95–101.

[13] J.P.M. Finberg, T. Takeshima, J.M. Johnston, J.W. Commissiong, Increased survival of dopamine neurons by rasagiline, a monoamine oxidase B inhibitor, NeuroReport 9 (1998) 703–707.

[14] H. Hara, R.M. Friedlander, V. Galgliardini, C. Ayata, K. Fink, Z. Huang, M. Shimizu-Sasamata, J. Yuang, M.A. Moskowitz, Inhibition of interleukin 1b converting enzyme family proteases reduced ischemic and excitotoxic neuronal damage, Proc. Natl. Acad. Sci. U. S. A. 94 (1997) 2007–2012.

[15] A. Hartmann, S. Hunot, P.P. Michel, M.P. Muriel, S. Vyas, B.A. Faucheux, H. Mouatt-Pigenet, H. Turmel, A. Srinivasan, M. Ruberg, G.I. Evan, Y. Agid, E.C. Hirsch, Caspase 3: A vulnerability factor and final effector in apoptotic death of dopaminergic neurons in Parkinson's disease, Proc. Natl. Acad. Sci. U. S. A. 97 (2000) 2875–2880.

[16] R.E. Heikkila, L. Manzino, F. Cabbat, R.C. Duvoisin, Protection against the dopaminergic neurotoxicity of 1-methyl-4-phenyl-1,2,3,6-tetrahydropyridine by monoamine oxidase inhibitors, Nature 311 (1984) 467–469.

[17] W. Huang, Y. Chen, E. Shohami, M. Weinstock, Neuroprotective effect of rasagiline, a selective monoamine oxidase B inhibitor, against closed head injury in the mouse, Eur. J. Pharmacol. 366 (1999) 127–135.

[18] R. Ishitani, D.-M. Chuang, Glyceraldehyde-3-phosphate dehydrogenase antisense oligodeoxynucleotides protect against cytosine arabinonucleoside-induced apoptosis in cultured cerebellar neurons, Proc. Natl. Acad. Sci. U. S. A. 93 (1996) 9937–9941.

[19] K.A. Jellinger, Cell death mechanism in Parkinson's disease, J. Neural Transm. 107 (2000) 1–29.

[20] N. Katsube, K. Sunaga, H. Aishita, D. Chuang, R. Ishitani, ONO-1603, a potential antidementia drug, delays age-induced apoptosis and suppresses overexpression of glyceraldehydes-3-phosphate dehydrogenase in cultured central nervous system neurons, J. Pharmacol. Exp. Ther. 288 (2000) 6–13.

[21] K. Kieburtz, Parkinson Study Group, Efficiency and safety of rasagiline as monotherapy in the early Parkinson's disease, Abstract of the XIV International Congress on Parkinson's disease, Parkinsonism Relat. Disord. vol. 7, Elsevier, 2001, p. S60.

[22] E. Koutsilieri, T.-S. Chen, W.-D. Rausch, P. Riederer, Selegiline is neuroprotective in primary brain cultures treated with 1-methyl-4-phenypyridinium, Eur. J. Pharmacol. 306 (1996) 181–186.

[23] E. Kragten, I. Lalande, K. Zimmermann, S. Roggo, P. Schindler, D. Müller, J. van Oostrum, P. Waldmeier, P. Fürst, Glyceraldehyde-3-phosphate dehydrogenase, the putative target of the antiapoptotic compounds CGP 3466 and R-(−)-deprenyl, J. Biol. Chem. 273 (1998) 58121–58128.

[24] G. Kroemer, B. Dallaporta, M. Resche-Rigon, The mitochondrial death/life regulation in apoptosis and necrosis, Annu. Rev. Physiol. 60 (1998) 619–642.

[25] I. Lamensdorf, M.B.H. Youdim, J.P.M. Finberg, Effect of long-term treatment with selective monoamine oxidase A and B inhibitors on dopamine release from rat striatum in vivo, J. Neurochem. 67 (1996) 1532–1539.

[26] I. Mahmood, Clinical pharmacokinetics and pharmacodynamics of selegiline. An update, Clin. Pharmacokinet. 33 (1997) 91–102.

[27] W. Maruyama, M. Naoi, T. Kasamatsu, Y. Hashizume, T. Takahashi, K. Kohda, P. Dostert, An endogenous dopaminergic neurotoxin, N-methyl(R)-salsolinol, induces DNA damage in human dopaminergic neuroblastoma SH-SY5Y cells, J. Neurochem. 69 (1997) 322–329.

[28] W. Maruyama, T. Takahashi, M. Naoi, (−)-Deprenyl protects human dopaminergic neuroblastoma SH-SY5Y cells from apoptosis induced by peroxynitrite and nitric oxide, J. Neurochem. 70 (1999) 2510–2515.

[29] W. Maruyama, M. Naoi, Neuroprotection by (−)-deprenyl and related compounds, Mechan. Age Dev. 111 (1999) 189–200.

[30] W. Maruyama, Y. Akao, M.B.H. Youdim, M. Naoi, Neurotoxins induce apoptosis in dopamine neurons: Protection by N-propargyl-1(R)- and (S)-aminoindan, rasagiline and TV 1022, J. Neural Transm., Suppl. 60 (2000) 171–186.

[31] W. Maruyama, A.A. Boulton, B.A. Davis, P. Dostert, M. Naoi, Enantio-specific induction of apoptosis by an endogenous neurotoxin, N-methyl(R)salsolinol, in dopaminergic SH-SY5Y cells: Suppression of apoptosis by N-(2-heptyl)-N-methylpropargylamine, J. Neural Transm. 108 (2001) 11–24.

[32] W. Maruyama, Y. Akao, M.B.H. Youdim, G.A. Davis, M. Naoi, Transfection-enforced Bcl-2 overexpression and an antiparkinson drug, rasagiline, prevent nuclear accumulation of glyceraldehydes-3-phosphate dehydrogenase induced by an endogenous dopaminergic neurotoxin, N-methyl(R)salsolinol, J. Neurochem. 78 (2001) 727–735.

[33] W. Maruyama, T. Takahashi, M. Youdim, M. Naoi, The anti-Parkinson drug, rasagiline, prevents apoptotic DNA damage induced by peroxynitrite in human dopaminergic neuroblastoma SH-SY5Y cells, J. Neural Transm. 109 (2002) 467–481.

[34] E.F. McCord, I. Friedovich, Superoxide dismutase. An enzymatic function for erythrocuprein (hemocuprein), J. Biol. Chem. 244 (1969) 6049–6055.

[35] M. Mogi, A. Togari, T. Kondo, Y. Mizuno, O. Komure, S. Kuno, H. Ichinose, T. Nagatsu, Caspase activities and tumor necrosis factor receptor R1 (p55) are elevated in the substantia nigra from parkinsonian brain, J. Neural Transm. 107 (2000) 335–341.

[36] C. Mytilineou, G. Cohen, Deprenyl protects dopamine neurons from the neurotoxic effect of 1-methyl-4-phenylpyridinium ion, J. Neurochem. 45 (1985) 1951–1953.

[37] M. Naoi, W. Maruyama, T. Kasamatsu, P. Dostert, Oxidation of N-methyl (R)salsolinol: Involvement to neurotoxicity and neuroprotection by endogenous catechol isoquinolines, J. Neural Transm., Suppl. 52 (1998) 125–138.

[38] M. Naoi, W. Maruyama, A. Akao, J. Zhang, H. Parvez, Apoptosis induced by an endogenous neurotoxin, N-methyl(R)Salsolinol, in dopamine neurons, Toxicology 153 (2000) 123–141.

[39] M. Naoi, W. Maruyama, T. Takahashi, Y. Akao, Y. Nakagawa, Involvement of endogenous N-methyl(R)salsolinol in Parkinson's disease: Induction of apoptosis and protection by (−)deprenyl, J. Neural Transm., Suppl. 58 (2000) 111–121.

[40] M. Narita, S. Shimizu, T. Ito, T. Chittenden, R.J. Litz, H. Matsuda, Y. Tsujimoto, Bax interacts with the permeability transition pore to induce permeability transition and cytochrome c release in isolated mitochondria, Proc. Natl. Acad. Sci. U. S. A. 95 (1998) 14681–14686.

[41] Parkinson Study Group, Effect of deprenyl on the progression of disability in early Parkinson's disease, N. Engl. J. Med. 321 (1998) 1369–1371.

[42] Parkinson Study Group, Effect of tocopherol and deprenyl on the progression of disability in early Parkinson's disease, N. Engl. J. Med. 328 (1993) 176–183.

[43] J.G. Pastorino, G. Shimbula, K. Yamamoto, P.A. Glascott Jr., R.J. Rothman, J. Faber, The cytotoxicity of tumor necrosis factor depends on induction of the mitochondrial permeability transition, J. Biol. Chem. 271 (1996) 29792–29798.

[44] P. Riederer, E. Sofic, W.D. Rausch, B. Schmidt, G.P. Reynolds, K. Jellinger, M.B. Youdim, Transition metals, ferritin, glutathione, and ascorbic acid in parkinsonian brains, J. Neurochem. 52 (1989) 515–520.

[45] I. Semkova, P. Wolz, M. Schilling, J. Krieglstein, Selegiline enhances NGF synthesis and protects central nervous system neurons from excitotoxic and ischemic damage, Eur. J. Pharmacol. 315 (1996) 19–30.

[46] S. Shimizu, M. Narita, Y. Tsujimoto, Bcl-2 family proteins regulate the release of apoptogenic cytochrome c by the mitochondrial channel VDAC, Nature 399 (1999) 483–487.

[47] Z. Speiser, O. Katzir, M. Rehavi, T. Zabarski, S. Cohen, Sparing by rasagiline (TVP-1012) of cholinergic functions and behavioral in the postnatal anoxia rat, Pharmacol. Biochem. Behav. 60 (1998) 387–393.

[48] N.A. Tatton, Increased caspase 3 and Bax immunoreactivity accompany nuclear GAPDH translocation and neuronal apoptosis in Parkinson's disease, Exp. Neurol. 116 (2000) 29–43.

[49] W.G. Tatton, W.J.H. Ju, D.P. Holland, C. Tai, M. Kwan, (−)-Deprenyl reduces PC12 cell apoptosis by inducing new protein synthesis, J. Neurochem. 63 (1994) 1572–1575.

[50] W.G. Tatton, R.M.E. Chalmers-Redman, Modulation of gene expression rather than monoamine oxidase inhibition: (−)-Deprenyl-related compounds in controlling neurodegeneration, Neurology 47 (Suppl. 3) (1996) S171–S183.

[51] C.B. Thompson, Apoptosis in the pathogenesis and treatment of disease, Science 267 (1995) 1456–1462.

[52] Y. Tsujimoto, S. Shimizu, Bcl-2 family: Life–death switch, FEBS Lett. 466 (2000) 6–10.

[53] J.S. Wadia, R.M.E. Chalmers-Radma, W.J.H. Ju, G.W. Carlile, J.L. Phillips, A.D. Fraser, W.G. Tatton, Mitochondrial membrane potential and nuclear changes in apoptosis caused by serum and nerve growth factor withdrawal: Time course and modification by (−)-deprenyl, J. Neurosci. 18 (1998) 932–947.

[54] P.C. Waldmeier, A.A. Boulton, A.R. Cools, A.C. Kato, W.G. Tatton, Neurorescuing effects of the GAPDH ligand CGP 3466B, J. Neural Transm., Suppl. 60 (2000) 197–214.

[55] M.B.H. Youdim, J.P.M. Finberg, R. Levy, J. Sterling, D. Lerner, T. Berger-Paskin, H. Yellin, R-enantiomers of N-propargyl-aminoindan compounds. Their preparation and pharmaceutical composition containing them. United States Patent (5) (1995) 457, 133.

[56] M.B. Youdim, A. Wadia, W. Tatton, M. Weinstock, The anti-Parkinson drug rasagiline and its cholinesterase inhibitor derivatives exert neuroprotection unrelated to MAO inhibition in cell culture and in vivo, Ann. N.Y. Acad. Sci. 939 (2001) 450.

[57] P. Yu, B. Davis, A. Boulton, Aliphatic propargylamines: Potent, selective, irreversible monoamine oxidase B inhibitors, J. Med. Chem. 35 (1992) 3705–3713.

[58] P.H. Yu, B.A. Davis, D.A. Durden, A. Barber, I. Terleckyj, W.G. Nicklas, A.A. Boulton, Neurochemical and neuroprotective effects of some aliphatic propargylamines: New selective nonamphetamine-like monoamine oxidase B inhibitors, J. Neurochem. 62 (1994) 697–704.

[59] N. Zamzami, P. Marchetti, M. Castedo, T. Hirsch, S.A. Susin, B. Masse, G. Kroener, Inhibitors of permeability transition interfere with the disruption of the mitochondrial transmembrane potential during apoptosis, FEBS Lett. 383 (1996) 53–57.

[60] X. Zhang, D. Zuo, P.H. Yu, Neuroprotection by R(−)-deprenyl and N-2-hexyl-N-methylpropargylamine on DSP-4, a neurotoxin, induced degeneration of noradrenergic neurons in the rat locus coeruleus, Neurosci. Lett. 186 (1995) 45–48.

ELSEVIER

Neurotoxicology and Teratology 24 (2002) 683–693

NEUROTOXICOLOGY
AND
TERATOLOGY

www.elsevier.com/locate/neutera

Review article

Docosahexaenoic acid attenuated hypertension and vascular dementia in stroke-prone spontaneously hypertensive rats

S. Kimura[a], H. Saito[b,*], M. Minami[c], H. Togashi[d], N. Nakamura[e], K. Ueno[d], K. Shimamura[a], M. Nemoto[b], H. Parvez[f]

[a]Department of Clinical Pharmacology, Faculty of Pharmaceutical Sciences, Health Sciences University of Hokkaido, Ishikari-Tobetsu, Hokkaido, 061-0293, Japan
[b]Department of Basic Sciences, Japanese Red Cross Hokkaido College of Nursing, Kitami, Hokkaido, 090-0011, Japan
[c]Department of Pharmacology, Faculty of Pharmaceutical Sciences, Health Sciences University of Hokkaido, Ishikari-Tobetsu, Hokkaido, 061-0293, Japan
[d]Department of Pharmacology, Hokkaido University Graduate School of Medicine, Sapporo, Hokkaido, 060-8638, Japan
[e]Department of Laboratory Technology, College of Medical Technology, Hokkaido University, Sapporo, Hokkaido, 060-8638, Japan
[f]Unit of Neuroendocrinology and Neuropharmacology of Development, Institute Alfred Fessard of Neurosciences, Gif Sur Yvette CNRS, France

Received 25 June 2001; accepted 24 January 2002

1. Introduction

The Japanese enjoy the longest average life span in the world: in 1999, the life expectancy for Japanese women and men was 83.99 and 77.10 years, respectively [14]. The aging of Japan's population is increasing rapidly. Among age-related diseases, hypertension, stroke, and vascular dementia are preponderant.

According to Yamori's autopsy study on 178 persons aged between 65 and 97 years in Shimane Prefecture, one third of those studied suffered from senile dementia before they died, and cerebral infarction was macroscopically evident in half of the entire group. In Japan, half of all cases of senile dementia are due to vascular dementia. This preponderant tendency for vascular dementia seems to be related to the prevalence of hypertension and stroke in Japan [30].

Yamori's recent 15-year epidemiological survey of dietary conditions of 60 centers in 25 countries throughout the world suggested that stroke can be prevented by various improvements in diet such as reduction of salt intake, and increases in the intake of fish oil, soybean-protein, and dietary fiber obtained from seaweed. Yamori reported that improved diet is a key factor for transforming long life spans into healthy long life spans. He also demonstrated that the traditional Japanese diet is one of the healthiest in the world. This traditional diet contains rice, fish, soybean, and seaweed. These foods are a factor in the prevention of hypertension, stroke, arteriosclerosis, and diabetes mellitus. In line with the

above, we have been probing the effects of docosahexaenoic acid (DHA) derived from fish oil on blood pressure and stroke-related behavior in stroke-prone spontaneously hypertensive rats (SHRSP) [19,31]. Since 1979, we have used genetic rat models of hypertension and stroke, such as SHRSP. We suggested that SHRSP might be an animal model useful in the study of vascular dementia [20,25]. Four years ago, our group started studying whether DHA obtained from fish oil helps to inhibit hypertension and to prevent vascular dementia. DHA is $n - 3$ unsaturated fatty acid derived from fish oil (Fig. 1) [2]. DHA's various physiological properties produce increases in learning ability [29], reflex activity of the retina [5], antithrombotic effect [27], anti-inflammatory action [1], and decreases serum lipid levels [3]. The effect of DHA on blood pressure, however, has not been established.

Using SHRSP, we studied whether DHA has an anti-hypertensive effect. An attempt was also made to elucidate the effects of DHA on learning–memory impairment and abnormal behavior, which are typically observed in SHRSP after the occurrence of stroke.

An attempt was also made to clarify the mechanism of DHA actions by studying the histopathological findings of the brains of DHA-treated SHRSP. The effects of DHA on plasma catecholamine concentrations and brain acetylcholine levels were also evaluated in SHRSP.

2. Characteristics of SHRSP

The blood pressure of SHRSP begins to increase at the age of 6–7 weeks and their systolic blood pressure rises to

---

* Corresponding author. Fax: +81-157-61-3125.

docosahexaenoic acid
(DHA ; C22 : 6, n-3)

Fig. 1. Chemical structure of DHA.

200 mmHg at 20 weeks. Almost all SHRSP suffer stroke and die at the age of approximately 35 weeks. This life span of SHRSP is less than half of that of WKY. Our behavioral study revealed that SHRSP underwent impairment of spontaneous motor activity similar to that demonstrated in patients with vascular dementia. More specifically, desynchronization of the light and dark alternation cycles was observed in SHRSP. Spontaneous motor activity decreased during the dark period after the occurrence of stroke and increased ambulatory activity was observed during the light period. Abnormal behavior, such as remarkable increases in ambulatory activity, was observed just before death in SHRSP [13]. These phenomena appear to be similar to nocturnal wandering in patients with vascular dementia. Moreover, SHRSP showed decreased ability in the passive avoidance response, a criteria connected with learning–memory function [25]. Multiple cerebral infarction is associated with increased incidence of vascular dementia in humans. Recently, histopathological findings have revealed multiple cerebral infarction in SHRSP [9].

3. Duration of DHA administration

In order to clarify the pharmacokinetics, toxic effects, blood pressure, and passive avoidance response of DHA-treated SHRSP, DHA was administered over a period of 14 weeks to animals between the ages of 6 and 20 weeks. To study the effects of DHA on spontaneous motor activity and life span, SHRSP were treated with DHA from the age of 6 weeks to death.

4. Changes in pharmacokinetics of long-term administration of DHA

We studied the changes in plasma and brain levels of DHA and its unsaturated fatty acid concentrations after long-term administration of DHA. Plasma DHA concentrations increased dose-dependently after DHA treatment (1% DHA and 5% DHA) as compared with concentrations in nontreated SHRSP. Similarly, DHA levels in the cerebral cortex and the hippocampus also rose significantly in accordance to the dose of DHA. These findings clarified

that DHA was distributed to the blood and brain. On the other hand, after long-term administration of DHA (1% DHA and 5% DHA), arachidonate concentrations decreased significantly in the plasma, cerebral cortex, and hippocampus. The ratio between $n-3$ and $n-6$ unsaturated fatty acids rose significantly after SHRSP were treated with DHA (1% DHA and 5% DHA) as compared with values in nontreated SHRSP. It was hypothesized that SHRSP might have a genetically caused increase in $n-6$ unsaturated fatty acids accompanied with a decrease in $n-3$ unsaturated fatty acids. It was presumed that abnormal ratios of $n-3$ and $n-6$ may be associated with the pathogenesis of hypertension and stroke.

5. Does DHA have toxic effects?

We studied whether DHA has toxic effects. Administration of DHA did not cause any changes in body weight or blood chemical data, including white blood cells, red blood cells, Hb, total protein, s-GOT, and s-GPT. The 14-week administration of DHA did not produce any toxic effects in SHRSP.

6. Effects of long-term administration of DHA on blood pressure in SHRSP

SHRSP were divided into three groups of similar blood pressure. In nontreated SHRSP, systolic blood pressure was 120.7 ± 2.2 mmHg at the age of 6 weeks, which increased to 203.2 ± 5.6 mmHg at 20 weeks. Systolic blood pressure in 1% DHA-treated SHRSP rose to 168.1 ± 7.1 mmHg (mean $\pm$ S.E., $P < .01$ vs. nontreated SHRSP), while that in 5% DHA-treated SHRSP increased to 150.2 ± 3.2 mmHg (mean $\pm$ S.E., $P < .001$ vs. nontreated SHRSP). Thus, systolic blood pressure decreased significantly in accordance to the dose of DHA (Fig. 2).

The administration of a mixture of DHA and EPA produced significant decreases in blood pressure in SHR and SHRSP [11]. It has been reported that antihypertensive action induced by fish oil was observed in SHRSP. In patients with hypertension, the administration of unsaturated fatty acid including EPA and DHA decreased blood pressure [15]. Although 4 weeks of continuous administration of DHA induced changes in blood lipid levels, DHA did not induce a change in blood pressure in Dahl-salt-sensitive rats [12]. The mechanism of hypertension in Dahl-salt-sensitive rats is due to kidney impairment induced by salt-load. Therefore, the mechanism of hypertension in Dahl-salt-sensitive rats is different from that in essential hypertension. The DHA effect on blood pressure had not been clarified due to the difference of purity, duration of administration, route of administration, doses, and formula of administration. The present study, however, clarified that 1% DHA and 5% DHA produced an anti-

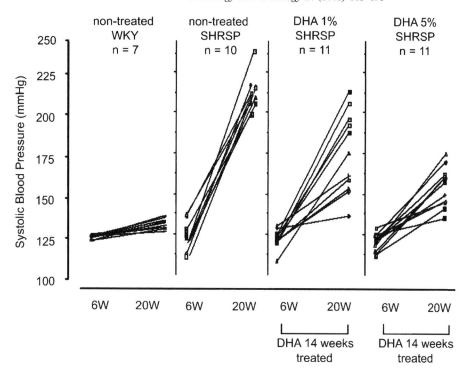

Fig. 2. Effect of DHA on blood pressure in SHRSP.

hypertensive action in SHRSP, a genetic animal model of essential hypertension.

7. Effects of DHA on passive avoidance response in SHRSP

The objective evaluation of memory function is difficult in animal experiment. In this study, we used the passive avoidance test because it is a relatively simple method for short-term experiments. Using the passive avoidance test, we studied the effect of DHA on learning–memory function in SHRSP. Four groups were compared: WKY, nontreated SHRSP, 1% DHA-treated SHRSP, and 5% DHA-treated SHRSP. Passive avoidance responses showed that the response latency of nontreated SHRSP decreased significantly during the first and seventh days after the acquisition trials, indicating the impairment of learning–memory function. There were no significant differences in response latency among the 1% DHA-treated SHRSP, the 5% DHA-treated SHRSP, and the WKY groups. At the same time, the response latency of both the 1% DHA-treated and the 5% DHA-treated SHRSP increased in comparison to the retention trials of the nontreated SHRSP. Therefore, decreased passive avoidance response in SHRSP might reflect impairment of memory function in the vascular dementia. Yamamoto et al. [29] reported that administration of $n - 3$ unsaturated fatty acid improved the learning function in rats. It was also reported that decreased $n - 3$ unsaturated fatty acid accompanied with increased $n - 6$ unsaturated fatty acid produced an impairment of learning

ability [28]. These reports indicate that a change in the balance of $n - 3$ unsaturated fatty acid and $n - 6$ unsaturated fatty acid produced a decrease in cerebral DHA levels and that this change in balance produced a decrease in learning–memory function.

A decrease in learning–memory function may be associated with the impairment of the nervous system in the hippocampus. Therefore, decreased brain DHA might impair the nerve transmission system. Continuous administration of DHA attenuated the decreased passive avoidance results in SHRSP. These findings suggest that DHA inhibited the impairment of memory-retention function in SHRSP.

8. Effects of continuous administration of DHA on spontaneous motor activity in SHRSP

We studied the effects of long-term administration of DHA on spontaneous motor activity in SHRSP. The automatic ambulo-drinkometer was used to analyze stroke-related behavior. At 15 weeks, just before the occurrence of stroke in SHRSP, no differences were observed among nontreated SHRSP, 1% DHA treated SHRSP, and 5% DHA-treated SHRSP with respect to spontaneous motor activity and ambulatory activity patterns during the light and dark periods. However, at the age of 30 weeks when stroke usually occurs in approximately 100% of SHRSP, a decrease in spontaneously motor activity over a 24-h period was observed in nontreated SHRSP as compared with their motor activity at 15 weeks. Also, at 30 weeks, nontreated

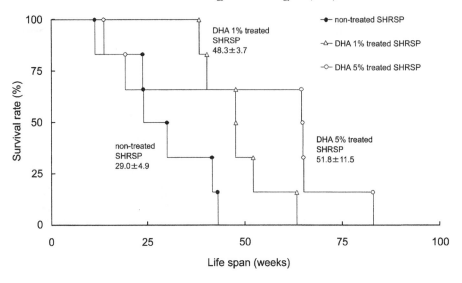

Fig. 3. Effect of DHA on the average survival time of SHRSP.

SHRSP showed an increase in ambulatory activity during the light period and a decrease in ambulation during the dark period.

On the other hand, at the age of 15 weeks, no difference was observed in spontaneous motor activity and ambulatory activity patterns during the light and dark periods in 1% DHA-treated and 5% DHA-treated SHRSP. Just before the death of SHRSP, the nontreated SHRSP group showed an increase in ambulatory activity accompanied with an abrupt desynchronization between light and dark alternation cycles reflecting an inversion of ambulatory activity during the light and dark periods. Neither of the DHA-treated SHRSP groups (1% and 5%) exhibited an increase in ambulatory activity or desynchronization of the light and dark periods. A power spectrum analysis revealed that all six cases of nontreated SHRSP showed a 24-h circadian rhythm accompanied with an abnormal 14-day cyclic rhythm. These findings suggest that the abnormal behavior observed in nontreated SHRSP seems to be similar to the abnormal behavior including nocturnal wandering in patients with vascular dementia. Our studies clarified that DHA produced decreases in spontaneous motor activity, alleviated desynchronization during light and dark periods, and reduced abnormal rhythms in SHRSP.

9. Effects of DHA on life span in SHRSP

Using the automatic ambulo-drinkometer, DHA was continuously administered to SHRSP from the age of 6 weeks until death. We compared the life span among nontreated SHRSP, 1% DHA-treated SHRSP, and 5% DHA-treated SHRSP. Our results showed that the life span of 1% DHA-treated SHRSP increased compared to that of the nontreated SHRSP group. Therefore, it was assumed that DHA might be connected with the prolongation of life span in SHRSP (Fig. 3).

10. Pathological changes in the SHRSP brain

The SHRSP brain shows slight atrophy of the cerebral cortex with focal discoloration of the parenchyma and dilation of the lateral ventricles (coronal sections) (Fig. 4). The stained coronal slices exhibit focal rarefaction of the cerebral cortex and pallor at the periventricular white matter (Fig. 5).

The SHRSP brain was compared to that obtained from the autopsy of a patient with multiple cerebral infarction to determine whether similar pathological changes occurred. The cerebral cortex of the SHRSP demonstrated enlarge-

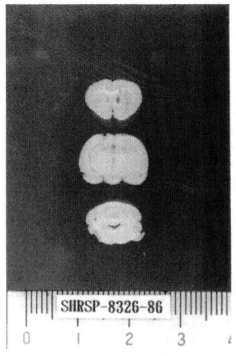

Fig. 4. Cut surfaces of the SHRSP brain showing slight dilatation of the lateral ventricles.

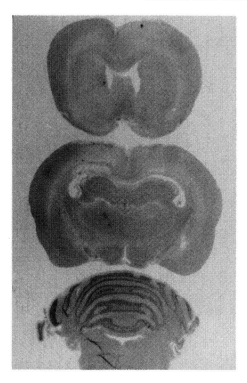

Fig. 5. The specimens from the slices shown in Fig. 4, stained by hematoxylin and eosin.

ment of the perivascular space and rarefaction of the parenchyma (Fig. 6) with scattered neuronal loss and astrocytic proliferation. As shown in Fig. 7, however, the neuronal depopulation of the SHRSP cerebral cortex was seen in SHRSP. Disintegration of the fibers in the white matter with scattered macrophages and reactive astrocytes is shown in Fig. 8.

This change in white matter is not uncommon in the SHRSP brain. The density of the pyramidal cell neurons of the hippocampus and the Ammon's horn of the SHRSP brain was initially lower than that of the human brain. Nearly half of the nerve cells of the pyramidal cell layer showed shrinkage, vacuolar formation, faintness of staining, or other degenerative changes (Fig. 9).

The so-called senile changes observed in the human case were not detected in the rat brain. This pathological study suggests that the cerebrovascular disorder in SHRSP is associated with lesions in the brain similar to those typically seen in humans with multiple cerebral infarction before the symptoms of dementia are apparent. The so-called senile changes (Hirano bodies and senile plagues), observed in the hippocampus of the human case, were not observed in SHRSP brains.

11. Histopathological findings in the brains of DHA-treated SHRSP

We studied the effect of long-term administration of DHA on the pathological changes in SHRSP with stroke. SHRSP were treated orally with DHA from 6 weeks of age until death. After death, histopathological examinations were performed. The histopathological findings of non-treated SHRSP show both congestion and subarachnoidal hemorrhage in the cerebral cortex (Fig. 10A); depopulation of nerve cells is also observed (Fig. 10B). Fig. 11A illustrates the depopulation of the nerve cells observed in the cerebral cortex of 1% DHA-treated SHRSP. The histopathological findings in Fig. 11B show infarction of the cerebral cortex due to vasculitis in 5% DHA-treated SHRSP.

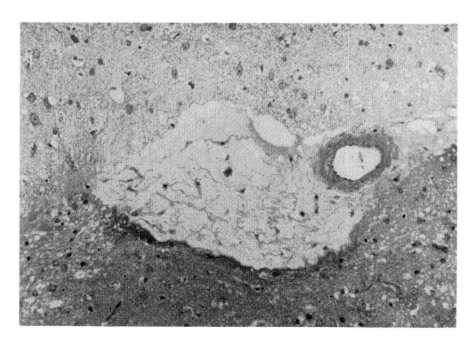

Fig. 6. Perivascular cystic change found at the border zone between the cortex and the white matter.

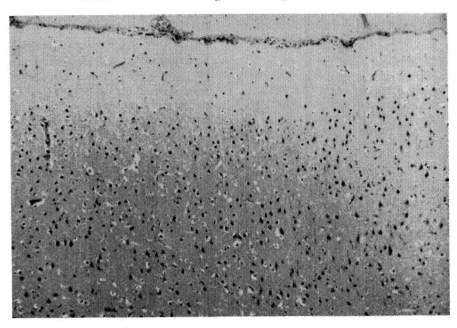

Fig. 7. The cerebral cortex of the rat showing depopulation of nerve cells.

Histopathological examinations revealed local ischemia in nontreated SHRSP, 1% DHA-treated SHRSP, and 5% DHA-treated SHRSP. No differences were observed in the histopathological changes between the brains of nontreated SHRSP and DHA-treated SHRSP, although DHA produced a significant increase in the life span of SHRSP.

Kashiwagi et al. [7] reported that the survival rate of SHRSP receiving EPA was prolonged significantly compared with that of nontreated SHRSP at the age of 23 to 28 weeks. Slight changes in the rate of stroke were also observed in SHRSP [7]. The present study of DHA-treated SHRSP coincided well with Kashiwagi's results.

On the other hand, Yasugi et al. [32] reported that salt-loaded SHRSP displayed cerebral bleeding after administration of EPA and that both the frequency and the severity of the cerebral bleeding exceeded that of the control group. Therefore, we investigated whether DHA induces cerebral bleeding in SHRSP. According to our histopathological findings, no cerebral bleeding resulted from long-term administration of DHA. The fact that DHA prolonged the

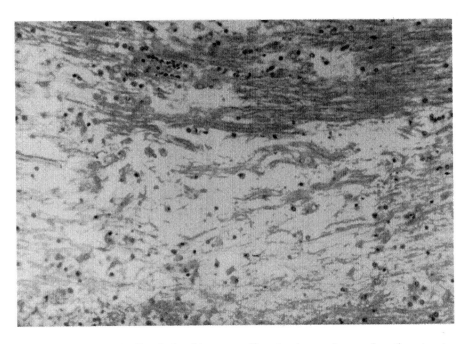

Fig. 8. Disintegration of the fibers in the white matter with scattered macrophages and reactive astrocytes.

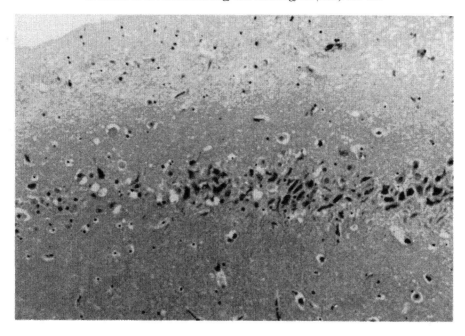

Fig. 9. The hippocampus of the rat showing depopulation and vacuolar degeneration of the pyramidal nerve cells.

life span of SHRSP suggests that even if DHA may not inhibit the occurrence of stroke, it might prolong its onset.

12. Mechanism of DHA-induced antihypertensive action

It has been reported that both SHRSP and patients with essential hypertension display increased sympathetic nerve activity [4]. The present study was undertaken to clarify whether the sympathetic nervous system is associated with

the antihypertensive action induced by DHA. Plasma norepinephrine, epinephrine, and dopamine concentrations were used as the criteria for sympathetic nerve activity. As shown in Table 1, plasma catecholamine concentrations in nontreated SHRSP rose significantly as compared with those in WKY. On the other hand, no difference was noted between nontreated SHRSP and DHA-treated SHRSP (1% DHA and 5% DHA) (Table 1).

The fact that plasma norepinephrine and epinephrine levels in nontreated SHRSP increased as compared with

(A)   (B)

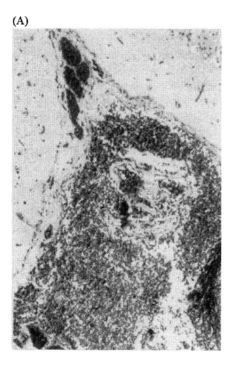
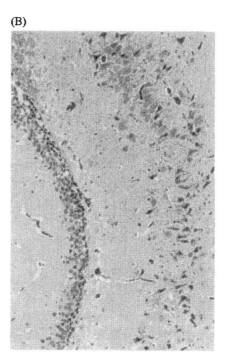

Fig. 10. (A, left), (B, right) Histopathological findings of nontreated SHRSP cerebral cortex.

(A) (B)

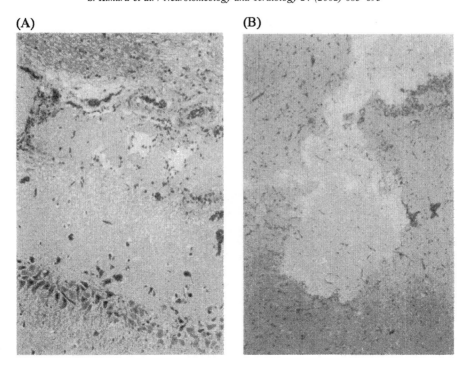

Fig. 11. (A, left) Histopathological findings of 1% DHA-treated SHRSP cerebral cortex. (B, right) Histopathological findings of 5% DHA-treated SHRSP cerebral cortex.

those in WKY may indicate that SHRSP is a suitable animal model for high-catecholamine [4] and high-renin [8] type essential hypertension in humans. Singer et al. [21] reported that the blood norepinephrine concentrations of healthy humans who ingested DHA- and EPA-containing foods showed a significant decrease. On the other hand, decreased blood pressure accompanied with unchanged urinary norepinephrine and epinephrine contents was observed in healthy people who ingested $n-3$ unsaturated fatty acid [10].

Nagatsu et al. [17] reported that, in SHR, dopamine-beta hydroxylase activity (DBH) rose at the age of 3–5 weeks, and that after the development of hypertension, DBH activity decreased [17]. Moreover, it has been reported that at the onset of hypertension, the activity of the sympathetic nervous system of hypertensive patients increased, indicating elevated blood catecholamine concentrations. These reports suggest a different pathophysiology between the initial and the maintenance stages of hypertension. Therefore, after the development of hypertension, blood norepi-

nephrine and epinephrine concentrations may not be influenced by DHA. On the other hand, the ingestion of DHA produced a tendency towards decreased plasma dopamine concentrations. More specifically, large amounts of plasma dopamine were observed in its conjugated form, but only small amounts were present as free dopamine. Therefore, the level of plasma dopamine as such is not significant. Accordingly, the tendency of decreased plasma dopamine does not explain the antihypertensive action of DHA.

Our results clarified that DHA did not influence catecholamine concentrations after the development of hypertension. However, we could not rule out a possible role of the sympathetic nervous system in DHA-induced antihypertensive action. Increased sympathetic nerve activity at an early stage might relate to the onset of hypertension in SHRSP. Therefore, further studies are needed to elucidate the effect of DHA on the blood pressure at the initial stage of hypertension in SHRSP.

13. Mechanism of the ameliorating effect of DHA on decreased passive avoidance response in SHRSP

In order to clarify the mechanism of the ameliorating effect of DHA on decreased passive avoidance response in SHRSP, we measured brain acetylcholine and choline concentrations in both the cerebral cortex and the hippocampus of SHRSP.

In the cerebral cortex of nontreated SHRSP, acetylcholine and choline concentrations decreased significantly as compared with those in WKY. Acetylcholine and choline levels

Table 1
Effects of DHA on catecholamine levels

	WKY control	SHRSP	DHA 1% SHRSP	DHA 5% SHRSP
Norepinephrine	0.67 ± 0.09	1.46 ± 0.24*	1.61 ± 0.22*	1.68 ± 0.33*
Dopamine	0.07 ± 0.01	0.17 ± 0.04*	0.14 ± 0.06	0.13 ± 0.04
Epinephrine	0.45 ± 0.06	0.68 ± 0.08*	0.48 ± 0.05*	0.45 ± 0.03**

Mean (ng/mg protein) ± S.E.
* $P < .05$ vs. WKY.
** $P < .05$ vs. SHRSP control.

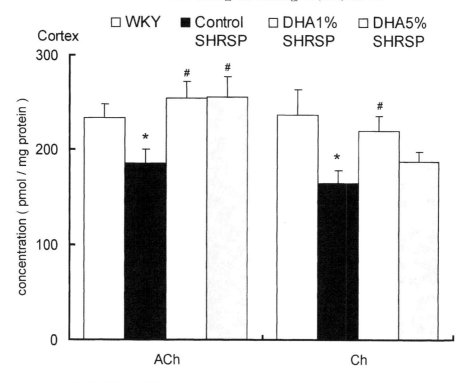

Fig. 12. Effects of DHA on cerebral cortex levels of acetylcholine and choline.

in 1% DHA-treated SHRSP increased significantly as compared with levels in nontreated SHRSP. Acetylcholine concentration in 5% DHA-treated SHRSP rose significantly as compared with values in nontreated SHRSP, however, DHA did not produce any significant effect on the choline levels in the cerebral cortex (Fig. 12).

In the hippocampus of nontreated SHRSP, acetylcholine and choline concentrations decreased significantly as compared to those in WKY. Acetylcholine and choline concentrations in 1% DHA-treated SHRSP were not significantly different from those in nontreated SHRSP. However, 5% DHA-treated SHRSP produced significant increases in both

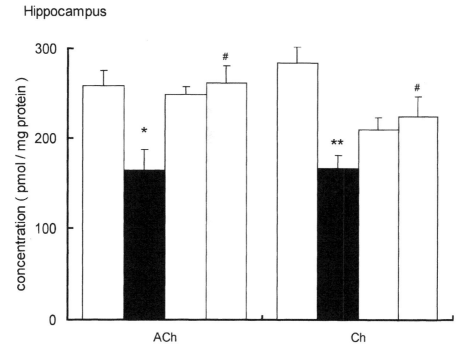

Fig. 13. Effects of DHA on hippocampal levels of acetylcholine and choline.

acetylcholine and choline concentrations in nontreated SHRSP (Fig. 13).

We studied the relationship between the response latency of passive avoidance response and the acetylcholine concentration of the cerebral cortex. A significant positive correlation was observed. Numerous reports have been presented hitherto linking cholinergic nerve deficiency and memory impairment of patients with senile dementia of the Alzheimer type [6,16]. In animal experiments, after intraventricular injection of AF64A, a substance that impairs the cholinergic nervous system, both choline acetyltransferase activity and passive avoidance response decreased. The choline acetyltransferase catalyzes the biosynthesis of acetylcholine in the hippocampus, thus bringing about the decreases [26]. It was reported that decreased passive avoidance response induced by AF64A was recovered by muscarinic agonists and that the muscarinic receptor agonists increased acetylcholine release from the hippocampus [18]. In Takagi et al.'s [22] experiment using models for multiple cerebral infarction, decreases in both active and passive avoidance responses were observed. Decreases in both acetylcholine levels and choline acetyltransferase activity in the cerebral cortex were also observed. In this study, the acetylcholine concentrations in the hippocampus and the cerebral cortex in SHRSP decreased significantly as compared with those in WKY. Togashi et al. [24] reported that acetylcholine and choline levels in the cerebrospinal fluid of SHRSP were lower than those in WKY. Our previous study of SHRSP also revealed decreases in choline levels accompanied with decreases in acetylcholine release as compared with levels in the WKY hippocampus [23]. In this situation, choline is a substrate for acetylcholine synthesis. Moreover, brain vitamin B_{12} concentration, which acts as coenzyme for choline biosynthesis, decreased significantly as compared with WKY [13]. These findings suggest that decreased acetylcholine concentration in the hippocampus and the cerebral cortex may be associated with decreased cholinergic nerve activity in SHRSP.

The fact that DHA treatment produced increases in acetylcholine and choline levels suggests that decreased cholinergic nerve activity was ameliorated by DHA administration. It has been reported that $n - 3$ unsaturated fatty acid ameliorated learning–memory function. No reports, however, are available in the literature concerning the ameliorating effect of DHA on decreased cholinergic function. The present study confirmed that DHA had an ameliorating effect on decreased cholinergic nerve activity.

14. Summary

According to the epidemiological studies, DHA, which is an $n - 3$ unsaturated fatty acid, was discovered in fish oil about 20 years ago and its numerous physiological and pharmacological activities were observed. Around the same time, prostanoids, such as prostaglandin, and eicosanoids were discovered. Attention has focused on the relationship between unsaturated fatty acids in the body and various diseases. Whereas EPA, which is also an $n - 3$ unsaturated fatty acid, is frequently used in clinical medicine as an antiplatelet drug, DHA has not been approved as a therapeutic drug.

The present study was undertaken to elucidate the effects of DHA on hypertension and stroke-related behavior in a genetic animal model, SHRSP.

The distribution of DHA was studied and its effect on biological functions was analyzed. Our results showed that administration of DHA increased DHA concentrations in the blood and brain, while decreasing their arachidonic acid levels. Moreover, long-term administration of DHA suppressed the development of hypertension in SHRSP. At the same time, decreased passive avoidance response, which is a criterion for learning–memory function, was inhibited by DHA in SHRSP. DHA also alleviated desynchronization between the light and dark alternation cycle and the occurrence of abnormal rhythms in SHRSP. It was clarified that DHA prolonged the life span of SHRSP.

These findings suggest that DHA might exert antihypertensive and antagonistic actions against vascular dementia in SHRSP. We presumed that endogenous DHA might be associated with the pathophysiology of hypertension and cerebrovascular diseases. We also clarified that exogenous continuous administration of DHA produced a decrease in the occurrence of defects in normal physiology.

We studied the mechanism of DHA-induced antihypertensive action and antagonistic actions against vascular dementia in SHRSP. We found that acetylcholine concentrations in the hippocampus and the cerebral cortex of SHRSP were significantly decreased as compared with those in WKY. Long-term administration of DHA produced an ameliorating effect on the decreased acetylcholine levels observed in nontreated SHRSP. These findings suggest that DHA may ameliorate the decreased function of the central nervous system through improvement of the decreased acetylcholine levels in the SHRSP brain.

References

[1] W. Christian, A. Martin, L. Inge, Z. Brigitte, C.W. Peter, Docosahexaenoic acid inhibits PAF and LTD4 stimulated $[Ca^{2+}]$i-increase in differentiated monocytic U937 cells, Biochim. Biophys. Acta 1133 (1991) 38–45.

[2] J. Dyerberg, H.O. Bang, N. Hjorne, Fatty acid composition of the plasma lipids in Greenland Eskimos, Am. J. Clin. Nutr. 28 (1975) 958–966.

[3] L. Froyland, H. Vaagenes, D.K. Asiedu, A. Garras, O. Lie, R.K. Berge, Chronic administration of eicosapentaenoic acid and docosahexaenoic acid as ethyl esters reduced plasma cholesterol and changed the fatty acid composition in rat blood and organs, Lipids 31 (1996) 169–178.

[4] P.R. Howe, P.F. Rogers, M.J. Morris, J.P. Chalmers, R.M. Smith, Plasma catecholamines and neuropeptide-Y as indices of sympathetic nerve activity in normotensive and stroke-prone spontaneously hypertensive rats, J. Cardiovasc. Pharmacol. 8 (1986) 1113–1121.

[5] J.P. Infante, Docosahexaenoate-containing phospholipids in sarcoplasmic reticulum and retinal photoreceptors. A proposal for a role in Ca^{2+}-ATPase calcium transport, Mol. Cell. Biochem. 74 (1987) 111–116.

[6] T.C. Joseph, L.P. Donald, R.D. Mahlon, Alzheimer's disease: A disorder of cortical cholinergic innervation, Science 219 (1983) 1184–1189.

[7] F. Kashiwagi, Y. Katayama, H. Memezawa, A. Terashi, Effect of long term administration of eicosapentaenoic acid (EPA) on survival, lipid metabolism, endothelium of middle cerebral (MCA) and experimental cerebral ischemia in stroke-prone spontaneously hypertensive rats, Jpn. J. Stroke 12 (1990) 326–333.

[8] K. Kawashima, K. Shiono, H. Sokabe, Variation of plasma and kidney renin activities among substrains of spontaneously hypertensive rats, Clin. Exp. Hypertens. 2 (1980) 229–245.

[9] S. Kimura, H. Saito, M. Minami, H. Togashi, N. Nakamura, M. Nemoto, H.S. Parvez, Pathogenesis of vascular dementia in stroke-prone spontaneously hypertensive rats, Toxicology 16 (2000) 167–178.

[10] R. Lorenz, U. Spengler, S. Fischer, J. Duhm, P.C. Weber, Platelet function, thromboxane formation and blood pressure control during supplementation of the Western diet with cod liver oil, Circulation 67 (1983) 504–511.

[11] T.M. Mark, S. Bexis, M.Y. Abeywardena, E.J. McMurchie, R.A. King, R.M. Smith, R.J. Head, Fish oils modulate blood pressure and vascular contractility in the rat and vascular contractility in the primate, Blood Pressure 4 (1995) 177–186.

[12] M. Matsuoka, Y. Watanabe, M. Soma, Y. Izumi, T. Yasugi, Effect of docosahexaenoic acid on blood pressure, fatty acid composition of kidney and renal prostaglandins in Dahl salt sensitive rats, in: T. Yasugi (Ed.), Advances in Polyunsaturated Fatty Acid Research, Elsevier, New York, 1993, pp. 138–185.

[13] M. Minami, S. Kimura, K. Ueno, T. Endo, Y. Monma, H. Togashi, M. Matsumoto, M. Yoshioka, H. Saito, H.S. Parvez, Role of vitamin B_{12} on behavioral changes in SHRSP, H. Saito, Y. Yamori, M. Minami, S.H. Parvez (Eds.), Progress in Hypertension, vol. 3, VSP, Netherlands, 1995, pp. 173–188.

[14] Ministry of Health, Labour and Welfare, Abridged Life Tables for Japan, 1999. Vital Statistics, 2000.

[15] M.C. Morris, F. Sacks, B. Rosner, Does fish oil lower blood pressure? A meta-analysis of controlled trials, Circulation 88 (1993) 523–533.

[16] J.L. Muir, Acetylcholine, aging, and Alzheimer's disease, Pharmacol., Biochem. Behav. 56 (1997) 687–696.

[17] T. Nagatsu, T. Kato, Y. Numata, I. Keiko, H. Umezawa, Serum dopamine beta-hydroxylase activity in developing hypertensive rats, Nature 251 (5476) (1974) 630–631.

[18] N. Ogane, Y. Takada, Y. Iga, G. Kawanishi, F. Mizobe, Effects of a M1 muscarinic receptor agonist on the central cholinergic system, evaluated by brain microdialysis, Neurosci. Lett. 114 (1990) 95–100.

[19] K. Okamoto, Y. Yamori, A. Nagaoka, Establishment of the stroke-prone spontaneously hypertensive rat (SHR), Circ. Res. 34/35 (Suppl. I) (1974) 143–153.

[20] H. Saito, H. Togashi, M. Yoshioka, N. Nakamura, M. Minami, S.H. Parvez, Animal models of vascular dementia with emphasis on stroke-prone spontaneously hypertensive rats, Clin. Exp. Pharmacol. Physiol. 22 (Suppl. 1) (1995) 257–259.

[21] P. Singer, W. Jaeger, M. Wirth, S. Voigt, E. Naumann, S. Zimontkowski, I. Hajdu, W. Goedicke, Lipid and blood pressure-lowering effect of mackerel diet in man, Atherosclerosis 49 (1983) 99–108.

[22] N. Takagi, K. Miyake, T. Taguchi, H. Tamada, K. Takagi, N. Sugita, S. Takeo, Failure in learning task and loss of cortical cholinergic fibers in microsphere-embolized rats, Exp. Brain Res. 114 (1997) 279–287.

[23] H. Togashi, S. Kimura, M. Matsumoto, M. Yoshioka, M. Minami, H. Saito, Cholinergic changes in the hippocampus of stroke-prone spontaneously hypertensive rats, Stroke 27 (1996) 520–525.

[24] H. Togashi, M. Matsumoto, M. Yoshioka, M. Hirokami, M. Minami, H. Saito, Neurochemical profiles in cerebrospinal fluid of stroke-prone spontaneously hypertensive rats, Neurosci. Lett. 116 (1994) 117–120.

[25] H. Togashi, M. Matsumoto, M. Yoshioka, M. Minami, H. Saito, Behavioral and neurochemical evaluation of stroke-prone spontaneously hypertensive rats, in: T. Nagatsu (Ed.), Basic and Clinical Therapeutic Aspects of Alzheimer's and Parkinson's Diseases, vol. 1, Plenum, New York, 1990, p. 473.

[26] T.J. Walsh, H.A. Tilson, D.L. Dehaven, R.B. Mailman, A. Fisher, I. Hanin, AF64A, a cholinergic neurotoxin, selectively depletes acetylcholine in hippocampus and cortex, and produces long-term passive avoidance and radial-arm maze deficits in the rat, Brain Res. 321 (1984) 91–102.

[27] S. Watanabe, E. Suzuki, N. Kojima, R. Kojima, Y. Suzuki, H. Okuyama, Effect of dietary alpha-linolenate/linoleate balance on collagen-induced platelet aggregation and serotonin release in rats, Chem. Pharm. Bull. 37 (1989) 1572–1575.

[28] N. Yamamoto, Y. Okaniwa, S. Mori, M. Nomura, H. Okuyama, Effects of a high-linoleate and a high-alpha-linolenate diet on the learning ability of aged rats. Evidence against an autoxidation-related lipid peroxide theory of aging, J. Lipid Res. 29 (1988) 1013–1021.

[29] N. Yamamoto, Y. Okaniwa, S. Mori, M. Nomura, H. Okuyama, Effects of a high-linoleate and high-alpha-linolenate diet on the learning ability of aged rats, J. Gerontol. 46 (1991) 17–22.

[30] Y. Yamori, Predictive and preventive pathology of cardiovascular diseases, Acta Pathol. Jpn. 39 (1989) 683–705.

[31] Y. Yamori, Dietary intervention of hypertensive diseases—epidemiological and experimental studies. Satellite symposium to the 10th International Symposium on SHR and Molecular Medicine, May 2, 2001, Berlin-Buch, 2001.

[32] T. Yasugi, T. Fujioka, E. Saito, H. Kanno, T. Ueno, T. Matsumoto, Y. Takahashi, H. Watanabe, T. Yamada, N. Saito, The influence of docosahexaenoic acid loading on stroke-prone spontaneously hypertensive rats, Ann. N. Y. Acad. Sci. 676 (1993) 70–82.

ELSEVIER

Neurotoxicology and Teratology 24 (2002) 695–701

NEUROTOXICOLOGY
AND
TERATOLOGY

www.elsevier.com/locate/neutera

Review article

Diabetic neuropathies in brain are induced by deficiency of BDNF

A. Nitta[a,*], R. Murai[a], N. Suzuki[a], H. Ito[a], H. Nomoto[a], G. Katoh[b], Y. Furukawa[c], S. Furukawa[a]

[a]Laboratory of Molecular Biology, Gifu Pharmaceutical University, 5-6-1 Mitahora-Higashi, 502-8585 Gifu, Japan
[b]Department of Neuroanatomy, Chiba College of Health Science, Chiba, Japan
[c]Aichi Bunkyo Women's College, Nishi-Machi, Inazawa, Japan

Received 1 August 2001; accepted 24 January 2002

Abstract

Diabetes is known to be one of the risk factors for dementia; however, neuropathic changes in the brain of patients with the disease have not been completely revealed. So in the present study, we investigated the brain function of rats with diabetes induced by streptozotocin (STZ), one of the most commonly used animal models for diabetes. In the diabetic rats, immediately working memory performance was impaired in the Y-maze task and neuronal cytoskeleton proteins such as calbindin, synaptophysin, and syntaxin were reduced. Furthermore, morphological observation by Golgi staining showed a decrease in the number of basal dendrites and abnormality of spine structure. Next, we measured the content of brain-derived neurotrophic factor (BDNF) in the diabetic brain, because BDNF is one of the essential proteins for the maintenance of neuronal functions including synapse function and neuronal transmissions. In the diabetic brains, both protein and mRNA levels of BDNF were severely reduced. These results suggest that, in diabetes, synapse dysfunction is, at least in part, caused by a failure of BDNF synthesis in the brain. © 2002 Elsevier Science Inc. All rights reserved.

Keywords: Brain-derived neurotrophic factor; Diabetes; Learning memory; Y-maze task; Rats

1. Introduction

Neuropathy is one of the major complications contributing to morbidity in patients with diabetes mellitus. Diabetes leads to a wide range of peripheral neuronal deficits including reduced motor nerve conduction velocity, impaired sciatic nerve regeneration, axonal shrinkage in association with reduced neurofilament delivery, and deficient anterograde axonal transport [1,2,28]. In rats with diabetes experimentally induced by streptozotocin (STZ), the nerve damage observed parallels in many ways the nerve degeneration seen in human diabetic neuropathy.

A lack of neurotrophins is considered to be one of the important reasons for neuropathies induced by diabetes [21,25]. Neurotrophins, including nerve growth factor (NGF), brain-derived neurotrophic factor (BDNF), neurotrophin-3 (NT-3), and NT-4/5, are synthesized in the target neurons [15]. BDNF affects the survival and differentiation of cultured motor neurons [11], mesencephalic dopaminergic neurons [14], and septal cholinergic neurons [14]. In adult rats, BDNF mRNA is more widely distributed in the whole brain than are mRNAs of NGF and NT-3 [20] and is regulated by glutamate or γ-aminobutyric acid neurotransmission [22]. In addition, enhanced expression is seen following the establishment of long-term potentiation [22]. BDNF thus seems to participate in various activity-dependent events, including synapse plasticity. In the sciatic nerve of diabetic rats, expression of BDNF is reduced [4]. Previous studies of ours demonstrated that 4-methylcatechol, which is a stimulator for BDNF synthesis [18], protected against the peripheral neuropathy induced by diabetes [6,7]. These results suggest that BDNF helps in the neurogeneration in the diabetic nervous system. Here, we focus on BDNF in the diabetic brains.

Recently, pathological studies have suggested that diabetes is one of the risk factors for senile dementia of Alzheimer type [8,16]. Many studies about the relationship between diabetes and peripheral neuropathy have been done to date [3,5]. However, the effects of diabetes on the brain itself have not been studied and no abnormality has been found in the central nervous system of patients with diabetic neuropathy. In the present study, we demonstrated neurological dysfunction and morphological change in the central nervous system of diabetic rats and found that the a reduction in BDNF synthesis in the brain accompanied the symptoms.

* Corresponding author. Tel.: +81-52-744-2671.
E-mail address: a-nitta@med.nagoya-u.ac.jp (A. Nitta).

2. Materials and methods

2.1. Materials

STZ was purchased from Wako (Osaka, Japan). BDNF was generously donated by Sumitomo Pharmaceutical (Osaka, Japan) and anti-BDNF antibody was prepared as described previously [18]. Antibodies specific for calbindin, synaptophysin, and syntaxin were from Chemicon International (Temecula, CA). Antibody against glial fibrillary acidic protein (GFAP) was purchased from DAKO Japan (Kyoto, Japan). All other materials used were reagent grade. Rats were purchased from Nippon SLC (Shizuoka, Japan) and they were treated according to the Guideline of Experimental Animal Care issued from the Office of the Prime Minister of Japan.

2.2. Induction of diabetes

Diabetes was induced in male Wistar rats (Nippon SLC, Hamamatsu, Japan) by the intraperitoneal administration of STZ (65 mg/kg), which was dissolved in freshly prepared 0.05 M citrate buffer, pH 4.5, immediately before injection [17]. The rats were kept for 4 weeks prior to the experiment so that a severe diabetic state could be induced. Glucose concentrations in both plasma and urine were monitored once every week. The criteria for determining diabetic rats were as follows: minimum plasma glucose of over 250 mg/dl and presence of urinary glucose. Age-matched, nontreated (citrate buffer only) rats were used as the control.

2.3. Immunoblots [4]

Hippocampus from diabetic rats were disrupted by sonicating in PBS containing 0.5% sodium dodecyl sulfate (SDS, 10% w/v) and centrifuging at $10,000 \times g$ for 30 min. The supernatants were then subjected to SDS-polyacrylamide gel electrophoresis (PAGE) (5%) under reducing conditions. Protein in the gel were transferred onto nylon membranes. The membranes were then washed twice in PBS containing 0.05% Tween 20, reacted with anti-calbindin, -synaptophysin, -syntaxin, or -GFAP antibodies for 15 h at 4 °C, and then reaction with biotinylated peroxidase complexes (Vector Lab.). Peroxidase activity was visualized as described earlier [4]. Expressions of each proteins were quantitatively analyzed by an image analyzer (BAS 2000, Fuji Film).

2.4. Y-maze task [27]

Y-maze task was carried out by using the diabetic rats 4 weeks after the injection of STZ. Immediately working memory performance was assessed by recording spontaneous alternation behavior in a single session in a Y-maze. The maze was made of gray-pained wood. Each arm was 40 cm long, 30 cm high, and 15 cm wide. The arm converged in an equilateral triangular central area that was 15 cm at its longest axis. The procedure was basically the same as that described previously: each rat, naive to the maze, was placed at the end of one arm and allowed to move freely through the maze during an 8-min session. The series of arm entries was recorded visually. Arm entry was considered to be completed when the hind paws of the rat had been completely placed in the arm. Alternation was defined as successive entries into the three arms on overlapping triplet sets. The alternation percentage was calculated as the ratio of actual to possible alternations (defined as the total number of arm entries minus two).

2.5. Measurement of BDNF contents

BDNF content in the brain tissue was determined with a newly developed two-site enzyme immunoassay (EIA) that was described recently [19]. Each brain tissue was added to homogenizing buffer (0.1 M Tris–HCl, pH 7.4, containing 1 M NaCl, 2% BSA, 2 mM EDTA, 0.2% Na_3N) for a concentration of 1 g wet weight/19 ml buffer, pulse-sonicated for 30 s, and centrifuged at $10,000 \times g$ for 30 min. The supernatant was then mixed vigorously with 100 μl of chloroform and centrifuged at $20,000 \times g$ for 15 min, after which the aqueous phase was collected and used for the EIA measurement.

The EIA system for BDNF was based on the method originally developed for NGF [13]. In short, multiwell plates (Falcon 3910; Becton Dickinson, Franklin Lakes, NJ) were incubated with 5 μl of anti-BDNF antibody in 0.1 M Tris–HCl buffer (pH 9.0, 10 μg/ml) per well for 12 h, washed with washing buffer (0.1 M Tris–HCl buffer, pH 7.4, containing 0.4 M NaCl, 0.02% Na_3N 0.1% BSA, and 1 mM $MgCl_2$), and then blocked with washing buffer containing 1% (w/v) skim milk. Tissue extract or BDNF standard (30 μl) in washing buffer was then added to each antibody-coated well and incubation was carried out for 5 h at 25 °C. After three washes with washing buffer, 30 μl of biotinylated anti-BDNF antibody (10 ng/ml) in washing buffer was added to each well and the plate was then incubated for 12 to 18 h at 4 °C. The biotinylated secondary antibodies were then reacted with avidin-conjugated β-galactosidase (Boehringer Mannheim, Mannheim, Germany) for 1 h. Then, following thorough washing with washing buffer, the enzyme activity retained in each well was measured by incubation with fluorogenic substrate; 4-methylumbelliferyl-β-D-galactoside (100 μM) in the washing buffer, and the fluorescence was monitored with 360-nm excitation and 448-nm emission. The detection limit of the EIA was as low as 5 pg/ml. The recovery of BDNF (61.8 pg/ml) exogenously added to the homogenizing buffer following disruption of the rat hippocampus was 80.5%. The value of BDNF content thus obtained was expressed without correction. In sample of brain tissue from mutant mice lacking BDNF gene, no signal could be detected by the EIA system (A.N. and S.F., unpublished data).

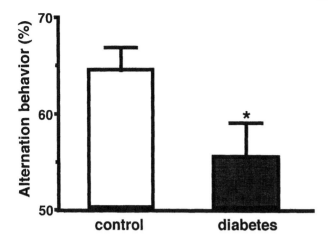

Fig. 1. Effects of diabetes on alternation behavior of rats in the Y-maze task. Y-maze task was carried out 4 weeks after the injection of STZ. *$P < 0.05$ compared with the control (Tukey's test).

2.6. Semiquantification of BDNF mRNA expression

The reverse transcription polymerase chain reaction (RT-PCR) was used to evaluate the BDNF and β-actin mRNA levels, as described earlier [18]. Total RNA was prepared from the cells by use of Isogen (Nippon Gene, Tokyo, Japan), which is basically composed of guanidine isothiocyanate. Oligonucleotide primers for the respective genes of rat BDNF and β-actin were used. RT-PCR was performed with a GeneAmp Thermostable rTth reverse transcriptase RNA PCR kit (Perkin-Elmer, Oak Brook, IL) according to the manufacturer's instruction. In short, 500 ng of total RNA was reverse-transcribed with 0.75 mM downstream primer by rTth polymerase in the presence of Mn^{2+} for 15 min at 60 °C. The synthesized cDNA was amplified by PCR in the presence of Mg^{2+} for 15 min with both up- and downstream primers. The thermal cycle profile used for amplification was 28 cycles of (1) 94 °C for 1 min, (2) 55 °C for 1 min, and (3) 72 °C for 1 min. A portion (10 μl) of the PCR products was resolved by PAGE and visualized by ethidium bromide staining. The density of the BDNF PCR products was analyzed by image analysis Macintosh system and was expressed as the ratio of the sample density to the density of the β-actin PCR products amplified from an identical RNA sample.

2.7. Golgi staining [24]

The Golgi staining technique was used in a separate series of three animals each to see morphological changes in dendrites and synapses 30 days after the injection of STZ. The brain was removed and quickly immersed in a mixture of solutions A and B: solution A contained potassium dichromate and mercury chloride and solution B contained potassium chromate and sodium tungstenate. Following

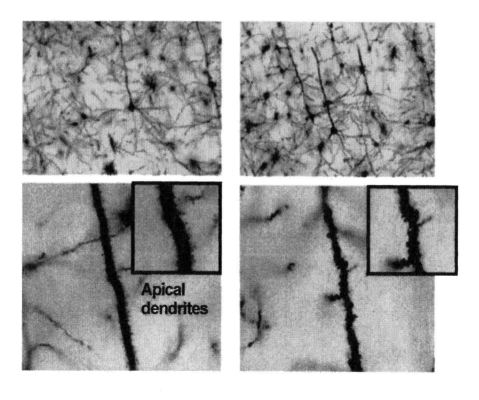

control   diabetes

Fig. 2. Chronological changes in the dendritic structure of the diabetic cortex. Golgi staining was carried out 4 weeks after the injection of STZ.

immersion for 1 month in this staining mixture, the brains were processed for microscopic examination.

2.8. Statistical analysis

The results of the Y-maze task and immunoblots were expressed as means (±S.E.) and analyzed by Tukey's test.

3. Results

The body weights of the diabetic and normal rats were 158±19.2 and 275±33.1 g, respectively, 4 weeks after the injection of STZ. General behavior and spontaneous activity were the same in both groups.

3.1. Alternation behavior in Y-maze

Fig. 1 shows the results for the Y-maze task, in which short-term memory performance can be examined. There was no significant difference in the total number of times the animal entered an arm. However, the alternation score of the diabetic rats was lower than that of the control.

3.2. Golgi staining

In the Golgi study, morphological changes appeared in dendrites and synaptic spines in the cortex of the diabetic rats. There was reduction in the number of both basal and apical dendrites (Fig. 2).

3.3. Immunoblots study

The expressions of calbindin, synaptophysin, and syntaxin were significantly reduced in the diabetic hippocampus (Fig. 3). These proteins are synaptosome-associated proteins and relate to the secretion of neuronal transmitters such as glutamate, dopamine, and acetylcholine. By contrast, the expression of GFAP was unchanged in the diabetic hippocampus.

3.4. Effects of diabetes on BDNF contents and BDNF mRNA expression in the hippocampus and cortex

Our EIA system for BDNF was sensitive enough to measure BDNF contents in the hippocampus and cortex. The BDNF contents were 10.04 and 5.6 ng/g in the

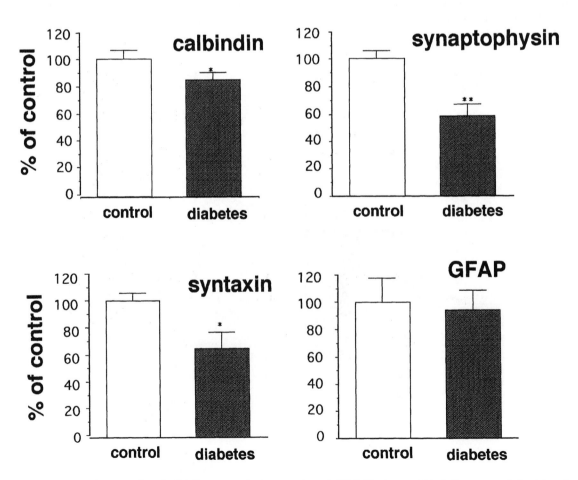

Fig. 3. Effects of diabetes on the expression of calbindin, synaptophysin, syntaxin, and GFAP. The immunoblot studies were carried out 4 weeks after the injection of STZ. *$P < 0.05$ compared with control (Tukey's test).

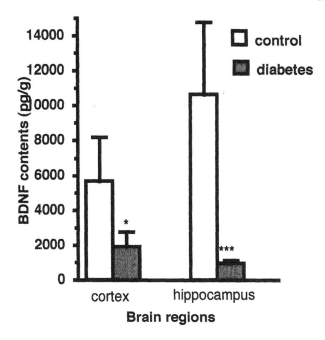

Fig. 4. Effects of diabetes on the BDNF contents in the hippocampus and cortex of the rat brain. BDNF contents were measured by our two-site EIA system. Each brain was removed 4 weeks after the STZ injection. *$P < 0.05$ compared with control (Tukey's test).

hippocampus and cortex, respectively, of the control animals. BDNF contents in both cortex and hippocampus were severely reduced in diabetic brain (Fig. 4). RT-PCR analysis showed that the ratio of BDNF to β-actin product was significantly reduced in the diabetic brain (Fig. 5).

4. Discussion

The diabetic brain has not been studied so much and its possible dysfunctions remain to be clarified; although severe peripheral neuropathy has been defined in diabetic patients. Very few etiological studies of relationship between diabetes and learning memory have been shown. One research group demonstrated that Alzheimer's patients had a relatively high frequency of diabetes mellitus [8]. However, another group found no significant differences between diabetes and control subjects with respect to severity of Alzheimer-type pathologies including both senile plaques and neurofibrillary tangles, although diabetics showed impaired cognitive performance relative to age-matched control subjects [9]. These previous results suggest that diabetes induces impairment of cognitive performance; however, the decrease of learning ability may not be restricted to only Alzheimer's disease.

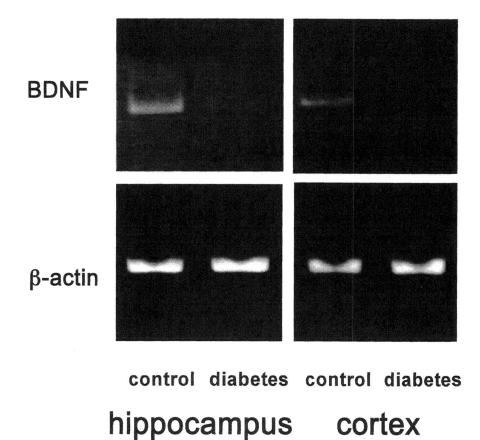

Fig. 5. Effects of diabetes on the expression of BDNF and β-actin mRNA in the hippocampus and cortex of rat brain. Total RNA from each culture was subjected to RT-PCR for BDNF of β-actin mRNA. Each brain was removed 4 weeks after the STZ injection.

Even in animal models of diabetes, no direct evidence of relationship between diabetes and memory impairment has been obtained. In our present study, the diabetic rats showed learning impairment in the Y-maze task, which is an indicator for the spontaneous alternation behavior. As was shown in Fig. 1, the score of alternation behavior of diabetic rats was significantly lower than that of the control. We had attempted other tasks to estimate the learning abilities of the diabetic rats. Such tasks requested some kinds of motivations, e.g., food, water, or swimming. Body weight, intake of food or water, and spontaneous motor activity of diabetic rats were significantly different from these of the control rats. The Y-maze task is a moderate task that does not require such motivations. Our finding that the alternation score of the diabetic rats was lower than that of control ones is the first evidence of memory impairment in a diabetic animal model.

Next, we investigated the cell number and possible morphological change in the brains of the diabetic rats. However, we failed to find any change in neuronal cell number when we used Nissl staining (data not shown). In Golgi staining, morphological changes were evident in the diabetic cortex. Similar morphology changes were observed in the hippocampus at the time of delayed neuronal death induced by the transient global cerebral ischemia in rodents [24]. The initiating event that leads to delayed neuronal cell death is possibly neuronal excitation caused by glutamate and subsequent calcium influx into the cell. Abnormal influxes of glutamate and/or calcium might also occur in diabetic brain, because the expressions of calbindin, synaptophysin, and syntaxin were reduced in such brains, which are proteins related to a Ca^{2+} binding protein or synaptic secretions. Their level indicate the degree of expression of glutamate, which initiates the release of other neuronal transmitters [10,12,26]. Unusual influxes of glutamate and/or calcium might cause the morphological changes in the diabetic brain. The neuronal cell death induced by ischemia is due to a failure of recovery process following excitatory damages to this particular neuronal circuitry. Lack of several neurotrophic factors such as BDNF or NGF is considered to cause the neuronal cell death after ischemia. A previous study of ours [26] demonstrated that BDNF was found only in the hippocampal CA1 dendritic field, in which the delayed neuronal damage occurs after transient cerebral ischemia. The intracerebroventricular injection of BDNF can protect against neuronal cell death after ischemia in rodents [23]. The reduced level of BDNF in the diabetic hippocampus in the present study may induce the neuronal dysfunction, i.e., impairment of cognition. Such dysfunction was prevented by the administration of a stimulator of BDNF synthesis (Nitta et al., unpublished data).

In conclusion, this present report provides the first evidence that diabetes induces a reduction in the BDNF level in the brain and poor cognitions. Further study is needed to investigate the detailed mechanisms that lead to these abnormalities.

Acknowledgments

The study was supported in part by a grant from the Smoking Research Foundation.

References

[1] J.D. Curb, B.L. Rodriguez, R.D. Abbott, H. Petrovitch, G.W. Ross, K.H. Masaki, D. Foley, P.L. Blanchette, T. Harris, R. Chen, L.R. White, Longitudinal association of vascular and Alzheimer's dementias, diabetes, and glucose tolerance, Neurology 52 (1999) 971–975.

[2] P.A. Ekstrom, D.R. Tomlinson, Impaired nerve regeneration in streptozotocin-diabetic rats is improved by treatment with gangliosides, Exp. Neurol. 109 (1990) 200–203.

[3] E.L. Feldman, M.J. Stevens, D.A. Greene, Pathogenesis of diabetic neuropathy, Clin. Neurosci. 4 (1997) 365–370.

[4] P. Fernyhough, L.T. Diemel, W.J. Brewster, D.R. Tomlinson, Altered neurotrophin mRNA levels in peripheral nerve and skeletal muscle of experimentally diabetic rats, J. Neurochem. 64 (1995) 1231–1237.

[5] D.A. Greene, M.J. Stevens, E.L. Feldman, Diabetic neuropathy: Scope of the syndrome, Am. J. Med. 30 (1999) 2S–8S.

[6] Y Hanaoka, T. Ohi, S. Furukawa, Y. Furukawa, K. Hayashi, S. Matsukura, The therapeutic effects of 4-methylcatechol, a stimulator of endogenous nerve growth factor synthesis, on experimental diabetic neuropathy in rats, J. Neurol. Sci. 122 (1994) 28–32.

[7] Y. Hanaoka, T. Ohi, S. Furukawa, Y. Furukawa, K. Hayashi, S. Matsukura, Effect of 4-methylcatechol on sciatic nerve growth factor level and motor nerve conduction velocity in experimental diabetic neuropathic process in rats, Exp. Neurol. 115 (1996) 292–296.

[8] Y. Harris, P.B. Gorelick, S. Freels, M. Billingsley, N. Brown, D. Robinson, Neuroepidemiology of vascular and Alzheimer's dementia among African-American women, J. Natl. Med. Assoc. 87 (1995) 741–745.

[9] J. Heitner, D. Dickson, Diabetics do not have increased Alzheimer-type pathology compared with age-matched control subjects. A retrospective postmortem immunocytochemical and histofluorescent study, Neurology 49 (1997) 1306–1311.

[10] A. Helme-Guizon, S. Davis, M. Israel, B. Lesbats, J. Mallet, S. Laroche, A. Hicks, Increase in syntaxin 1B and glutamate release in mossy fibre terminals following induction of LTP in the dentate gyrus: A candidate molecular mechanism underlying transsynaptic plasticity, Eur. J. Neurosci. 10 (1998) 2231–2237.

[11] C.E. Henderson, W. Camu, C. Mettling, A. Gouin, K. Poulsen, M. Karihaloo, J. Rullamas, T. Evans, S.B. McMahon, M.P. Armanini, L. Berkemeier, H.S. Phillips, A. Rosenthal, Neurotrophins promote motor neuron survival and are present in embryonic limb bud, Nature 363 (1993) 266–270.

[12] A. Jouvenceau, B. Potier, R. Battini, S. Ferrari, P. Dutar, J.M. Billard, Glutamatergic synaptic responses and long-term potentiation are impaired in the CA1 hippocampal area of calbindin D(28 k)-deficient mice, Synapse 33 (1984) 172–180.

[13] K. Kaechi, R. Ikegami, N. Nakamura, M. Nakajima, Y. Furukawa, S. Furukawa, 4-Methylcatechol, an inducer of nerve growth factor synthesis, enhances peripheral nerve regeneration across nerve gaps, J. Pharmacol. Exp. Ther. 272 (1995) 1300–1304.

[14] B. Knusel, J.W. Winslow, A. Rosenthal, L.E. Burton, D.P. Seid, K. Nikolics, F. Hefti, Promotion of central cholinergic and dopaminergic neuron differentiation by brain-derived neurotrophic factor but not neurotrophin-3, Proc. Natl. Acad. Sci. U. S. A. 88 (1991) 9621–9625.

[15] Y. Ming, E. Bergman, E. Edstrom, B. Ulfhake, Reciprocal changes in the expression of neurotrophin mRNAs in target tissues and peripheral nerves of aged rats, Neurosci. Lett. 273 (1999) 187–190.

[16] K.F. Mortel, S. Wood, M.A. Pavol, J.S. Meyer, J.L. Rexer, Analysis

of familial and individual risk factors among patients with ischemic vascular dementia and Alzheimer's disease, Angiology 44 (1993) 589–605.

[17] M. Nadai, H. Yoshizumi, T. Kuzuya, T. Hasegawa, I. Johno, S. Kitazawa, Effect of diabetes on disposition and renal handling of cefazolin in rats, Drug Metab. Dispos. 18 (1990) 565–570.

[18] A. Nitta, M. Ito, H. Fukumitsu, M. Ohmiya, H. Ito, A. Sometani, H. Nomoto, Y. Furukawa, S. Furukawa, 4-Methylcatechol increases brain-derived neurotrophic factor content and mRNA expression in cultured brain cells and in rat brain in vivo, J. Pharmacol. Exp. Ther. 291 (1999) 1276–1283.

[19] A. Nitta, M. Ohmiya, T. Jin-nouchi, A. Sometani, T. Asami, H. Kinukawa, H. Fukumitsu, H. Nomoto, S. Furukawa, Endogenous neurotrophin-3 is retrogradely transported in the rat sciatic nerve, Neuroscience 88 (1999) 679–685.

[20] H.S. Phillips, J.M. Hains, G.R. Laramee, A. Rosenthal, J.W. Winslow, Widespread expression of BDNF but not NT-3 by target areas of basal forebrain cholinergic neurons, Science (Washington, D.C.) 250 (1990) 290–294. •

[21] A. Rodriguez-Pena, M. Botana, M. Gonzalez, F. Requejo, Expression of neurotrophins and their receptors in sciatic nerve of experimentally diabetic rats, Neurosci. Lett. 200 (1995) 37–40.

[22] L.C. Rutherford, A. DeWan, H.M. Lauer, G.G. Turrigiano, Brain-derived neurotrophic factor mediates the activity-dependent regulation of inhibition in neocortical culture, J. Neurosci. 17 (1997) 4527–4535.

[23] W.R. Schabitz, C. Sommer, W. Zoder, M. Kiessling, M. Schwaninger, S. Schwab, Intravenous brain-derived neurotrophic factor reduces infarct size and counterregulates Bax and Bcl-2 expression after temporary focal cerebral ischemia, Stroke 31 (2000) 2212–2217.

[24] T. Shigeno, T. Mima, K. Takakura, D.I. Graham, G. Kato, Y. Hashimoto, S. Furukawa, Amelioration of delayed neuronal death in the hippocampus by nerve growth factor, J. Neurosci. 11 (1991) 2914–2919.

[25] D.R. Tomlinson, P. Fernyhough, L.T. Diemel, Role of neurotrophins in diabetic neuropathy and treatment with nerve growth factors, Diabetes 46 (1997) S43–S49.

[26] R. Tyzio, A. Represa, I. Jorquera, Y. Ben-Ari, H. Gozlan, L. Aniksztejn, The establishment of GABAergic and glutamatergic synapses on CA1 pyramidal neurons is sequential and correlates with the development of the apical dendrite, J. Neurosci. 19 (1999) 10372–10382.

[27] K. Yamada, T. Tanaka, L.B. Zou, K. Senzaki, K. Yano, T. Osada, O. Ana, X. Ren, T. Kameyama, T. Nabeshima, Long-term deprivation of oestrogens by ovariectomy potentiates beta-amyloid-induced working memory deficits in rats, Br. J. Pharmacol. 128 (1999) 419–427.

[28] M.A. Yorek, T.J. Wiese, E.P. Davidson, J.A. Dunlap, M.R. Stefani, C.E. Conner, S.A. Lattimer, M. Kamijo, D.A. Greene, A.A. Sima, Reduced motor nerve conduction velocity and Na(+)-K(+)-ATPase activity in rats maintained on L-fucose diet. Reversal by *myo*-inositol supplementation, Diabetes 42 (1993) 1401–1406.

Printed and bound by CPI Group (UK) Ltd, Croydon, CR0 4YY

03/10/2024

01040328-0016